INTRODUCTION TO ALGEBRAIC SYSTEM THEORY

This is Volume 151 in
MATHEMATICS IN SCIENCE AND ENGINEERING
A Series of Monographs and Textbooks
Edited by RICHARD BELLMAN, *University of Southern California*

The complete listing of books in this series is available from the Publisher
upon request.

INTRODUCTION TO ALGEBRAIC SYSTEM THEORY

MICHAEL K. SAIN

Department of Electrical Engineering
University of Notre Dame
Notre Dame, Indiana

 1981

ACADEMIC PRESS
A Subsidiary of Harcourt Brace Jovanovich, Publishers

New York   London   Toronto   Sydney   San Francisco

ACADEMIC PRESS, INC.
111 Fifth Avenue, New York, New York 10003

United Kingdom Edition published by
ACADEMIC PRESS, INC. (LONDON) LTD.
24/28 Oval Road, London NW1 7DX

Library of Congress Cataloging in Publication Data

Sain, Michael K.
    Introduction to algebraic system theory.

    (Mathematics in science and engineering)
    Bibliography: p.
    Includes index.
    1. System analysis. 2. Algebra. I. Title.
II. Series.
QA402.S34   003   80–522
ISBN 0–12–614850–3

PRINTED IN THE UNITED STATES OF AMERICA

81 82 83 84   9 8 7 6 5 4 3 2 1

With Love
to
My Happy Frances

CONTENTS

4. GROUP MORPHIC SYSTEMS

5. INVERSES OF GROUP MORPHIC SYSTEMS

6. INTERCONNECTED SYSTEMS

7. ASPECTS OF MODULE MORPHIC SYSTEMS

PREFACE

The material in this volume has been developed as part of a lecture course given to graduate students at the University of Notre Dame during the past five years. Basically, the purpose of this course has been to provide a bridge over which interested persons could pass from the classical and traditional concepts of introductory system theory to those of its rapidly developing extension, known today as algebraic system theory.

In order to make such a passage, then, it will be most efficient if the reader already has the intuition that comes from a first course in system theory, including at least some exposure to the ideas of input–output functions, state equations, and transforms. This knowledge might come from a junior or senior level course. Absence of such a course does not represent an impediment to reading the material. Indeed, we have endeavored to make the book as self-contained as possible. However, a finite volume represents a fixed total resource; and we have chosen to allot more of this resource toward carefully motivating the way in which the ideas of algebra mesh with the ideas of system theory than toward motivating system theory in itself.

The author's experience in presenting this material has suggested that this is one effective way to communicate the concepts to interested audiences, which often begin with very heterogeneous notions of algebra itself. Almost everyone has a firm impression of algebra. Sometimes these impressions differ enough from modern technical algebra that they may actually have to be *un*learned. A ready example is available from transform theory.

Few system theorists are without knowledge of the Laplace and Fourier transformations, and we suspect that the vast majority who have such exposure have at one time or another been confronted with the misleading statement that "transformation of an ordinary linear differential equation with constant coefficients converts the problem of solving the equation into a problem in *algebra.*'' Nothing, of course, could be further from the truth; the original equation is as algebraic in nature as the transformed equation, and the transform itself is quite distinctly algebraic as well. In a way, this common lack of insight suggests that system theorists in the past have, perhaps, failed to perceive fully the algebraic underpinnings of their discipline. This may have occurred in much the same way as did the story of various blind people describing the elephant, with one feeling the tail and seeing it as a rope, another feeling the trunk and regarding it as a hose, and yet another feeling the leg and guessing it to be like a tree.

This situation is now beginning to change. Many of the older, narrow concepts of algebra are giving way to newer, more broadly based viewpoints. The impetus for this development has come from many directions, particularly from singular advances which are not so well explained or understood in terms of aged and shopworn algebraic clichés. Noteworthy examples of such advances are the uses of polynomial modules for depicting dynamical action in linear systems, the employment of the subspace lattice for a geometric theory of control of linear multivariable systems, the mating of cyclic and convolutional ideas in coding for communication with reduced errors, and the development of new forms of nonlinear dynamical filters that take greater advantage of the bilinear and multilinear nature of many nonlinearities. Fortunately, such trends have, we believe, created a new and much more open attitude among systems personnel toward the uses for algebra in their work. These positive attitudes are now becoming quite apparent in the classroom, when compared with the situation in existence just a few years ago. Seniors and beginning graduate students in systems studies no longer see the use of algebra as a far-out activity. Moreover, similar interests appear to be surfacing in industry, where systems workers see the additional structural insights afforded by a judicious choice of algebra as a great help in understanding and troubleshooting complex technological systems.

Unfortunately, this new openness is not as yet very well aided by textual material. Personnel wishing to assess the new methods are often faced on one hand with highly technical papers which assume that they already know their algebra and on the other hand with algebra books which assume that they already know all their motivating systems examples.

It is to help fill this gap in the emerging literature that we have prepared this volume, which is intended for three types of use:

(1) Textbook for a Second Course in Systems. Such a course could take place at the senior, introductory graduate, or advanced graduate level, depending upon the curriculum, the school, and the size of the student body. It is assumed that the student has introductory knowledge of difference equations and matrix algebra. We have tried to write so that, on the average, the level is for an introductory graduate course. In this type of use, the material has been tested in class situations five times. After much consideration, we have decided not to use a

theorem–proof format. Instead, we have tried to emphasize examples and to control carefully the length of sections. Indeed, some entire sections are essentially examples. A considerable number of illustrative exercises have been provided. These should also give the instructor some ideas for useful examination questions. We have found it useful to generate additional examples from the students' current experience. One purpose of the text examples and section exercises is to aid in this task. Inasmuch as algebraic system theory remains a rapidly moving research area, many "natural" examples are in reality small research problems. For this reason, we have tried to exercise caution in the complexity of the exercises.

(2) Reference for a First Course in Systems. The use of selected chapters and ideas from the text can offer additional perspective to many introductory systems concepts. For example, students in our experience have reacted quite positively to a lecture or two on the generalization of linear dynamical systems to group morphic systems.

(3) Seminars and Self-Study. The book has been arranged so that a casual reading can proceed from beginning to end in a self-contained manner. Nonetheless, certain of the sections and chapters can easily be excerpted for seminar discussions. Chapters 4 and 5 on group morphic systems and their inverses are an illustration of this.

Our purpose—to help fill the gap between technical systems papers having little algebraic explanation and algebraic textbooks having no systems examples—has demanded that we be reasonably broad in topic selection and that we employ many examples. Necessarily, then, we often bypass proofs in favor of illustrations. No attempt has been made to be encyclopedic. We must apologize in advance to those whose favorite topic has been excluded.

The reader does not require prior or concurrent training in technical algebra; algebraic ideas are developed along the way as needed. In most cases, these ideas are strongly concentrated in one or two sections in each chapter, though definitions and properties are distributed among a wide variety of sections. Thus this is not a reference book in algebra. The reader should have no difficulty in making contact with such reference books, if algebraic system theory stimulates further interest. Generally, however, we have found that students tend to profit more from such references after they have developed notions of how algebraic ideas work and how they can be applied.

The idea of the book is to take successive looks at dynamical systems, with each look adding a bit more algebraic structure.

Chapter 1 contains some intuitive material which can be selected for introductory discussion or reading. Depending upon the background of the student, certain of these sections may have more initial meaning than others. We have had good success by beginning with Sections 1.1, 1.2, and 1.3, and returning later to sample the material in Sections 1.4 and 1.5. It should be emphasized that the "firm" material begins in Chapter 2, and that the ideas provided in Chapter 1 are to be regarded as suggested material for a motivational introduction. If a student has doubts about functions, it may be best to read Section 2.1 before sampling Chapter 1.

Chapters 2 and 3 deal with systems defined on sets. The only structures put in place on the sets are relations, though some operations on relations are considered. The main purpose here is to show that many systems ideas require very little in the way of algebraic assumptions. Chapters 4 and 5 continue the discussion by permitting binary operations on the basic system sets. In a way, Chapter 5 serves a dual purpose. On one hand, it is an extensive illustration of certain of the materials in Chapter 4, and on the other hand, it lays motivational background for the interconnection discussions of Chapter 6. Though the concept of ring is introduced first in Chapter 5 to support the inverse system calculations with morphisms in that chapter, the main illustrative application for the ring is found in Chapter 6, where a "connection" between interconnections and ring operations is developed. The last chapter is a natural consequence of Chapter 6, which used the action of ring elements on commutative group elements. Combining both the ring and group structures, Chapter 7 is of course the most technical. Selection of topics in Chapter 7 has been made in such a way that the volume ends at a point from which the reader will be able to proceed naturally to other available materials. The last section lays the groundwork for algebraic realization theory, which is available in other textbooks. Moreover, it is a relatively easy transition from Chapter 7 to algebra books on rings and modules. Finally, the chapter just touches on vector spaces. Our belief here is that most readers have already been exposed to vector spaces before they come to this volume. Moreover, there is textual material on algebraic system theory over vector spaces already available for further reading.

Finally, I wish to express thanks to those who have helped to make this book a reality. At the outset, I am indebted to Dr. W. M. Wonham, whose fascinating lectures during my visit to the University of Toronto in 1972 were the main force leading to my own interest in algebraic system theory. Also, I am appreciative of the encouragement offered by Dr. James L. Melsa, Chairman of the Department of Electrical Engineering at the University of Notre Dame. Dr. Bostwick F. Wyman of the Mathematics Department at The Ohio State University has shared innumerable hours of helpful conversation on the general subject and its potential as a research tool. Dr. Stanley R. Liberty of Old Dominion University has been of great help by teaching a course from the manuscript at Norfolk in fall 1978. Many students have listened and shared their thoughts and comments. Among these, I am especially thankful to K. Dudek, R. R. Gejji, R. M. Schafer, V. Seshadri, and S. Yurkovich. Finally, Mrs. Tamara Youngs has been a cheerful and cooperative assistant in preparing the typewritten manuscript.

1 INTUITIONS

Anyone coming to a study of algebraic system theory is the inevitable bearer of a certain amount of baggage containing preconceived notions of the subject. In some cases, it must be true that this is all for the good; in other cases, however, it is probable that a great deal of this baggage will have to be unloaded before the journey can proceed.

The purpose of this chapter is to review certain of the more common misconceptions about the subject of algebraic system theory and to develop on an intuitive basis some notions about both algebra and system theory.

Quite a number of the algebraic usages of later chapters have been quietly woven into the fabric of these sections. In this way, various motivations for the subsequent, more axiomatic, treatments should be established.

Readers having little experience with the use of algebra in system theory may be surprised at the stress on set structure. This is intended. Such a stress is crucial to a full appreciation of algebraic system theory, and yet is an ingredient often underplayed in introductory systems presentations. The approach is designed to be persuasive in nature and to provide a reasonably convincing argument in support of the logical necessity of addressing that structure in an adequate way.

A caveat may be in order. The process of selecting a structure for a set is not entirely deductive. For example, the number 2 may be regarded as a natural number (i.e., 0, 1, 2, . . .), an integer (i.e., 0, ± 1, ± 2, . . .), a ratio of integers (e.g., 6/3), a real number, or a complex number $(2 + j0)$. In subsequent paragraphs, therefore, the reader may decide that he or she might have made different choices. This is entirely to be expected and well within the rules of the game. Nonetheless, pitfalls do exist. Just as a tailor-made suit may fit the buyer much better than a suit purchased ready-made, so some algebraic structures seem to fit certain systems discussions much better than others. Eventually, designation of set structure becomes a matter of taste; but there is a community standard as to what constitutes poor taste!

1.1 MISCONCEPTIONS

To some extent, system theory has suffered from what might be described as less than judicious use of mathematics. During the burst of systems activity just after the midcentury, mathematics was often put in place as a sort of window dressing to make respectable the results which had been found without its explicit use. The last few years have witnessed, with thanks, the gradual dissipation of such activities. During the remainder of this century, it seems more likely that mathematical structures will be emplaced at the start of systems investigations and will be used routinely and explicitly in problem solution.

An illustration should help to clarify this point. Consider the differential equation

$$\begin{bmatrix} \dot{x}_1 \\ \dot{x}_2 \end{bmatrix} = \begin{bmatrix} 1 & 2 \\ 3 & 4 \end{bmatrix} \begin{bmatrix} x_1 \\ x_2 \end{bmatrix}, \qquad t \geq 0, \tag{1-1}$$

where $x_i(t)$, $i = 1, 2$, is a real-valued function of the real variable t on the interval indicated and where the dot signifies ordinary differentiation. It is quite common to point out that the coefficient array in (1-1) is just the matrix of a linear map $\mathbb{A}$ defined on the real space of real 2-tuples $\mathbb{R}^2$ to itself. It is then quite possible, and almost as common, to obtain the solutions and properties of the equation without ever referring again to the notion of space.

Such window-dressing usage has tended to create the rather debilitating impression that mathematics is a pleasant form of culture but is not really needed in the trenches. Without doubt, this has created delays in the advancement of system theory, because of the confusions which can occur when equations such as (1-1) are stated in a different light. For example, suppose that the coefficients are to be regarded as *integers* instead of real numbers. What are the consequences of such a stipulation? The *formal* matrix theory does not speak as conveniently to these more delicate issues, which are associated very naturally with generalizations of the concept of real space.

This is not to say, of course, that the remarkable systems advances of the last quarter-century have not used mathematics. Indeed they have, and sometimes with force and elegance. But in a great many instances it can be argued that these usages were of the bargain-basement variety and were only rarely tailor-made for the problem. Thus, when problems took on new and unforeseen dimensions, the standard bargains tended to be misfits.

One very well-worn example of bargain-basement usage should be quite familiar to most engineers. Suppose the equation

$$\dot{x} = x, \qquad t \geq 0, \tag{1-2}$$

is to be solved, subject to the condition

$$x(0) = 1. \tag{1-3}$$

It remains surprisingly common today for authors to suggest that a useful procedure to follow is to convert this initial-value problem in ordinary differential equations into a problem in *algebra* by means of the Laplace transformation. Everyone knows the procedure, which modifies (1-2) and (1-3) to achieve

$$s\mathbb{X} - 1 = \mathbb{X}, \tag{1-4}$$

which then solves for the Laplace transform $\mathbb{X}$ of x in the usual manner

$$\mathbb{X}(s) = 1/(s - 1). \tag{1-5}$$

The point is not that (1-4) fails to be algebraic—for it most certainly is—but rather that the original initial-value problem was *also* algebraic.

Thus it is *not* a question of using the Laplace technique to turn a non-algebraic problem into an algebraic one. But it *is* a question of turning one type of algebraic problem, namely, (1-2)–(1-3), into another type of algebraic problem, namely, (1-4).

Such continuing confusions say much about what algebraic system theory must bring to mind for the students and practitioners of the art as well as for personnel from other disciplines. Far too often, it would appear that algebra is conceived in terms of such activities as solving the quadratic equation

$$ax^2 + bx + c = 0 \tag{1-6}$$

or the simultaneous equations

$$ax + by = e, \qquad cx + dy = f. \tag{1-7a,b}$$

This is nowhere near the modern notion of algebra, and thus is not an adequate algebraic notion for this work.

The development of a more comprehensive notion of algebra in the system theory context is of course one of the main purposes of the following chapters. At this point, however, it is desirable to set into motion certain basic intuitions about the subject.

1.2   AN ASPECT OF ALGEBRA

Speaking about algebra begins with speaking about sets. In fact, it is immediately convenient to assume that there are two sets available, say S and T. For, once this point is reached, various relationships between the elements of S and the elements of T can be postulated. Among the most fundamental of such relationships is the idea of a *function* f which assigns to each element s in S an element $f(s)$ in T.

Pictorially, such a function is indicated in Fig. 1.1. A more compact depicting is achieved by condensing the figure to

$$f: S \to T. \tag{1-8}$$

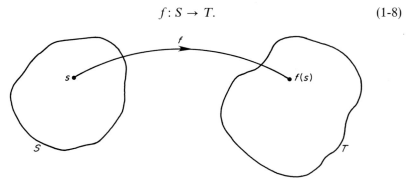

Fig. 1.1. Pictorial view of a function.

It should be remarked that sketches such as Fig. 1.1 have certain pedagogical disadvantages which result from their finite extent. For example, if S is a finite-dimensional real space, then it must be closed under arbitrarily large real scaling in any direction, a property which is at conceptual odds with the bounded picture. Because of this, it is very convenient to have available an alternative view of the function, of a type such as that shown in Fig. 1.2. Fortunately, this view is highly compatible with current algebraic practice; and it will be convenient in the sequel to adhere to it—except in certain special circumstances.

$$S \xrightarrow{\quad f \quad} T$$ **Fig. 1.2.** Alternative view of a function.

With only these two ideas, sets and functions, one aspect of algebra can be brought into sharp focus. Of course, every insight has its own price, and in this case it is a matter of an increment in notation. What is required is a symbol for the set of all functions f of the type indicated in Fig. 1.2. The desired symbol is

$$T^S \triangleq \{f: S \to T\}, \tag{1-9}$$

which often seems improbable on a first reading. Actually, however, T^S is a very natural generalization of symbols in common use throughout the range of applications. In fact, such a symbol, $\mathbb{R}^2$, has already been used in this chapter.

An additional remark may serve to clarify this issue. The symbol

$$\mathbb{Z} = \{0, 1, -1, 2, -2, \dots \} \tag{1-10}$$

denotes the set of integers; and the corresponding symbol

$$\mathbb{Z}_n = \{0, 1, 2, \dots, n - 1\} \tag{1-11}$$

describes integers from 0 to $n - 1$. Now consider

$$\mathbb{Z}_2 = \{0, 1\} \tag{1-12}$$

as the set S and $\mathbb{R}$ the set of real numbers as the set T. The function f is completely defined by specifying its action on the elements of S, namely

$$f(0), \quad f(1). \tag{1-13}$$

If

$$f(0) = r_1, \quad f(1) = r_2, \tag{1-14a,b}$$

for real numbers r_1 and r_2, it is clear that knowledge of f is the same thing as knowledge of

$$(r_1, r_2), \tag{1-15}$$

which is understood as an element of $\mathbb{R}^2$. To complete the illustration, simply observe that

$$T^S \doteq \mathbb{R}^{\mathbb{Z}_2} \tag{1-16}$$

and that $\mathbb{R}^2$ can be understood as an abbreviation for $\mathbb{R}^{\mathbb{Z}_2}$.

With an explicit notation T^S for sets of functions available, a focus can now be made on the algebraic aspect just hinted. It is not intended to define this aspect, but rather to illustrate it. Proceed, therefore, to assume the presence of two such sets of functions, denoted in straightforward manner by

$$T_i^{S_i}, \quad i = 1, 2. \tag{1-17}$$

As mentioned at the beginning of this section, it is always a convenience to assume that there are two sets available; for then it is possible to posit a function. Do so, and denote this higher-level function by

$$f_{12}: T_1^{S_1} \to T_2^{S_2}. \tag{1-18}$$

Note that, in grammatical terms, f_{12} is a function which assigns functions to functions.

As a brief digression, it can be seen that the matrix in (1-1), namely,

$$\begin{bmatrix} 1 & 2 \\ 3 & 4 \end{bmatrix}, \tag{1-19}$$

is a representation of just such a function, if

$$T_1 = T_2 = \mathbb{R}, \qquad S_1 = S_2 = \mathbb{Z}_2, \tag{1-20}$$

and this is entirely in accord with the preceding discussion of $\mathbb{R}^2$ as a set of functions.

Continuing, it is now possible to describe all functions of the form (1-18) by a direct application of the notation for sets of functions. The reader has probably already foreseen the construction

$$(T_2^{S_2})^{(T_1^{S_1})} \tag{1-21}$$

for this set. In due course, it would be convenient if another such object were made available, say in the manner

$$(T_4^{S_4})^{(T_3^{S_3})} \tag{1-22}$$

because functions

$$f_{12,34}: (T_2^{S_2})^{(T_1^{S_1})} \to (T_4^{S_4})^{(T_3^{S_3})} \tag{1-23}$$

are then an obvious constructive consequence. Nor is a function such as $f_{12,34}$ difficult to exhibit. Consider the matrix

$$\begin{bmatrix} 1 & 2 \end{bmatrix}, \tag{1-24}$$

which carries 2×2 matrices into 1×2 matrices in the manner suggested by

$$\begin{bmatrix} x & x \end{bmatrix} = \begin{bmatrix} 1 & 2 \end{bmatrix} \begin{bmatrix} x & x \\ x & x \end{bmatrix}, \tag{1-25}$$

and select

$$T_1^{S_1} = \mathbb{R}^2, \qquad T_2^{S_2} = \mathbb{R}^2, \qquad T_3^{S_3} = \mathbb{R}^2, \qquad T_4^{S_4} = \mathbb{R}^1. \tag{1-26}$$

By now, the suggested algebraic aspect should be beginning to make itself clear. The set of all functions of the form (1-23) can be symbolized immediately as

$$((T_4^{S_4})^{(T_3^{S_3})})^{((T_2^{S_2})^{(T_1^{S_1})})} \tag{1-27}$$

and discussion could automatically proceed to functions on such sets.

Further pursuit of this idea here would be unnecessarily cumbersome without compensating benefit. However, it should be evident that the repeated application of the notions

$$\text{sets,} \qquad \text{functions,} \qquad \text{sets of functions,} \ldots$$

would inevitably and eventually carry each reader beyond his present level of easy perception. It should also be evident that the concept of simple matrices, such as

$$[1 \quad 2], \qquad \begin{bmatrix} 1 & 2 \\ 3 & 4 \end{bmatrix},$$

can be used *on more than one level* of meaning. This important observation shows clearly that the understanding of a function f is not possible outside the context of (1-8), wherein a clear specification of the pair of sets associated with the function is provided.

Accordingly, an entrance to algebraic system theory is marked by a more decided attention to functions and their definitions than is usually evident in typical introductions to systems. At first, these attentions can cause impatience as familiar ideas are reconsidered with what often seems to be only added tedium. As experience is gained, however, the added distinctions begin to produce a resolution of issues almost completely invisible in less detailed treatments.

1.3 AN ASPECT OF SYSTEM THEORY

The elemental algebraic ideas of set and function provide immediate access to the notion of system.

To see this, let I be a set of inputs, O a set of outputs, and $f : I \rightarrow O$ a function. In this situation, it is typical to refer to f as a *system function*. The situation is indicated in Fig. 1.3, which is a sort of archetypal input–output system description.

Though remarkably general, Fig. 1.3 is overly simple from a systems viewpoint because it fails to provide for adequate notions of building up bigger and more complicated systems from smaller and simpler ones. The missing idea is that which a systems specialist would probably call *connection*. Connection, in a systems sense, turns out to be very closely related to the concept of a binary operation, in an algebraic sense.

The idea of a connection can seem deceptively simple. Consider Fig. 1.4, which seems at first glance to be an acceptable rendering of two systems

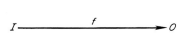

Fig. 1.3. Archetypal input–output system description.

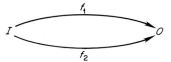

Fig. 1.4. Two systems in parallel?

working in parallel. System 1 defined by function f_1, produces from an input i in I an output o_1 in O; likewise, system 2, defined by function f_2, produces from the same input i in I an output o_2 in O. The basic issue revolves about what is to be done in O with outputs o_1 and o_2. A moment's reflection should convince the reader that there is as yet no relationship between f_1 and f_2 at all, even though they appear somehow related in Fig. 1.4. In fact, the figure merely says that f_1 and f_2 are elements of O^I. Nothing else is, or at this point can be, inferred.

What is missing? The missing idea is that of operating on two elements of the set O to produce another element of O. Because two elements are used in the operation, it is technically described as a binary operation. The reader is familiar with key examples of binary operations, including addition and multiplication of real numbers.

So as not to prejudice the concept at this point, use is made of a general symbol $\square$ for such an operation, a precise definition of which will be given in a following chapter. With the aid of such a symbol, however, it is now possible to combine o_1 and o_2 in O, a construction which is indicated in the manner

$$o = o_1 \,\square\, o_2. \tag{1-28}$$

It may be surprising to the reader to see that this one simple idea essentially solves the connection issue, for (1-28) induces a corresponding binary operation on the functions f_1 and f_2, which can be denoted by

$$f_1 \,\square\, f_2 : I \rightarrow O \tag{1-29}$$

and defined by

$$(f_1 \,\square\, f_2)(i) = (f_1(i)) \,\square\, (f_2(i)) \tag{1-30}$$

for each i in I. It is crucial to take cognizance of the fact that no restrictions whatever have been placed on the system functions, except that they be functions in the usual sense.

Equipped with an appropriate binary operation $\square$, then, the output set O provides possibility for a more meaningful version of Fig. 1.4. One way to indicate such a version is to make use of the binary operation symbol $\square$ as a joining point or junction in the sketch, as shown in Fig. 1.5. Note that $\square$ is not a symbol for a set but rather for a binary operation on a set, in this case O. The set label O can be written alongside the junction symbol, if there is need to avoid confusion.

Finally, a parallel equivalent of Fig. 1.5 can be visualized with the aid of (1-29)–(1-30), as in Fig. 1.6.

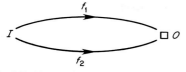

Fig. 1.5. More meaningful version of Fig. 1.4.

$$f_1 \ \Box \ f_2$$
$$I \xrightarrow{\hspace{4cm}} O$$

Fig. 1.6. Parallel equivalent of Fig. 1.5.

Illustrations abound. Select I and O to be the set of real numbers $\mathbb{R}$, f_1 to be sine, and f_2 to be cosine. If $\Box$ stands for the usual multiplication, then

$$(f_1 \ \Box \ f_2)(r) = \sin(r)\cos(r), \qquad r \in \mathbb{R}, \qquad (1\text{-}31)$$

whereas, if $\Box$ stands for the usual addition, then

$$(f_1 \ \Box \ f_2)(r) = \sin(r) + \cos(r), \qquad r \in \mathbb{R}. \qquad (1\text{-}32)$$

This is a flexible notion of connection whose meaning is intricately inter-twined with the nature of binary operations on the set O. Fundamentally, it is precisely these operations which determine what is usually known as the algebraic structure of the set. It is somewhat satisfying to be able to see at the very outset that system structure, in a connection sense, is going to be closely related to algebraic structure, in a mathematical sense.

In the case of series connection, there is a very natural concept available immediately without any need to install a binary operation on any of the sets.

Consider the two systems indicated in Fig. 1.7. To establish a favorable climate for series connection, simply arrange that

$$O_1 = I_2, \qquad (1\text{-}33)$$

a situation which is sketched in Fig. 1.8. The idea is basic. An input i_1 in I_1 produces an output o_1 in O_1 through f_1; o_1, by reason of (1-33), can be identified with an input i_2 in I_2, which then produces an output o_2 in O_2 through f_2. Overall, it follows that

$$o_2 = f_2(f_1(i_1)) \qquad (1\text{-}34)$$

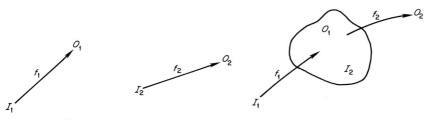

Fig. 1.7. Two systems.

Fig. 1.8. Two systems in series ($I_2 = O_1$).

for each i_1 in I_1, an equation which defines in algebraic terms the *composition*

$$f_2 \circ f_1 : I_1 \to O_2 \tag{1-35}$$

of the functions f_1 and f_2 according to the rule

$$(f_2 \circ f_1)(i_1) = f_2(f_1(i_1)), \qquad i_1 \in I_1. \tag{1-36}$$

The notion of composition then permits an equivalent for Fig. 1.8; such an equivalent is shown in Fig. 1.9.

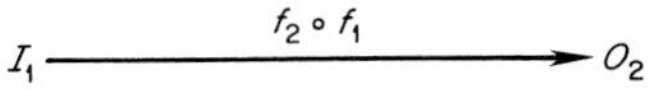

Fig. 1.9. Series equivalent of Fig. 1.8.

A comparison[†] may be drawn between the way in which the notion of parallel connection evolves from set structure and the way in which the alternative notion of series connection evolves. In both cases, these notions of connection depend upon the possibility of carrying out a certain type of activity on elements in the sets. The series connection is certainly the more fundamental, relying as it does only upon the concept of equating two elements of a set. The parallel connection is more advanced, requiring not only a notion of equality but also a binary operation. There are contrasts also, to be sure, with the series case involving in general four sets and the parallel case only two sets.

But the idea is seminal: connections of systems, in a reasonable sense, are related to algebraic structure on the sets upon which these systems act. In fact, they follow from it.

It is interesting to speculate on the basis of this premise. For example, it has been hinted in the preceding section that it is not very commonplace for system theoretic introductions to focus on the structure of sets. A possible conclusion might be that the everyday genre of systems concept might have some difficulty in coping with connections, especially prolific ones—large scale, in the current jargon.

Note also that, if the systems of interest are in T^S, then connections in the sense exposed above involve operations, such as $f_1 \,\square\, f_2$ and $f_2 \circ f_1$, on T^S. As will become clear in the following chapter, such operations also have the interpretation of functions, in an appropriate sense. Accordingly, in the hierarchy of the section preceding, it may be argued that connections exist on a level higher than the simple concept of system itself.

It is of some help when dealing with interconnections of systems to realize that the discourse takes place on more than one level. Indeed, the

---

[†] The comparison is tenuous at this point because it is predicated upon the concept of equality, which is so deeply entrenched that its absence may be difficult to visualize at this juncture.

phrase "constraints imposed by the connections" enjoys a certain amount of popularity and is quite compatible with the idea of connections being a higher-level entity.

Clearly, moreover, the surface has barely been scratched; and it is possible to conjecture about meaningful concepts of connecting connections.

1.4   AN ASPECT OF ALGEBRAIC SYSTEM THEORY

A substantial portion of what is discussed in algebraic system theory can be glimpsed rather suggestively by refining slightly the observations of the preceding section.

The idea is simple enough, and involves a process of interconnecting a number of identical systems.

Historically, of course, similar ideas abound. As examples, consider the many-body problem in theoretical mechanics or the resistive network problem in electricity. In fact, the science of statistics has a number of key aspects which are predicated upon large numbers of essentially identical activities, such as sampling. Moreover, ever increasing digitization has raised to an even higher level the worth of such considerations.

A much simpler illustration will serve adequately here. Begin by selecting a typical atomic system. Each reader can probably visualize such a system convenient to his or her own experience. Of interest at this time, for example, are transistors, which may be deposited by the thousands on thin wafers. Again, a timely example could be residential energy loads, of gas or electricity, arranged in the grids typical of established city neighborhoods or new suburban developments. Studies of physical systems have, because of the phrasing of physical laws in differential form, led to a special interest in the idea of an integrator as an atomic building block for larger systems. Readers who have been involved with simulations, of analog, hybrid, or even certain digital types, are certainly acquainted with such a concept of interconnected integrators. Yet another atomic system of pervasive practical merit is the delayor associated with systems predicated upon difference equations or growing out of numerical approaches to the solution of differential equations.

For simplicity, it is thus sufficiently general to assume that an atomic system **a** has been chosen, as in Fig. 1.10. Because a number of these elements

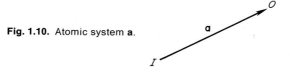

Fig. 1.10. Atomic system **a**.

are to be connected together, assume that the input set *I* and the output set *O* are identical. Also, to make use of the foregoing ideas on connections of parallel type, establish a binary operation □ on the input and output sets. Finally, adopt the approach of Fig. 1.5, wherein the operation symbol □ replaces the set symbol, and suppress the set symbols altogether, inasmuch as they are now a fixed quantity throughout a connection. The constructor of large systems, in such an instance, has available the equivalent of a box of atomic systems **a**, each identical to the other, from which he connects together a larger, more complicated structure (Fig. 1.11).

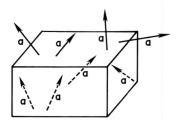

Fig. 1.11. Box of atomic systems.

Consider at the outset the two fundamental connections, parallel and series, discussed in this chapter on an intuitive basis. These are indicated in Fig. 1.12. The parallel connection, according to the earlier discussion, is equivalent to the expression

$$\mathbf{a} \,\square\, \mathbf{a}, \tag{1-37}$$

whereas the series connection is equivalent to

$$\mathbf{a} \circ \mathbf{a}. \tag{1-38}$$

To construct larger systems, of course, it will be necessary to hybrid these two basic connections in various ways. An interesting threesome of hybrids is sketched in Fig. 1.13. Begin with the leftmost, and calculate in a manner analogous to (1-37)–(1-38) that the equivalent must be

$$\mathbf{a} \circ (\mathbf{a} \,\square\, \mathbf{a}). \tag{1-39}$$

Now proceed to the center sketch and find the equivalent

$$(\mathbf{a} \,\square\, \mathbf{a}) \quad \mathbf{a}. \tag{1-40}$$

Fig. 1.12. Parallel and series connections.

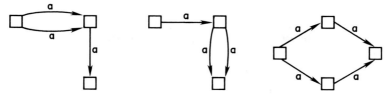

Fig. 1.13. Hybrid connections.

Finally, determine the equivalent of the rightmost case to be

$$(\mathbf{a} \circ \mathbf{a}) \,\square\, (\mathbf{a} \circ \mathbf{a}). \tag{1-41}$$

A key issue concerns whether any of the expressions (1-39)–(1-41) are themselves equivalent.

An example will illustrate the question. Suppose that $I = O = \mathbb{R}$ (the real numbers) and that $\square$ is just real addition $(+)$. Further, adopt the common convention of suppressing the symbol $\circ$. For the moment, moreover, write a for **a**. Then the question concerns

$$a(a + a) \overset{?}{=} (a + a)a \overset{?}{=} aa + aa, \tag{1-42}$$

which appears very similar to basic properties of polynomial algebra, with the variable being a.

It is, of course, no accident whatsoever that polynomic calculations resemble the manipulations attached to elemental interconnections of numbers of identical atomic systems, and in fact the resemblance is much more intrinsic than is apparent at first glance.

Later purposes will be quite well served if the resemblance is pursued more adequately at this juncture. Just how close does connection algebra, whatever that is, resemble polynomial algebra, with its own meaning not yet made precise? Intuitively, quite a few points can be established, though of course the axiomatic underpinnings must be delayed to later chapters.

As is typical in algebra, it is necessary to tie things down a bit better, even before proceeding on an intuitive basis. Begin by examining a familiar polynomial, such as

$$2.7 + 3.1a - 7.7a^2. \tag{1-43}$$

It is easy enough to identify the as with atomic systems, but what about 2.7? In fact, the nature of these coefficients is a crucial constituent of the definition of the polynomials, even to the extent that the usual symbol for them is taken to be

$$\text{coefficients}[a]. \tag{1-44}$$

For example, if the coefficients are real numbers, then (1-44) can be written $\mathbb{R}[a]$; or, if the coefficients are integers, then (1-44) becomes $\mathbb{Z}[a]$. There are thus three aspects to the traditional symbol for polynomials: the coeff-icient symbol, the square bracket, and the variable symbol. Algebraically, the variable is usually called an *indeterminate*, and this is the term which will be used in the sequel. Pictorially, the notation has been sketched in Fig. 1.14.

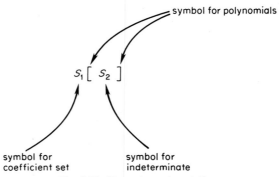

symbol for polynomials

$S_1 \begin{bmatrix} S_2 \end{bmatrix}$

symbol for coefficient set

symbol for indeterminate

Fig. 1.14. The polynomic notation.

Because, as mentioned in Section 1.2, it is possible to proceed to quite a distance in introductory systems topics without coming to grips with the idea of stating carefully what the sets are in an application, it is often easy to demean the importance of this notation. For example, is it not possible to deal just with polynomials of the type

$$a + a^2, \qquad a^2 + a^3, \tag{1-45}$$

and so forth? This seems to require little emphasis on the coefficients, which are then all the same. All right, but now add the polynomials in (1-45), in the usual manner

$$(a + a^2) + (a^2 + a^3) \tag{1-46}$$

to obtain the result

$$a + (1 + 1)a^2 + a^3. \tag{1-47}$$

There is now an issue to be resolved, namely, what is $(1 + 1)$? It is not un-common to approach this classical test of basic knowledge with, at least initially, an elegant disdain. One reader may advance the classical answer

$$1 + 1 = 2, \tag{1-48}$$

in which case he would create the sum polynomial

$$a + 2a^2 + a^3 \tag{1-49}$$

having coefficients which are not all the same. Such a decision might lead to the choice $\mathbb{Z}$ or $\mathbb{R}$ for the coefficient set, or even to $\mathbb{Z}_3$. Later chapters will show that such a choice has an important effect on the conceptual and numerical tools available to calculate in large-scale situations. In the hope of preserving the identical coefficients in (1-45), another reader might select

$$1 + 1 = 0. \tag{1-50}$$

Far from avoiding the coefficient issue, however, this choice suggests strongly that the coefficients are in $\mathbb{Z}_2$. Thus there is very little way to bypass the coefficient issue, and so it is best to face up to it at the very beginning.

Of what significance are these coefficients for interconnections of atomic systems **a**? Much more is here than meets the eye. Note, for example, that (1-42) already indicates the need for a unity coefficient of some sort, corresponding to the use of just one atomic element **a**. Let the typical symbol 1 stand for such a coefficient. Then (1-49) would in more tedious form be written

$$1a + 2a^2 + 1a^3. \tag{1-51}$$

This idea leads to a very important system theoretic side effect, which is suggested to some extent by (1-43); this is the idea of the straight-through connection, which is in the polynomial sense a consequence of the ordinary agreement

$$a^0 = 1 \tag{1-52}$$

and in the system sense a result of using the identity function

$$1 : I \to O, \tag{1-53}$$

which assigns to each element i of the input set I the same element i, but now regarded as a member of the output set O. Such an assignment is meaningful within the context of this section, in which I and O are the same sets.

It is certainly of interest to observe at this point that the simple connections of Fig. 1.13 have, under introductory study, suggested rather strongly that the box of Fig. 1.11 was packed incompletely at the factory, and really should logically contain some wire, which can be indicated in the manner of Figs. 1.15 and 1.16.

Next return to (1-43) for yet another glance, which should be focussed on the nature of the coefficients 2.7, 3.1, and 7.7. Clearly, these coefficients would require in a polynomial sense the selection of a suitably rich set

Fig. 1.15. The straight-through connection.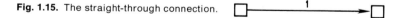

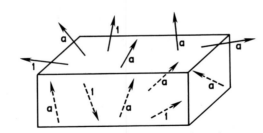

Fig. 1.16. Completed box of atomic systems.

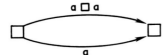

Fig. 1.17. Extended parallel connection.

S_1, such as the real numbers $\mathbb{R}$. But what do they imply in the system connection sense?

The reader may possibly believe that the use of nonintegral numbers such as 2.7 is a complication which might well have been avoided at this stage by remaining in a coefficient set such as $\mathbb{Z}$. But an elementary investigation of such a position shows that, although in principle feasible, this leads to a rather ponderous situation. What follows is prefigured by (1-49).

Consider the leftmost, or parallel, connection of atomic systems **a** in Fig. 1.12. It has already been indicated that this connection is just the same as one equivalent system of type (1-37). Now form the extended parallel connection of Fig. 1.17, in which a third atomic system has been added in parallel with the original two such systems. Straightforwardly, Fig. 1.17 is the same as an equivalent system

$$(\mathbf{a} \,\square\, \mathbf{a}) \,\square\, \mathbf{a}, \tag{1-54}$$

where it is reasonable in most cases to drop the parentheses in the manner

$$\mathbf{a} \,\square\, \mathbf{a} \,\square\, \mathbf{a} \tag{1-55}$$

inasmuch as all the atomic systems are identical.[†] Continuing in this manner to build up extended parallel connections leads inexorably to

$$\mathbf{a} \,\square\, \mathbf{a} \,\square\, \mathbf{a} \,\square\, \mathbf{a},$$
$$\mathbf{a} \,\square\, \mathbf{a} \,\square\, \mathbf{a} \,\square\, \mathbf{a} \,\square\, \mathbf{a}, \tag{1-56}$$
$$\vdots$$

[†] Technically, this is an assumption on the nature of the connection represented by $\square$, as will become clear at the beginning of the next chapter. Nonetheless, it is certainly the choice appropriate to analog systems and almost always appropriate to digital systems.

Fig. 1.18. Equivalent for multiple parallel connections.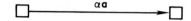

and the wisdom of writing the compact forms

$$2\mathbf{a},\quad 3\mathbf{a},\quad 4\mathbf{a},\quad 5\mathbf{a},\quad \ldots \tag{1-57}$$

for the expressions (1-37), (1-55), (1-56) becomes apparent. But (1-57) has implications of its own, which are familiar to students of the systems folklore. Write α for an arbitrary coefficient of $\mathbf{a}$ in the manner

$$\alpha\mathbf{a}. \tag{1-58}$$

The depiction of this entity (1-58) introduces interesting issues. Basically, of course, Fig. 1.18 could serve the purpose. The trouble with Fig. 1.18 is that it presents two aspects, the one immensely practical and the other decidedly the opposite. On the practical side, there are numerous real-life situations in which this is exactly how things have been done. For example, an electronics store does not stock all unit resistors $\mathbf{a}$; rather, it stocks a large number of different sizes of resistors, namely $\alpha\mathbf{a}$, $\beta\mathbf{a}$, $\gamma\mathbf{a}$, and so forth. The same is true of weights, tuning forks, and numerous other physical building blocks. The economic reasons for this should be apparent; they would lead to yet another modification of the basic box of elements, say in the manner of Fig. 1.19. On the other hand, it turns out that such an aspect is from an algebraic point of view very uneconomical. Preferable, from this aspect, is the use of the original box of Fig. 1.16 together with an additional box of coefficients, as in Fig. 1.20. Before the reader discards this approach as too

Fig. 1.19. Another box of atomic systems.

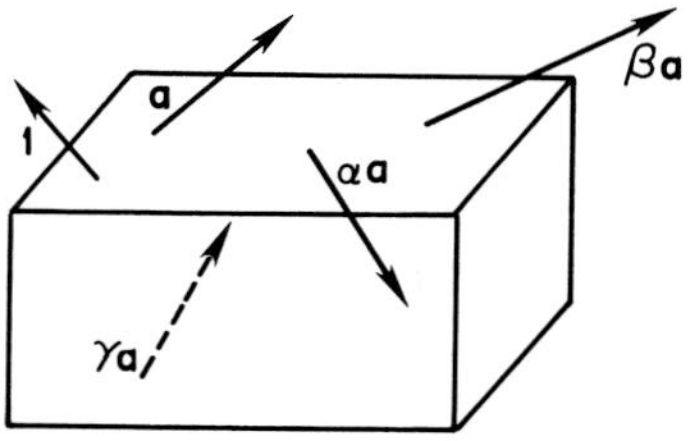

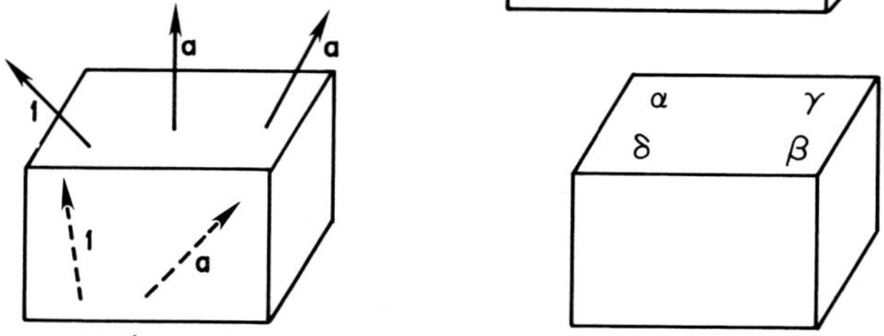

atomic systems

coefficients

Fig. 1.20. The two-box approach.

unrealistic, it is worthwhile to examine briefly the manner in which the contents of these two boxes are put together. Algebraically, such a melding process is used many times in the sequel, and so is appropriately introduced on an intuitive basis here.

In a somewhat lighter vein, the goal is indicated in Fig. 1.21. The pieces from which the desired result is to be attained are indicated again in Fig. 1.22.

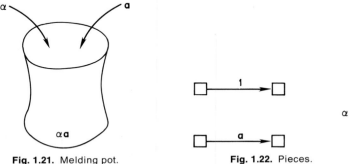

Fig. 1.21. Melding pot.   **Fig. 1.22.** Pieces.

What is the potential meaning of $\alpha\mathbf{a}$? Recall that $\mathbf{a}: I \to O$ is a function from an input set I to an output set O. In this section, I and O are, as previously agreed, identical sets. Indeed, at least in the context of this early chapter, there is little choice except to suggest that $\alpha\mathbf{a}$ should also be such a function from I to O. Thus, let it be agreed that

$$\alpha\mathbf{a}: I \to O \qquad (1\text{-}59)$$

is the issue at hand, with the question of how the boxes of Fig. 1.20 are to be melded reducing to the question of how the function (1-59) should be defined. Suppose α is, for illustrative purposes, 3; here S_1 could be taken as $\mathbb{Z}$. Moreover, write

$$\mathbf{a}(i) = o \qquad (1\text{-}60)$$

for an appropriate i and o in I and O, respectively. Then one very natural interpretation of (1-59) and (1-60) would be

$$3\mathbf{a}(i) = o \,\square\, o \,\square\, o. \qquad (1\text{-}61)$$

This is in essence the classical construction "three times o" and suggests strongly a sort of multiplicative structure. Such structure would work as in (1-61) with $S_1 = \mathbb{Z}$ and without requiring further refinement of the nature of I and O. Suppose, however, that $S_1 = \mathbb{R}$ and α is 2.7. Because the statement

$$2.7\mathbf{a}(i) = \underbrace{o \,\square\, o \,\square\, \cdots \,\square\, o}_{2.7 \text{ times}} \tag{1-62}$$

is rather unsatisfactory, it becomes necessary to say something further about the nature of I, O, and $\square$.

The obvious resolution of the situation is to be able to write

$$2.7 \text{ times } o \tag{1-63}$$

in an appropriate sense. This in turn suggests a two-box structure for the output set O, in a manner corresponding to Fig. 1.21. Such a structure could be indicated as in Fig. 1.23. There are a number of quite common types of output sets which could be suggested to meet such requirements. The most familiar is undoubtedly the real vector space. A related and very useful structure which will be used frequently in the later chapters is that of a module.

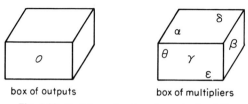

box of outputs   box of multipliers

Fig. 1.23. Two-box structure for output set O.

In a module, the multipliers α would not usually be taken from a set quite so rich algebraically as the real numbers $\mathbb{R}$. Suppose, instead, that α is an element of $\mathbb{R}[\mathbb{D}]$, the set of polynomials in an indeterminate $\mathbb{D}$ with coefficients which are real numbers. For O select the set of real-valued functions of a real variable having an arbitrary number of continuous derivatives. Such a set can be dubbed C^∞ and is a subset of $\mathbb{R}^{\mathbb{R}}$. Now, for

$$e^t \cos 3t \tag{1-64}$$

in C^∞, and for $5\mathbb{D} - 2$ in $\mathbb{R}[\mathbb{D}]$, the multiplication by $5\mathbb{D} - 2$ is indicated by

$$(5\mathbb{D} - 2)e^t \cos 3t, \tag{1-65}$$

which if $\mathbb{D}^i$ has the interpretation of ith ordinary derivative, leads to the result

$$3e^t \cos 3t - 15e^t \sin 3t. \tag{1-66}$$

This example illustrates the idea that ordinary differentiation is a candidate for an atomic system also, inasmuch as $\mathbb{D}$ can serve as an indeterminate.

Even more subtle, perhaps, is the fact that $\mathbb{D}$ has not been used here in the role of atomic system **a**, but rather in the role of a coefficient in the sense of the rightmost box of Fig. 1.20. Thus, it can be seen that one atomic system structure can be used as a set of coefficients for another atomic system interconnection. This layering effect is entirely within the spirit of Section 1.2 and is typical of the algebraic penchant for compounding structures.

It remains to indicate how the pieces of Fig. 1.22 are melded together. This amounts essentially to naming a convention for the meaning of Fig. 1.18. Probably the most widely used way of doing this is that sketched in Fig. 1.24, where advantage is taken of the fact that

$$\mathbf{1} \circ (\alpha \mathbf{a}) = (\alpha \mathbf{1}) \circ \mathbf{a} \qquad (1\text{-}67)$$

and the notation $\alpha \mathbf{1}$ is simplified to α. In an interconnection of this type, it is very important to recognize that when the symbol α stands alone, as it does in the last of the three pictures in Fig. 1.24, it stands not for itself but for $\alpha \mathbf{1}$.

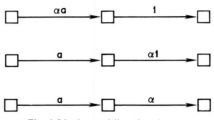

Fig. 1.24. Assembling the pieces.

In concluding this section, it should be remarked once again that the discussion of systems has again led back inexorably to the sets upon which they are defined. This case happened to involve the notion of interconnecting a large number of atomic systems **a** and relating the connection structure to that of the polynomials $S_1[S_2]$. A particularly striking feature of the close relationship between systems **a** and their input and output sets has been illustrated by addressing the meaning of $\alpha \mathbf{a}$. Here it has been found that the natural meaning of $\alpha \mathbf{a}$ is, as indicated in (1-63), intimately tied to the natural meaning of a corresponding calculation on the output set O.

Usually, discussion starts with the explanation of set structures and then proceeds to the description of function structures. This intuitive section proceeds inversely, in the belief that most readers encountering the material for the first time will be starting with a more extensive appreciation of the action of functions than of the sets upon which the functions are defined.

1.5 A GLIMPSE OF NEGATIVE FEEDBACK

The purpose of this section, which is of considerably greater brevity than its predecessor, is to introduce one or two of the implications of the minus sign which appears in (1-43).

With the background so far accumulated, the reader can quickly associate the -7.7 in that expression with a system, in the manner of Fig. 1.25, where the system, as explained above, is actually

$$-7.71. \tag{1-68}$$

From the earlier definition of the meaning of a function such as (1-68), it follows that

$$(-7.71)(i) = (-7.7)(\mathbf{1}(i)) = (-7.7)i \tag{1-69}$$

for each i in I. Now, the two-box structure imposed on O is also in place on I, as a result of the agreement that I and O are the same set for the current discussion. Thus the right member of (1-69) is easily enough understood as the multiplication of i by the real number -7.7.

Fig. 1.25. A negative system.

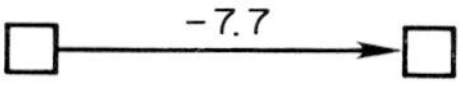

But there is a commonly held alternative meaning for the right member of (1-69), and that is,

$$-(7.7i) \tag{1-70}$$

as the negative of $7.7i$. The implications of this alternative viewpoint do, as might be suspected by now, run deeper than a surface inspection indicates. In fact, consideration of (1-70) can lead rapidly to a discovery of useful new structure for I. Within the context of (1-43), it is clear from Section 1.3 that $\square$ must be understood as a $+$, and so it is permissible to speak of

$$(7.7i) + (-7.7i), \tag{1-71}$$

which, by virtue of the interpretation (1-70), suggests the need for a zero element in I.

Such a zero is an example of what will be introduced later as a *unit e* for the binary operation $\square$, in the sense that

$$i\,\square\,e = i \tag{1-72}$$

for each i in I. The counterpart of

$$-(i), \tag{1-73}$$

in a manner analogous to (1-70), is denoted by $\hat{i}$ and defined by the equation

$$i \,\square\, \hat{i} = e. \tag{1-74}$$

A given set may fail to have a unit e. If it has e, it may be able to provide elements $\hat{i}$ for only a subset of elements i. In a general situation, then, there is no firm assurance that negatives make sense.

The careful reader has probably observed that (1-74) says

$$i + (-i) = \text{zero} \tag{1-75}$$

for the addition case, instead of

$$i - (i) = \text{zero}. \tag{1-76}$$

Algebraically, it is not common to define a symbol, say $\hat{\square}$, such that $i \,\hat{\square}\, i = e$. This means that there is a tendency to avoid the term subtraction in favor of the notion of adding negatives.

Various interpretations of $(-i)$ are possible. Suppose, for example, that the input set I has the two-box structure with αs taken from $\mathbb{Z}$. Then the interpretation

$$(-i) = (-1)i \tag{1-77}$$

is straightforward enough. A similar interpretation would work if the αs belong to $\mathbb{R}$. But consider the situation in which the αs come from $\mathbb{Z}_2$. Then the (not unreasonable) calculation

$$i + i = (1)i + (1)i = (1 + 1)i$$
$$= (0)i = \text{zero} \tag{1-78}$$

suggests that there is a certain amount of wisdom in using $\hat{i}$ in place of $(-i)$, inasmuch as the latter may not always be defined as a separate, nonzero entity. Yet another example will help to drive home this point. Let I be the real numbers $\mathbb{R}$ except for 0, which is deleted. More formally,

$$I = \{i \,|\, i \in \mathbb{R} \text{ and } i \neq 0\}; \tag{1-79}$$

a sketch is provided in Fig. 1.26. On the set I in (1-79), $\square$ can be chosen as real multiplication, denoted by $\cdot$, as in the calculation

$$2 \cdot 2 = 4. \tag{1-80}$$

It is clear enough that e may be taken as the real number 1, and

$$\hat{i} = i^{-1} \tag{1-81}$$

in the usual sense. In this case there is no idea of subtraction at all, but rather a different type of inverse which corresponds to division.

Nonetheless, when $\square$ is $+$, and when $\hat{i}$ is defined for each i, there is certainly a notion of negative, and a conception of subtracting junctions, as

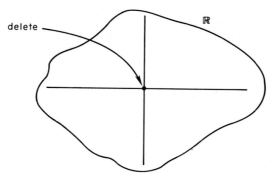

Fig. 1.26. The real numbers with zero deleted.

Fig. 1.27. A concept of subtracting junction.

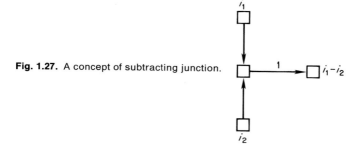

in Fig. 1.27, can be formed. There are, however, nontrivial difficulties with this apparently simple idea. Not the least of these is the way in which the result of junction action depends upon the order in which the inputs are processed. In more algebraic terms,

$$i_1 \square i_2 \neq i_2 \square i_1. \tag{1-82}$$

This is more than an inconvenience; it is a nuisance. For this reason, it is rather uncommon to encounter the idea of Fig. 1.27 in system theory. Instead, the more straightforward approach of Fig. 1.28 is in use. Note that according to the preceding discussion, the ± 1s appearing in Fig. 1.28 are to be understood as $(\pm 1)1$s. Sometimes, a more heuristic version of Fig. 1.28 is encountered, as in Fig. 1.29.

An important classical example of such a junction is the negative feedback scheme of Fig. 1.30. It is desired to have the output of a plant resemble a reference, and the key construction is a subtraction to form an error which excites the plant. Modern ways of looking at the feedback control problem are, of course, much more sophisticated. But they still use an error-forming scheme.

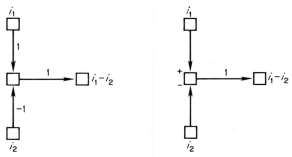

Fig. 1.28. A better idea.   **Fig. 1.29.** Heuristic version of Fig. 1.27.

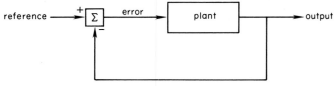

Fig. 1.30. Classical negative feedback scheme.

The junction shown in Fig. 1.30 is in the same spirit as Fig. 1.29. Note carefully that, although a subtraction is being performed, the Σ sign on the junction makes it a summing junction. While recent thought could reject Fig. 1.27 on the basis of (1-82), it is certainly reassuring to see that pioneers in system theory reached essentially the same conclusion, probably on the basis of a less axiomatic insight. In this treatment, the classical negative feedback scheme would be viewed as in Fig. 1.31.

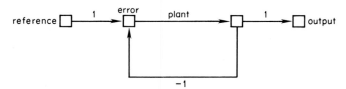

Fig. 1.31. Present view of Fig. 1.30.

This section shows yet another time how a simple system concept, in the present instance negative feedback, depends for its existence on the structure of the sets upon which systems act.

1.6 DISCUSSION

There can scarcely be a systems specialist who is unaware of the way in which his or her specialty makes use of functions. A glance at the keyboard of any recent electronic calculator is adequate confirmation of this assertion.

On the other hand, it appears to be generally true that there is less cognizance of the role in which sets relate to functions. For example, squaring or cubing often seems to make sense without any mention of *what* is squared or cubed. But $(2)^2$ is, in reality, an unknown quantity until this issue is settled. It could be 0, if 2 is in $\mathbb{Z}_4$; it could be 4, if 2 is in $\mathbb{R}$. Algebraic system theory begins with making these issues clear. To do so, it stresses *three* parts to a function: the set of elements upon which the function may act; the set of elements within which the results produced by the function reside; and the production rule or action of the function. All three of these should be understood in the symbolism $f : S \to T$. From the pictorial point of view, it is often very helpful to depict the function with f on the arrow between S and T, as in Fig. 1.2. Various misconceptions and half-truths can arise through a process of stressing less than all three parts, to the exclusion of the others.

The layered aspect of modern algebra can be foreseen through the simple process of following a few steps in the sequence: sets, functions, sets of functions, ... Not only does this algebraic feature suggest certain new types of activities in system theory, but it also makes vivid the necessity of being aware of the level on which any given activity is taking place. Surprising, perhaps, is the observation that a given numerical matrix, which seems concrete enough, can "live" on more than one level, which makes careful set designation all the more important.

In its simplest guise, system theory involves functions. On an introductory basis, a system can be regarded as a function. Once this is done, it turns out that the issues raised by interconnected systems are closely related to operations on the sets which are part of the system definition. This suggests immediately that complicated system interconnections require increased refinement of the sets involved, from an algebraic point of view. Such an insight can easily be overlooked by carelessly and needlessly over enriching these sets at the outset. To prevent this effect, the subsequent chapters proceed gradually, adding a bit of structure at each step. In fact, the following chapter places no structure whatever on the sets, so as to see what can be said on a ground-level basis.

The polynomial is certainly a workhorse of algebraic system theory. It can mean many things in many situations. One very basic system idea associated with a polynomial is that of an interconnection of identical, atomic systems. In fact, it is of course quite common to use such connections in systems studies, with integrators, delayors, and so forth being prime illustrations. The burgeoning electronic chip industry with thousands of transistors on small wafers is conceptually in the same spirit. A number of portentous observations can be drawn from a brief examination of atomic interconnections. Examples are the need for an identity function, the place of the scalar coefficient, and the role of units for binary operations on the sets.

The reader may advance whatever hypothesis seems most appropriate concerning which of these aspects is the least understood at introductory system levels. A good candidate for such speculation might be the scalar coefficient, the two-box structure, and the inferences which this logic imposes on the sets. Probably the most basic confusion which tends to occur is the confusion of systems and coefficients. This is why a careful understanding of Fig. 1.24 can be beneficial.

The two-box idea shows that different algebraic layers can be combined into multilayer structures. Like the idea of layers itself, such a possibility can rapidly stretch the imagination.

The purpose of this chapter has been to introduce, within the intuitive context, various representative notions of what algebraic system theory is all about. An effort has been made to draw upon familiar notions from the systems folklore to bring out these points. Wherever possible, definitions have been studiously avoided because they would have burdened the exposition. This, of course, has made it necessary to trek a very tenuous path of algebraic credibility.

In the following chapters, however, the path will be more adequately landscaped.

2 SET-DYNAMICAL SYSTEMS

It has been seen in Chapter 1 that a careful use of sets is essential to the understanding of functions and therefore to the understanding of systems, which are defined in terms of functions. Seen also was the fact that the richness of interconnection possibility was related to richness of the set structure upon which the functions are predicated. Very often, as discussed in Section 1.1, choice of set structure is a hastily chosen window dressing; and a not unusual result is set structure so rich that it conveys little insight. In terms of another metaphor, a size 48 suit may fit many wearers—in the sense that they can get inside—but it may reveal very little about the features of the wearer, once within.

This chapter strikes a blow to combat such tendencies. It does so by refusing to admit the usually assumed structures on sets. No binary operations are allowed. Such familiar activities as adding, multiplying, and so forth, are not permitted. In this way, it is hoped to set the stage for a more insightful later introduction of these activities. The reader may, in fact, be somewhat surprised at what can already be done in this chapter, from the systems point of view.

2.1 SETS, FUNCTIONS

A set is defined by giving its elements. Thus, there is the set of books on system theory, defined perhaps by the presence of the root "system" in the title, the set of authors of books on system theory, the set of readers of these books, and on and on. The symbolism $s \in S$ when s is an element of the set S, and $s \notin S$ when s is not, is in common use.

If every element of the set S is also an element of the set T, then S is a *subset* of T, denoted in the usual manner

$$S \subset T. \tag{2-1}$$

Two sets U and V are *equal*, denoted

$$U = V, \tag{2-2}$$

if

$$U \subset V \qquad \text{and} \qquad V \subset U. \tag{2-3a,b}$$

A set is empty if it has no elements. The empty set is denoted by $\varnothing$. $\varnothing$ is a subset of every set. A subset S of T is *proper* if it is not empty and it is not equal to T.

The *power set* $\mathbb{P}(T)$ of a set T is the set of all subsets of T, in the manner

$$\mathbb{P}(T) = \{S \mid S \subset T\}, \tag{2-4}$$

where (2-4) indicates the usual convention for defining a set in terms of a specified property.

It is useful here to establish the notation for a few basic sets. Some of these have already been introduced in Chapter 1: the real numbers $\mathbb{R}$, the integers $\mathbb{Z}$, and the integers from zero to $n - 1$, $\mathbb{Z}_n$. Others which are quite useful are the natural numbers

$$\mathbb{N} = \{0, 1, 2, \ldots\}, \tag{2-5}$$

the rational numbers $\mathbb{Q}$, and the complex numbers $\mathbb{C}$. It will later be seen that $\mathbb{Q}$ can be understood in terms of pairs of integers. For that matter, $\mathbb{C}$ can be understood in terms of pairs of real numbers. This suggests the very important notion of pairing, which is crucial to this chapter.

Let S and T be two sets. If $s \in S$ and $t \in T$, then there is a concept of *ordered pair*, denoted

$$(s, t), \tag{2-6}$$

of these elements. The concept derives its importance from the fact that, through (2-6), s and t can become related to each other. "A person is known by the company that he or she keeps." Consider, for example, a case in which s is 15.7 meters and t is -3.14 meters/second. If s and t happen to refer to the position and velocity of a particle in rectilinear motion, then their pairing says a great deal about the state of affairs with regard to that particle.

The set of such ordered pairs,

$$S \times T = \{(s, t) \mid s \in S \text{ and } t \in T\} \tag{2-7}$$

is the *Cartesian product* of S and T. In the sequel, the adjective Cartesian is typically suppressed. Relative to the earlier assertion about $\mathbb{Q}$, then, it can be shown that

$$\mathbb{Q} \subset \mathbb{Z} \times \mathbb{Z}, \tag{2-8}$$

in the sense that

$$\mathbb{Q} = \{(z_1, z_2) \,|\, z_1 \in \mathbb{Z}, z_2 \in \mathbb{Z}, \text{ and } z_2 \neq 0\}. \tag{2-9}$$

A similar statement establishes

$$\mathbb{C} \subset \mathbb{R} \times \mathbb{R}. \tag{2-10}$$

Once the product of two sets has been defined, it can be extended to more than that number. Thus,

$$S \times T \times U = (S \times T) \times U, \tag{2-11}$$
$$S \times T \times U \times V = (S \times T \times U) \times V, \tag{2-12}$$

and on in a recursive manner for higher fold products. Special interest attaches to the situation in which all the sets are the same. Thus the n-fold product, denoted

$$T \times T \times T \times \cdots \times T = T^n, \tag{2-13}$$

has already been used in Chapter 1, under the familiar guise $\mathbb{R}^2$.

For given sets S and T, a *function* assigns to each element s in S an element $f(s)$ in T. S is the *domain* of the function f, and T is its *codomain*. As already indicated in the first chapter,

$$f : S \rightarrow T \tag{2-14}$$

is a very useful notation for the function in prosaic situations, whereas

$$S \xrightarrow{f} T \tag{2-15}$$

is equally useful in diagrams. Neither (2-14) nor (2-15) defines, however, the *action* of a function, which must be indicated in an elementwise manner such as

$$f(s) = t. \tag{2-16}$$

Clearly, (2-16) is a form of pairing.

It should be carefully noted that changes in the domain or codomain change the function. Thus the absolute value function

$$f(s) = |s| \tag{2-17}$$

can be defined with $S = T = \mathbb{Z}$, or with $S = T = \mathbb{R}$; and each of these choices is understood to result in a different function. There are thus three aspects to a function: domain, codomain, and action. Two functions are *equal* if they are alike in all three aspects.

A certain amount of function terminology is unavoidable. The *image* of f, denoted

$$\text{Im } f = \{t \,|\, t = f(s), s \in S\}, \qquad (2\text{-}18)$$

is a subset of the codomain. f is *surjective*[†] if $\text{Im } f$ is T. The Greek prefix sur has the meaning onto. f is *injective* if

$$s_1 \neq s_2 \Rightarrow f(s_1) \neq f(s_2), \qquad (2\text{-}19)$$

where the arrow $\Rightarrow$ is read implies. Injective is the term for one-to-one. f is *bijective* if it is both injective and surjective, that is, if it is both one-to-one and onto.

The algebraic propensity to construct new levels of sets from old ones has been discussed to a considerable extent in Chapter 1. A key concept in this regard is the *function set* T^S, which is the set

$$T^S = \{f \,|\, f : S \rightarrow T\} \qquad (2\text{-}20)$$

of all functions having domain S and codomain T. The earlier intuitive relationship between $\mathbb{R}^2$ and $\mathbb{R}^{\mathbb{Z}_2}$ can now be expressed by observing that there is a bijection $b : \mathbb{R}^2 \rightarrow \mathbb{R}^{\mathbb{Z}_2}$ with the action

$$b(r_1, r_2) = f, \qquad (2\text{-}21)$$

where

$$f(0) = r_1, \qquad f(1) = r_2. \qquad (2\text{-}22a,b)$$

A very important function from the system theoretic point of view is the *identity function*

$$1_S : S \rightarrow S \qquad (2\text{-}23)$$

whose action is, as expected,

$$1_S(s) = s. \qquad (2\text{-}24)$$

Identities 1_S are available on every set S, regardless of what assumptions are in force on S. Do not confuse the identity with the *insertion*

$$i : S \rightarrow T \qquad (2\text{-}25)$$

of a subset $S \subset T$ into its parent set T, according to the action

$$i(s) = s, \qquad (2\text{-}26)$$

but with the right member of (2-26) now regarded as an element of T. The difference between identity and insertion is clear when it is remembered that a function has the three aspects of domain, codomain, and action; for

---

[†] Later, in other chapters, related words are used. For example, f can also be called *epic*. But in such situations, the sets have certain operational structures. The adjectives in this section apply when the only thing known is that S and T are sets.

then it is seen that the identity and the insertion do not have the same codomain.

Without making any assumption on T, U, and V, other than the fact that they are sets, there is in force the *composition function*

$$\circ: V^U \times U^T \to V^T, \tag{2-27}$$

whose action is traditionally represented by

$$(f \circ g)(t) = f(g(t)). \tag{2-28}$$

If T, U, and V are equal, this effectively provides an operation on V^T even though V and T have no similar feature. The utility of this feature for series connections has been illustrated in the previous chapter.

Inverse systems play a fundamental role in system theory. Thus, some remarks on inverse functions are in order. A function $f: S \to T$ with $S \neq \emptyset$ has *left inverses* $f^{-L}: T \to S$ satisfying

$$f^{-L} \circ f = 1_S, \tag{2-29}$$

if it is an injection, *right inverses* $f^{-R}: T \to S$ satisfying

$$f \circ f^{-R} = 1_T \tag{2-30}$$

if it is a surjection, and a unique, bijective, *two-sided inverse* $f^{-1}: T \to S$ satisfying

$$f^{-1} \circ f = 1_S, \qquad f \circ f^{-1} = 1_T \tag{2-31}$$

if it is a bijection.

Finally, if $f: S \to T$, it is customary to define functions $f_*: \mathbb{P}(S) \to \mathbb{P}(T)$ and $f^*: \mathbb{P}(T) \to \mathbb{P}(S)$ according to

$$f_*(U) = \{t \mid t = f(s), s \in U\}, \tag{2-32}$$
$$f^*(V) = \{s \mid f(s) = v, v \in V\}. \tag{2-33}$$

By widely accepted abuse of notation, $f_*(U)$ and $f^*(V)$ are often written $f(U)$ and $f^{-1}(V)$, respectively.

If a function $f: S \to T$ is defined, and if W is a subset of S, then it is possible to define another function $f: W \to T$ whose action on W is the same as that of f. This new function f is called the *restriction* of f to W, and is sometimes denoted $f \mid W$.

Exercises

2.1-1. In connection with later discussions of reachability in dynamical systems, it is important to understand how the functions f_* and f^* behave on nested subsets. Suppose that

$$S_0 \subset S_1 \subset S_2 \subset \cdots \subset S,$$

show that

$$f_*(S_0) \subset f_*(S_1) \subset f_*(S_2) \subset \cdots \subset T.$$

Analogously, if

$$T_0 \subset T_1 \subset T_2 \subset \cdots \subset T,$$

show that

$$f^*(T_0) \subset f^*(T_1) \subset f^*(T_2) \subset \cdots \subset S.$$

2.1-2. Let $f: S \to T$ and $g: T \to V$ be two bijections. Establish that their composite $g \circ f$ is also a bijection. Determine the inverse of the composite in terms of the inverses of f and g.

2.1-3. List all functions

$$\{a, b, c\} \to \{c, d\}.$$

Determine whether or not each function in your list is surjective or injective.

2.1-4. Weaken the assumption of Exercise 2.1-2 above so that f and g are merely injections. Is the composite an injection? Can you determine a left inverse of the composite in terms of left inverses of f and g?

2.1-5. Show that there is a bijection

$$b: (S \times T) \times U \to S \times (T \times U).$$

2.2 LOCAL DYNAMICAL FUNCTIONS

The intuitions of Chapter 1 include the idea of a system being defined by a function which acts upon a set I of inputs and produces results within a set O of outputs.

Such an input–output view of systems is indeed one of the most basic and appealing. Hidden within it, however, is a fascinating subtlety. It is not hard to locate this fine point, moreover, because the only technical term that has been employed is that of a function.

Of course, the key notion in the definition of a function $f: S \to T$ is its assignment to each element s in S of exactly one element $f(s)$ in T. This means that the input–output view of a system, as described in Chapter 1, would require each input in I to produce a *unique* output in O. This turns out to be a very restrictive way to look at general systems, a way which is really not flexible enough to encompass the accepted applications.

What is needed is a sort of parameterization of the input–output function. To see this, consider the following example.

EXAMPLE 2.2-1

A student in a course on psychokinesis is given the task of manipulating a set of levers so as to cause certain colored lights to be activated. The situation is sketched in Fig. 2.1. The student can be regarded as having access

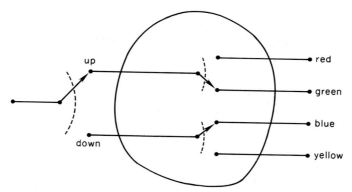

Fig. 2.1. Psychokinesis experiment.

to an input set

$$I = \{\text{up}, \text{down}\} \tag{2-34}$$

for the system, which in turn has an output set

$$O = \{\text{red}, \text{green}, \text{blue}, \text{yellow}\}. \tag{2-35}$$

The goal is to be able to produce a specified output upon request, through an appropriate choice of input. Unknown to the student, however, radio-active decay of an atomic element is being used to position randomly a pair of internal switches. This means that the student is not facing a problem of selecting elements in the domain of a function. For a given input, say up in I, the system does not produce a unique output in O. In fact, it may produce either red or green according to the position of the internal switches. Accordingly, the system rule $I \rightarrow O$ does not qualify as a function.   ∎

This example indicates that the sets I and O might be well augmented by an additional set S containing such knowledge as may be required to properly establish a function on I. For Example 2.2-1, a reasonable choice might be

$$S = V^2 \tag{2-36}$$

for

$$V = \{\text{up}, \text{down}\}. \tag{2-37}$$

In (2-36), the fact that $V^2 = V \times V$ is a set of ordered pairs permits an easy accounting of the pair of internal switches. The elements in the set S have come to be called the *states* of the system.

As will already be apparent, the creation of a state set S provides a new mechanism

$$S \times I \rightarrow O \tag{2-38}$$

for producing the system output. Moreover, provision is traditionally made for tracking any changes in the state set which may be dependent upon the input selected. This leads to another mechanism

$$S \times I \to S. \qquad (2\text{-}39)$$

The domain and codomain structure made explicit in (2-38) and (2-39) may be used to specify pairs of functions. Such a pair of functions is typical of those which can be used to establish axiomatic descriptions of dynamical systems.

Such a pair can be called the *local dynamical functions* of a system. More will be said about the use of these functions in Section 2.4.

A word about notation is in order. The notation I for sets of inputs has a pleasing sort of connotation for intuitive introductions. Eventually, however, it begins to clash with indices in standard summations, so that it is best to make a modification now. Though S and O could in principle be retained, the change in I suggests a complete change to another collection of set designations. The following are in common use in dynamical system theory.

Let U be a set of *inputs*, X be a set of *states*, and Y be a set of *outputs*. Passing to the next level of construction, establish functions

$$f: X \times U \to X, \qquad (2\text{-}40)$$
$$g: X \times U \to Y. \qquad (2\text{-}41)$$

The function f is called the *local transition function*, and the function g is called the *local output function*. A local transition function and a local output function comprise a set of local dynamical functions for a system.

Operation of the system is understood as follows. Suppose that the system is in state x, and that input u is applied. Then the system produces an output $g(x, u)$ in Y according to the function g and executes a transition to state $f(x, u)$ in X according to the function f.

The reader has certainly already noted that the choice (2-40) and (2-41) for local dynamical functions does not easily embody Example 2.2-1 because of randomness, which is not of primary interest in the sequel. It is therefore appropriate to close with a second example.

EXAMPLE 2.2-2

Widgets, Inc., is a national firm which manufactures three colors of widgets: red, white, and blue. Each color widget comes in two sizes: very large and extra large. Because widgets are hermetically sealed in packages of a dozen each, orders must be placed in multiples of 12. As a retailer, your mission, if you accept, is to set up an ordering system for widgets.

One way to do this is as follows. Define sets

$$V_1 = \{\text{red}, \text{white}, \text{blue}\}, \tag{2-42}$$
$$V_2 = \{\text{very large}, \text{extra large}\}, \tag{2-43}$$
$$U = V_1 \times V_2, \tag{2-44}$$
$$X = (V_1 \times V_2 \times \mathbb{N})^6, \tag{2-45}$$
$$V_3 = \{\text{order}, \text{do not order}\}, \tag{2-46}$$
$$Y = (V_1 \times V_2 \times V_3)^6. \tag{2-47}$$

The local output function $g: X \times U \rightarrow Y$ is then arranged so that, if a customer makes a selection u in U given by (blue, very large), then y in Y is (blue, very large, order) if the count of the item in X is 12 or less; otherwise y is (blue, very large, do not order). Similarly the local transition function $f: X \times U \rightarrow X$ is arranged to decrement the count of that item by one. ∎

The use of (2-45) for a state set X in the example is somewhat of an overkill, in the sense that $(\mathbb{Z}_{24})^6$ might have been used in its place.

Also, it is quite pertinent to examine the total implication of the statement, "decrement the count of that item by one." One way to think of this is in terms of subtraction. Another way is to place the standard ordering on $\mathbb{N}$. This latter idea leads into the material of the next section.

Exercises

2.2-1.   A certain systems expert rejects the definitions (2-40) and (2-41) for the local dynamical functions. Instead the expert chooses to define local transition and local output functions in the manner

$$\tilde{f}: U \times X \rightarrow X, \qquad \tilde{g}: U \times X \rightarrow Y.$$

Resolve any ambiguity by exhibiting a bijection

$$b: X \times U \rightarrow U \times X.$$

Express f and g in terms of $\tilde{f}$ and $\tilde{g}$.

2.2-2.   Let

$$X = \{0, 1, 2\}, \qquad U = \{a, b, c\}, \qquad Y = \{\triangle, \square\};$$

and define a local output function g. Show that $g^*(\triangle)$ and $g^*(\square)$ have no elements in common. Moreover, establish that $g^*(\triangle)$ and $g^*(\square)$ together comprise all elements in $X \times U$. $g^*(\triangle)$ and $g^*(\square)$ are known as a *partition* of $X \times U$.

2.2-3.   Discuss the possibility of defining insertion functions

$$U \rightarrow X \times U \qquad \text{and} \qquad X \rightarrow X \times U.$$

2.2-4.   Reconsider Example 2.2-2 from the viewpoint of replacing (2-45) with $(\mathbb{Z}_{24})^6$.

2.3 RELATIONS

This section introduces several types of relations, which will become the workhorse concepts later in this chapter.

The basic idea is that of a *binary relation B* on *S* to *T*, which is defined to be a subset

$$B \subset S \times T. \tag{2-48}$$

Examples of binary relations abound. Suppose that *S* and *T* are chosen to be $\mathbb{P}(X)$ and $\mathbb{P}(Y)$, respectively, for arbitrary sets *X* and *Y*. Then a binary relation *B* on $\mathbb{P}(X)$ to $\mathbb{P}(Y)$ can be specified by demanding some requisite property of pairs

$$(U, V) \in \mathbb{P}(X) \times \mathbb{P}(Y), \tag{2-49}$$

as for instance $U \subset V$ or $U \cap V = \varnothing$. And in $\mathbb{R} \times \mathbb{R}$, the well-known symbols $<, \leq, >, \geq$, and $=$ are binary relations.

Most basic of all relations is perhaps the binary relation of equality. It is available on *any* set to itself. Given such a set *S*, and a pair (s_1, s_2) of elements from the product $S \times S$ of *S* with itself, one way to see this is to argue that (s_1, s_2) is an element of the binary relation of equality on *S* to *S* if

$$\{s_1\} = \{s_2\}, \tag{2-50}$$

where $\{s_i\}$, $i = 1, 2$, is the singleton subset of *S* containing just the element s_i.

Now is a convenient time to point out one of the notational features associated with binary relations. If $(s, t) \in B$, then it is common to write $s \, B \, t$. Thus, if the binary relation of equality is denoted by $=$, then

$$s_1 = s_2 \tag{2-51}$$

is written in place of $(s_1, s_2) \in \, =$.

Moreover, if the set *T* is the same as the set *S*, as in the case of $=$, then the "binary relation *B* on *S* to *S*" is often replaced by the "binary relation *B* on *S*." This convention permits such convenient usages as "relations on *S*" in the following pages.

Certainly the idea of equality is so fundamental a relation, and is so available, that it is difficult to think in terms which do not include it.

Some properties of binary relations have become so frequently used that they receive special nomenclature.

A binary relation *B* on *S* is said to be *reflexive* if

$$(s, s) \in B \tag{2-52}$$

for all *s* in *S*; *symmetric* if

$$(s_1, s_2) \in B \Rightarrow (s_2, s_1) \in B \tag{2-53a}$$

for all (s_1, s_2) in B; *antisymmetric* if

$$(s_1, s_2) \in B \qquad \text{and} \qquad (s_2, s_1) \in B \qquad\qquad \text{(2-53b,c)}$$

imply

$$s_1 = s_2; \qquad\qquad \text{(2-53d)}$$

and *transitive* if

$$(s_1, s_2) \in B \qquad \text{and} \qquad (s_2, s_3) \in B \qquad\qquad \text{(2-54a,b)}$$

imply

$$(s_1, s_3) \in B \qquad\qquad \text{(2-54c)}$$

for all (s_1, s_2) and (s_2, s_3) in B.

A binary relation B on a set S is a *partial order relation* if it is reflexive, antisymmetric, and transitive. Partial order relations are frequently denoted by the symbol $\le$. In addition to these three properties, a fourth property is sometimes specified: given s_1 and s_2 in S, then either

$$s_1 \le s_2 \qquad \text{or} \qquad s_2 \le s_1. \qquad\qquad \text{(2-55a,b)}$$

When (2-55) is also satisfied, the relation is usually termed a *simple order relation*.

The next relation is in a certain sense the generalization of the binary relation of equality. Instead of using the idea that two elements are identical, it proceeds from the notion that two elements are the same, "at least insofar as anyone cares." A crucial notion, it relates to the real-world fact that everyday objects often possess a greater richness of description than the user would wish to utilize, or can be regarded from a great many more aspects than the user has any need to examine.

A binary relation B on a set S is an *equivalence relation* if it is reflexive, symmetric, and transitive. Equivalence relations are frequently denoted by $\equiv$, and will probably be the most frequently used concept in this volume.

EXAMPLE 2.3-1

An important way to obtain an equivalence relation is through the use of a function. Let $f : S \to T$ be a function, and consider Fig. 2.2. If f assigns the same value t to more than one value of s, then these values are somehow linked together in the process. This notion can be formalized as an equivalence relation on S by defining

$$s_1 \equiv s_2 \qquad \text{if} \qquad f(s_1) = f(s_2). \qquad\qquad \text{(2-56a,b)}$$

The facts that f is a function and that equality is an equivalence relation then establish that (2-56) defines an equivalence relation. For simplicity,

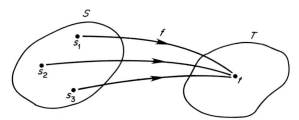

Fig. 2.2. A function assigning *t* to more than one *s*.

(2-56) can be replaced by

$$s_1 \equiv s_2(f). \tag{2-57}$$

Later, it will be shown that every equivalence relation can be understood as arising in this way, for some function f. ∎

EXAMPLE 2.3-2

    The notion of binary relation is the formalization of the intuitive statements about pairing in Section 2.1. In applications, binary relations arise through tabulating data, as in Fig. 2.3, or in drawing graphs, as in Fig. 2.4. Tabulations are common enough to rate special keys on machines with very limited space, such as the ubiquitous typewriter. Few typists, however, would be aware that pressing the tab key creates a relation! ∎

Student	Grade
Jones	B
Schaefer	A
Smith	C
Cosenta	A
Geddi	A

Fig. 2.3. A typical tabulation.

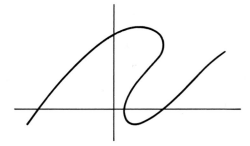

Fig. 2.4. A typical graph.

EXAMPLE 2.3-3

    Graphs have a special interest here, because of the way in which they relate to functions. Suppose $B \subset \mathbb{R} \times \mathbb{R}$. Both the graphs of Fig. 2.4 and Fig. 2.5 could define B, but only the binary relation of Fig. 2.5 could qualify as the graph of a function. To see this, assume that the horizontal axis represents the domain of the function and that the vertical axis represents

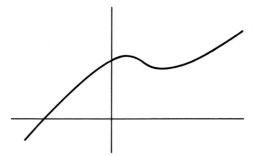

Fig. 2.5. Graph of a function.

the codomain. Then the fact that some domain elements in Fig. 2.4 correspond to more than one codomain element prevents function interpretation. More generally, a binary relation B on S to T defines a function if, whenever $(s_1, t_1) \in B$ and $(s_2, t_2) \in B$, the equality of s_1 and s_2 implies that of t_1 and t_2. ∎

EXAMPLE 2.3-4

Define a binary relation on $\mathbb{R}$ in the manner

$$B = \{(2, 2), (2, 3), (3, 4), (2, 4)\}. \tag{2-58}$$

B is transitive, but fails to be reflexive or symmetric. ∎

In the next section, the use of relations begins.

Exercises

2.3-1. Following the guidelines of Example 2.3-1, use the tabulation of Fig. 2.3 to establish an equivalence relation on the set of students shown.

2.3-2. Consider $\mathbb{R}$, the set of real numbers. Define a binary relation B on $\mathbb{R}$ to $\mathbb{R}$ by

$$(r_1, r_2) \in B \qquad \text{if} \qquad r_1 - r_2 = r\pi$$

for $r \in \mathbb{R}$. Study B for the properties of reflexivity, symmetry, and transitivity.

2.3-3. Let $\mathbb{P}(S)$ be the power set of a set S, and define a binary relation B on $\mathbb{P}(S)$ to itself by means of

$$(S_1, S_2) \in B \qquad \text{when} \qquad S_1 \subset S_2.$$

Show that B is a partial order relation. Is B a simple order relation?

2.3-4. In Example 2.3-1, it is asserted that the facts that f is a function and that equality is an equivalence relation establish (2-56) as an equivalence relation. Complete the missing steps in this argument.

2.3-5. Suppose that B_1 and B_2 are two binary relations on the set S to itself. If B_1 and B_2 are reflexive, what can be said of

$$B_1 \cap B_2?$$

Can your conclusions be extended to the case in which B_1 and B_2 are symmetric?

2.3-6. Repeat Exercise 2.3-5, but with focus on

$$B_1 \cup B_2.$$

2.4 DEFINITION OF A SET-DYNAMICAL SYSTEM

Intuitively, one could characterize a set-dynamical system as a dynamical system which is defined on sets. If the reader is already equipped with some preconceived notion of what a dynamical system means, then this approach seems plausible enough on the surface. Actually, however, the statement says very little. Every notion of dynamical system involves functions, and every function involves sets. Thus the intuitive euphemism above really amounts only to saying that "a system is a system is a system is a . . . ."

What distinguishes the term set-dynamical system, as used in the present context, from arbitrary dynamical systems is a convenant to admit only binary relations on the sets involved. This excludes, for example, such every-day constructions as addition and multiplication, in their various guises. In effect, then, a set-dynamical system is a dynamical system stripped of just about all of its algebraic structure. An illustration of the meaning of such a statement can be built upon Section 1.3. The binary operation $\square$ is part of the most elementary algebraic structures on sets, but it is excluded in this chapter because it is not a binary relation. Thus, adjectives such as linear or nonlinear have no meaning here, inasmuch as there is no addition operation upon which they could be predicated.

The intuitive discussion of the first chapter began treating systems in an input–output manner—in Section 1.3. Later in that chapter, however, Section 1.4 prefigured a different way of looking at systems through the medium of interconnecting numbers of identical systems. Prime examples would be integrators or delayors. It is this second, atomic way of looking at systems which is typically associated with the term dynamical. It is also the description which tends to occur when system equations are written down on the basis of scientific principle. In Section 2.2, the foundation for defining dynamical systems was laid by bringing forward the notion of local dynamical functions: the local transition function

$$f : X \times U \to X \qquad\qquad (2\text{-}59a)$$

and the local output function

$$g : X \times U \to Y. \tag{2-59b}$$

Operation of the system was to be understood as follows. Suppose that the system is in state $x \in X$ and that an input $u \in U$ is applied. Then the system produces an output $g(x, u)$ in Y according to the function g and executes a transition to state $f(x, u)$ according to the function f. Quite often, such activities are portrayed with a diagram such as the one indicated in Fig. 2.6. Here, within a given circle, the state is shown to the left above the slant line. The input is shown on an arrow leading to another circle, and the output resulting from a given state and input is shown to the right below the slant line.

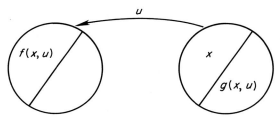

Fig. 2.6. Basic operation of system (2-59).

Though single transitions, such as that indicated in Fig. 2.6, give at least a ground-level feeling for dynamical action, the total dynamical scope is best perceived by means of a repetitive use of the local dynamical functions.

To bring in this aspect of system behavior, what is needed is a way to *relate* various usages of the local functions. To accomplish this, introduce a fourth set J, and equip it with a simple order relation denoted by $\leq$. Many readers will tend to think of J as a set of times, while others may prefer to regard it as a set of places. Either interpretation can be useful. Note that $\mathbb{N}$, $\mathbb{Z}$, and $\mathbb{R}$ could be candidates for the set J, whereas $\mathbb{C}$ could not. Note, moreover, that a simple order relation $\leq$ on $\mathbb{N}$ is not the same as a simple order relation $\leq$ on $\mathbb{Z}$ or $\mathbb{R}$, because in the first instance $\leq$ is a subset of $\mathbb{N}^2$ whereas in the second it is a subset of $\mathbb{Z}^2$ or $\mathbb{R}^2$. J will be called the *index set* of the dynamical system. In accordance with the convention mentioned in the section preceding, the cumbersome $(j_1, j_2) \in \leq$ will be replaced by $j_1 \leq j_2$.

Now consider functions

$$\alpha : \mathbb{N} \to J. \tag{2-60}$$

Such a function α in $J^{\mathbb{N}}$ is said to be *admissible* if it is injective and *order preserving*, which means that

$$n_1 \le n_2 \Rightarrow \alpha(n_1) \le \alpha(n_2), \tag{2-61}$$

where $\le$ on $\mathbb{N}$ is the usual natural simple order relation. An admissible function α in $J^{\mathbb{N}}$ will be called an *index function*.

Index functions permit the establishment of a notion of successive operations of the system.

Reconsider, then, the idea of Fig. 2.6, and associate the state x in X with an index $j(n)$ in Im α for some index function α in $J^{\mathbb{N}}$. Similarly, associate the input u in U with that same index. Then the output $g(x, u)$ can be associated with $j(n)$ as well, and the transition to state $f(x, u)$ can be visualized as leading to the next state, with next being quantified by the index function α in the manner $j(n + 1)$. These associations can be pictured conveniently in the tabular style of Fig. 2.7. Then the notations $x(j(n))$ and $u(j(n))$ are natural; and a local transition equation can be written in the manner

$$x(j(n + 1)) = f(x(j(n)), u(j(n))), \tag{2-62a}$$

as well as a local output equation

$$y(j(n)) = g(x(j(n)), u(j(n))), \tag{2-62b}$$

for $n \in \mathbb{N}$.

$j(n)$	x	u	$g(x, u)$
$j(n+1)$	$f(x, u)$	.	.

Fig. 2.7. Tabular view of dynamical operation.

The classical streamlining of (2-62) can be a useful introductory algebraic exercise. A focus on the left member of (2-62a) indicates the composition of two functions, one being $\alpha \in J^{\mathbb{N}}$ and the other being an element of X^J. In terms of functions, then, the situation in the left member has the potentiality

$$X^J \times J^{\mathbb{N}} \to X^{\mathbb{N}}, \tag{2-63}$$

which is usually accomplished by the replacement of $x(j(n + 1))$ with $x(n + 1)$. In the right member of (2-62a), the same type of simplifications can be achieved by writing $x(n)$ and $u(n)$ in place of $x(j(n))$ and $u(j(n))$. It is clear, therefore, that (2-62) can be understood also in the manner

$$x(n + 1) = f(x(n), u(n)), \qquad y(n) = g(x(n), u(n)). \tag{2-64a,b}$$

However, in the sequel, it is normally the custom to use symbols other than n for the domain $\mathbb{N}$ of an index function. Thus, as in Section 2.2, it is convenient to sideslip to the more eventually satisfactory form

$$x_{k+1} = f(x_k, u_k), \qquad y_k = g(x_k, u_k) \qquad \text{(2-65a,b)}$$

for $k \in \mathbb{N}$. Note especially the notational convenant which replaces $x(k)$ by x_k, primarily for visual acuity.

Figure 2.7 can then be replaced by the more revealing Fig. 2.8. Further, the preliminary Fig. 2.6 gives way to the more interesting Fig. 2.9.

0	x_0	u_0	y_0
1	x_1	u_1	y_1
2	x_2	u_2	y_2
3	x_3	u_3	y_3
⋮	⋮	⋮	⋮

Fig. 2.8. Extension of the tabular view.

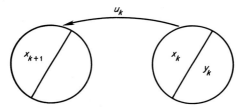

Fig. 2.9. Extended version of Fig. 2.6.

In summary, then, a *set-dynamical system* (SDS) consists of a number of sets, functions, and relations. Typical of the sets are the

(1) index set J,
(2) input set U,
(3) state set X,
(4) output set Y.

Typical of the functions are the local dynamical functions, which are

(5) the local transition function

$$f : X \times U \to X,$$

(6) and the local output function

$$g : X \times U \to Y.$$

Representative of the relations is

(7) the simple ordering $\leq$ on J.

The adjectives typical and representative have been used so as to indicate that added functions and sets are occasionally useful, and that variations can be made on the present definitions. As an example of the latter, consider the possibility of a local output function $g: X \rightarrow Y$ which is not dependent upon the input.

It is convenient to have a sort of shorthand notation for reference to the sets, functions, and relations which describe an SDS. This is very often accomplished by writing

$$(J, \leq; U, X, Y; f, g)$$

or some variant of it. In most problems, the index set J and its relation $\leq$ can be suppressed without loss, as for example

$$(U, X, Y; f, g).$$

On the other hand, it may well happen that one of the other sets has a relation on it, in which case it is possible to write

$$(U; X, B; Y; f, g),$$

in this case denoting a relation B on X.

Exercises

2.4-1. A function

$$\alpha: \mathbb{N} \rightarrow \mathbb{N}$$

is defined by

$$\alpha(n) = n^2 + 5.$$

Is α an index function? Suppose instead that the action of α is specified by

$$\alpha(n) = n^3 - 3n^2 + n + 7.$$

Do your conclusions change?

2.4-2. Consider the possibility of generalizing the notion used in defining an index function. Let B_1 be a binary relation on S and B_2 be a binary relation on T. Further, let $\alpha: S \rightarrow T$ be a function. Suggest a version of (2-61) to fit this case.

2.4-3. Two set-dynamical systems are given, namely,

$$(U_i, X_i, Y_i; f_i, g_i), \qquad i = 1, 2.$$

Arrange things so that

$$Y_1 = U_2,$$

so that

$$u_2 = g_1(x_1, u_1)$$

is sensible. This is a natural notion of series connection for SDSs. Is such a series connection another SDS?

   2.4-4. Sometimes the local output equation (2-65b) can be solved for

$$u_k = w(y_k, x_k)$$

as a function $Y \times X \rightarrow U$. When this can be done, (2-65a) can be expressed in terms of x and y instead of x and u. What is the meaning of such a substitution in (2-65a)?

2.5 ORDER REDUCTION

Practical considerations in dynamical systems often lead to methods of order reduction. To be explicit, consider the local transition equation

$$x_{k+1} = f(x_k, u_k). \tag{2-66}$$

An elementary notion of order can be obtained in the following way. Let

$$X = S^p, \tag{2-67}$$

for S some basic set of state variables. A natural notion of order is available in the context (2-67) by specifying the order of X to be the natural number p of repeated factors in the products of S with itself.

   Order reduction can then be approached by writing

$$p = n + m, \tag{2-68}$$

and defining

$$X_1 = S^n \quad \text{and} \quad X_2 = S^m. \tag{2-69a,b}$$

It is now straightforward to establish a bijection

$$b: X \rightarrow X_1 \times X_2 \tag{2-70}$$

according to the action

$$b(s_1, \ldots, s_n, s_{n+1}, \ldots, s_{n+m}) \rightarrow ((s_1, \ldots, s_n), (s_{n+1}, \ldots, s_{n+m})). \tag{2-71}$$

The bijection b induces another bijection

$$\tilde{b}: X \times U \rightarrow X_1 \times X_2 \times U \tag{2-72}$$

in the obvious way

$$\tilde{b}(x, u) = (b(x), u). \tag{2-73a}$$

Now (2-73a) can be rewritten with the aid of the identity function 1_U on U to itself, so as to give

$$\tilde{b}(x, u) = (b(x), 1_U(u)). \tag{2-73b}$$

Structures of the type (2-73b) occur frequently enough to receive their own appellation.

Specifically, suppose that two functions

$$h_i: S_i \to T_i, \qquad i = 1,2, \tag{2-74}$$

are given. Then a new function can be defined by

$$h_1 \times h_2: S_1 \times S_2 \to T_1 \times T_2 \tag{2-75}$$

with the action

$$(h_1 \times h_2)(s_1, s_2) = (h_1(s_1), h_2(s_2)). \tag{2-76}$$

The entity $h_1 \times h_2$ is called a (*Cartesian*) *product function*.

An interesting diagram is associated with product functions. As a prelude to considering it, however, examine Fig. 2.10, which introduces functions

$$p_S: S \times T \to S \qquad \text{and} \qquad p_T: S \times T \to T \tag{2-77a,b}$$

whose actions are given by

$$p_S(s, t) = s \qquad \text{and} \qquad p_T(s, t) = t. \tag{2-78a,b}$$

These functions are called *projections*. Next consider Fig. 2.11. This is an example of what is known as a *commutative diagram*, many of which occur in the sequel. The idea in a commutative diagram is that the same result is obtained by following parallel or alternative paths through the diagram from a fixed initial or originating set to a fixed final or ending set. In Fig. 2.11, the only originating set which can initiate alternative paths to the same ending set is $S_1 \times S_2$. Two ending sets are possible: T_1 and T_2. Each ending set can be reached along two alternate paths. Thus, in this example, com-

$$S \xleftarrow{\quad p_S \quad} S \times T \xrightarrow{\quad p_T \quad} T$$

Fig. 2.10. Product projections.

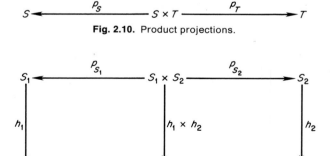

Fig. 2.11. Product function diagram.

mutation in the diagram amounts to commutation of the "left window" framed by the two paths to T_1 and of the "right window" framed by the two paths to T_2. To verify that the diagram does indeed commute, select an arbitrary element (s_1, s_2) in $S_1 \times S_2$. Application of p_{S_1} to this element yields s_1; and then h_1 leads to $h_1(s_1)$. Alternatively, application of $h_1 \times h_2$ to (s_1, s_2) gives $(h_1(s_1), h_2(s_2))$; and then p_{T_1} provides $h_1(s_1)$. Thus $h_1(s_1)$ is obtained by either path. A similar argument works on the right window.

Now return to (2-73b). In light of the discussion on product functions, it is clear that the bijection $\tilde{b}$ is given by

$$\tilde{b} = b \times 1_U. \tag{2-79}$$

If $b^{-1}: X_1 \times X_2 \to X$ is the inverse of b, then it follows immediately that

$$\tilde{b}^{-1} = b^{-1} \times 1_U. \tag{2-80}$$

A basic step in one very common order reduction procedure is to use $\tilde{b}$ to redefine (2-66) on $X_1 \times X_2 \times U$. To do this, apply the bijection b to both members of (2-66) and obtain

$$b(x_{k+1}) = (b \circ f)(x_k, u_k). \tag{2-81}$$

Note that

$$b(x_{k+1}) = (x_1(k+1), x_2(k+1)), \tag{2-82}$$

where a temporary reversal of last section's subscript notation is here helpful for clarity. To complete the transformation of the local transition equation, observe that

$$\tilde{b}^{-1} \circ \tilde{b} = 1_{X \times U}, \tag{2-83}$$

so that (2-81) can become

$$b(x_{k+1}) = (b \circ f \circ \tilde{b}^{-1})(\tilde{b}(x_k, u_k)). \tag{2-84}$$

Define

$$\tilde{f}: X_1 \times X_2 \times U \to X_1 \times X_2 \tag{2-85}$$

by

$$\tilde{f} = b \circ f \circ \tilde{b}^{-1}, \tag{2-86}$$

and note that

$$\tilde{b}(x_k, u_k) = (x_1(k), x_2(k), u(k)). \tag{2-87}$$

Then (2-84) becomes

$$(x_1(k+1), x_2(k+1)) = \tilde{f}(x_1(k), x_2(k), u(k)), \tag{2-88}$$

which completes the separation of the state x into two parts.

Moving on, then, to fulfilling the idea of order reduction, a common procedure is to assume that x_2, say, is not changing very much, so that

$$(p_{X_2} \circ \tilde{f})(x_1(k), x_2(k), u(k)) = x_2(k), \tag{2-89}$$

an equation which undergoes remarkable conceptual improvement when k is suppressed, namely,

$$(p_{X_2} \circ \tilde{f})(x_1, x_2, u) = x_2. \tag{2-90}$$

Now suppose that it is possible to solve (2-90) for x_2 in terms of x_1 and u in the manner

$$x_2 = h(x_1, u), \tag{2-91}$$

for $h: X_1 \times U \to X_2$. Then (2-88) can be reduced to an equation on X_1, thereby reducing the order of the problem from p to n, by writing

$$x_1(k + 1) = (p_{X_1} \circ \tilde{f})(x_1(k), h(x_1(k), u(k)), u(k)). \tag{2-92}$$

Finally, let

$$\tilde{h}: X_1 \times U \to X_1 \times X_2 \times U \tag{2-93}$$

be defined by

$$\tilde{h}(x_1, u) = (x_1, h(x_1, u), u); \tag{2-94}$$

then the reduced local transition equation is

$$x_1(k + 1) = (p_{X_1} \circ \tilde{f} \circ \tilde{h})(x_1(k), u(k)). \tag{2-95}$$

Note that x_2 can be recovered from (2-91).

The local output equation can be addressed in a similar way. Begin with

$$
\begin{align}
y_k &= g(x_k, u_k) \tag{2-96a} \\
&= (g \circ \tilde{b}^{-1})(\tilde{b}(x_k, u_k)) \tag{2-96b} \\
&= \tilde{g}(x_1(k), x_2(k), u(k)), \tag{2-96c}
\end{align}
$$

where $\tilde{g}: X_1 \times X_2 \times U \to Y$ has been defined as

$$\tilde{g} = g \circ \tilde{b}^{-1}. \tag{2-97}$$

Though there is no need to regress from y_k to $y(k)$ in (2-96), it does add notational symmetry to the equations. Proceed, then, to obtain

$$
\begin{align}
y(k) &= \tilde{g}(x_1(k), h(x_1(k), u(k)), u(k)) \tag{2-98a} \\
&= (\tilde{g} \circ \tilde{h})(x_1(k), u(k)). \tag{2-98b}
\end{align}
$$

This converts the original local output equation, but it may be desirable to add x_2 as an additional output. To do so, define

$$\tilde{\tilde{g}} : X_1 \times U \rightarrow Y \times X_2 \tag{2-99}$$

by the action

$$\tilde{\tilde{g}}(x_1, u) = ((\tilde{g} \circ \tilde{h})(x_1, u), h(x_1, u)). \tag{2-100}$$

In the notation, then, of the section preceding, the original SDS was represented by

$$(J, \leq ; U, X, Y; f, g), \tag{2-101a}$$

whereas the reduced SDS is represented by

$$(J, \leq ; U, X_1, Y \times X_2 ; p_{X_1} \circ \tilde{f} \circ \tilde{h}, \tilde{\tilde{g}}). \tag{2-101b}$$

In practice the variables in X_2 are often those which change very rapidly with k, while those in X_1 are more dominant. The key, of course, is whether h can be defined.

A goal of this section has been to show a nontrivial example of the algebraic treatment of a practical issue associated with set-dynamical systems. The pace has purposely been set so as to be as detailed as possible.

Figure 2.12 is a commutative diagram which is helpful in visualizing some of the constructions which have been carried out. Interpretation of the figure is aided by the observation that an arrow bearing a bijection, such as b or $\tilde{b}$, can be reversed if the functions are replaced by their inverses.

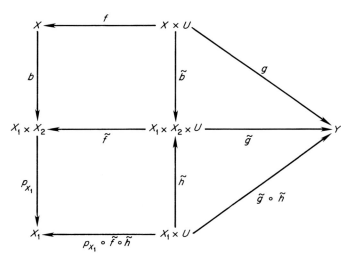

Fig. 2.12. Order reduction diagram.

One very important consequence of the discussion in this section is that, even if the SDS originally had a local output function $g: X \to Y$ which did not depend on inputs explicitly, the reduced SDS will probably exhibit explicit input dependence through $\tilde{\tilde{g}}: X_1 \times U \to Y \times X_2$. In fact, in many applications, such reductions can explain the original presence of u in $g(x, u)$.

Exercises

2.5-1. Show that the action (2-71) makes (2-70) a bijection.

2.5-2. Let S and T be sets. Show that

$$1_S \times 1_T = 1_{S \times T}.$$

Notice that to establish an equality of functions requires showing that they have the same domains, codomains, and actions.

2.5-3. Recall Example 2.2-2, and suppose that red, extra large widgets are not moving very well. Explain in detail how Widgets, Inc., should determine whether the methods of this section can be applied to streamline the operations of the corporation.

2.6 THE KEY TRIANGLE

An equivalence relation on a set S is a binary relation which is reflexive (r), symmetric (s), and transitive (t). When a binary relation B on S satisfies these r-s-t properties, it is often convenient to use the symbol E in place of B.

An equivalence relation E on a set S breaks S up into subsets in a natural way. These subsets are just those whose elements are equivalent under the relation E. The idea can be formalized. Begin by defining the *equivalence class* $\bar{s}$ of an element s in S as the set

$$\bar{s} = \{\tilde{s} \mid \tilde{s} \in S \text{ and } \tilde{s} E s\}. \tag{2-102}$$

In view of the fact that E is reflexive, it follows that $s \in \bar{s}$, so each element s in S is always a member of at least one equivalence class. Can it be a member of two distinct equivalence classes? Suppose so, then

$$s \in \bar{s}_1, \qquad s \in \bar{s}_2, \qquad \bar{s}_1 \neq \bar{s}_2. \tag{2-103}$$

By definition (2-102) of $\bar{s}$, it follows that

$$s E s_1, \qquad s E s_2 \tag{2-104}$$

and thus, by symmetry and transitivity, that

$$s_1 E s_2, \tag{2-105}$$

which is at odds with the assumption $\bar{s}_1 \neq \bar{s}_2$ of (2-103). Thus the contradiction establishes that each s in S can be a member of but one equivalence class. In total, the arguments using r-s-t properties provide that each s in S is a member of one and only one equivalence class.

It is therefore possible to define a function

$$\tilde{p}_E : S \rightarrow \mathbb{P}(S) \tag{2-106}$$

whose action is given by

$$\tilde{p}_E(s) = \bar{s}. \tag{2-107}$$

Im $\tilde{p}_E$ is called the *quotient set* of S by E and is denoted by S/E. It is then traditional to make a revision of (2-106) in such a way that the function becomes surjective. To achieve this, just change the codomain from $\mathbb{P}(S)$ to S/E, and supply a new symbol

$$p_E : S \rightarrow S/E \tag{2-108}$$

for the function, which is now called the *projection* of S onto S/E.

EXAMPLE 2.6-1

For the set S, select the subset

$$\{1, 2, 3\} \tag{2-109}$$

of integers $\mathbb{Z}$. To establish an equivalence relation E on (2-109), it is necessary to begin with reflexivity, which demands that

$$(1,1), \quad (2,2), \quad (3,3) \tag{2-110}$$

be elements of E. While this would be adequate to establish an equivalence relation, namely, that of equality, on (2-109), it is not very interesting. Therefore, add the pairs

$$(2,3), \quad (3,2) \tag{2-111}$$

to the relation. Then

$$E = \{(1,1), (2,2), (3,3), (2,3), (3,2)\}, \tag{2-112}$$

and the function p_E has action

$$p_E(1) = \{1\}, \qquad p_E(2) = \{2,3\}, \qquad p_E(3) = \{2,3\}. \tag{2-113a,b,c}$$

Moreover,

$$S/E = \{\{1\}, \{2,3\}\}, \tag{2-114}$$

so that there are two equivalence classes. A pictorial version of the relation (2-112) is seen in Fig. 2.13. Note that, in contexts of this type, reflexivity can be "seen" in terms of points along a line bisecting the angle between abscissa and ordinate axes. And symmetry can be expressed relative to the same line. ▮

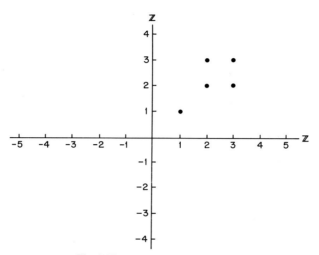

Fig. 2.13. An equivalence relation.

A brief word on the notation S/E may be appropriate at this point. Some readers may find it difficult to support the use of a division-type symbol in order to denote subsets. Actually, however, this difficulty has a workable solution. Simply regard the construction $p_E: S \rightarrow S/E$ as one of dividing S into subsets. This intuitive ploy has been helpful in the past, and is offered here in the same spirit. It is important to develop a sound intuitive sense for the notation S/E, which is pivotal in much of what follows.

Now recall from Section 2.3 that one way to create equivalence relations on S is with the aid of a function $f: S \rightarrow T$ having domain S. Then the procedure, described in Example 2.3-1, was to arrange

$$s_1 E s_2 \qquad \text{if} \qquad f(s_1) = f(s_2), \qquad (2\text{-}115\text{a,b})$$

the last equality occurring in T. Equations (2-115) are often replaced by the single statement

$$s_1 E s_2(f). \qquad (2\text{-}116)$$

It is possible to use the projection $p_E:S \rightarrow S/E$ associated with an *arbitrary* equivalence relation in much the same way. Simply select S/E for T, and note that

$$s_1\, E\, s_2(p_E). \tag{2-117}$$

This means that *every* equivalence relation E on S can be understood as arising from at least one function $f:S \rightarrow T$, namely, $p_E:S \rightarrow S/E$.

   The connection of equivalence relations with functions leads to the possibility of numerous notational proliferations. Already in Section 2.3, the symbol $\equiv$ has been used in place of E. In many cases, it is useful to be able to attach the symbol f to E in some more direct way. One way to do this is to regard E as an element in the image of a function which can assign an equivalence relation to f. For obvious reasons, it is convenient to use the symbol E for this function. Thus

$$E:T^S \rightarrow \mathbb{P}(S^2) \tag{2-118}$$

is a possibility, and $E(f)$ can be written for the equivalence relation created by f.

   Possibly the chief fact concerning equivalence relations is that they can be used to simplify the description of functions, provided that the functions to be considered are compatible with the equivalence relation. These thoughts can be put on a firm foundation as follows.

   Consider Fig. 2.14. The presence of the dashed line asks the question, "Does there exist a function $\bar{f}:S/E \rightarrow T$ so that the diagram commutes?" To understand the question, observe that $E(f)$ and E may not be the same relation. If they were, of course, $\bar{f}$ could be chosen as an appropriate injection.

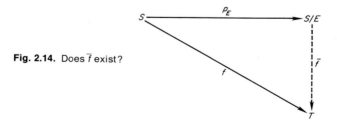

Fig. 2.14. Does $\bar{f}$ exist?

   The general answer to the question is negative. To see this, suppose that $\bar{f}:S/E \rightarrow T$ does exist. Then commutativity of the diagram requires

$$f = \bar{f} \circ p_E. \tag{2-119}$$

Now p_E assigns the same value, namely $\bar{s}$, to every element which is equivalent to s. Therefore the composition $\bar{f} \circ p_E$ must do likewise. If $\bar{f}$ exists, then, it is necessary that

$$s_1 \, E \, s_2 \Rightarrow f(s_1) = f(s_2). \tag{2-120}$$

Inference (2-120) can be taken as a precise statement of the earlier, more intuitive one that f is compatible with E. Actually, (2-120) is sufficient as well for the construction of $\bar{f}$. For let s be any element in $\bar{s}$; simply define

$$\bar{f}(\bar{s}) = f(s). \tag{2-121}$$

From (2-120) and (2-121), $\bar{f}(\bar{s})$ can change its value only if s changes its equivalence class. So $\bar{f}$ is well defined. There is a bonus, for this is the only way to define $\bar{f}$.

Accordingly, with reference to Fig. 2.14, there exists a function $\bar{f} : S/E \to T$ which makes the diagram commute when and only when f is compatible with E in the sense (2-120). Moreover, if $\bar{f}$ exists, it is unique. The situation is summarized in Fig. 2.15.

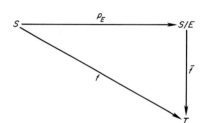

Fig. 2.15. Yes, if $p_E(s_1) = p_E(s_2) \Rightarrow f(s_1) = f(s_2)$.

EXAMPLE 2.6-2

An important application of this idea can be studied in Fig. 2.16. Given is a function $\alpha : S \to S$, and an equivalence relation E on S. Intuitively, p_E

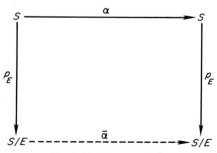

Fig. 2.16. Does $\bar{\alpha}$ exist?

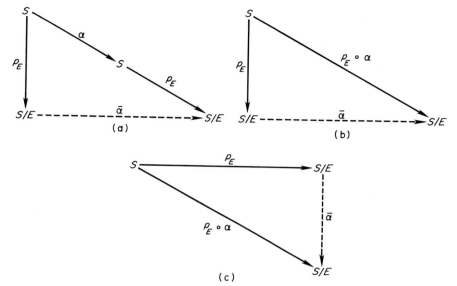

Fig. 2.17. Converting Fig. 2.16 to Fig. 2.15.

coarsens the set S, and the diagram inquires concerning whether α might be able to follow along. Technically, of course, the inquiry is about the existence of a function $\bar{\alpha}: S/E \to S/E$ which would make the diagram commute. This question can be answered by referring it back to the key triangle diagram of Fig. 2.15. The procedure for doing this is sketched in Fig. 2.17. In (a), the upper and right sides of the square have been coalesced into one side of a triangle; in (b) this new triangle side has been expressed in terms of one composite function $p_E \circ \alpha$; finally, in (c), the triangle has been rotated in such a way as to resemble Fig. 2.15. A comparison of Figs. 2.15 and 2.17c now shows that the function $\bar{\alpha}$ will exist if and only if

$$s_1 \, E \, s_2 \Rightarrow (p_E \circ \alpha)(s_1) = (p_E \circ \alpha)(s_2). \qquad (2\text{-}122)$$

Moreover, when (2-122) is satisfied, the function $\bar{\alpha}$ is unique. Now the right member of (2-122) will hold when and only when

$$\alpha(s_1) \, E \, \alpha(s_2), \qquad (2\text{-}123)$$

as a result of the nature of the projection p_E. Thus α takes equivalent elements into equivalent elements. In summary, relative to Fig. 2.16, $\bar{\alpha}: S/E \to S/E$ exists so as to make the diagram commute if and only if

$$s_1 \, E \, s_2 \Rightarrow \alpha(s_1) \, E \, \alpha(s_2). \qquad (2\text{-}124)$$

The situation is summarized in Fig. 2.18.   ∎

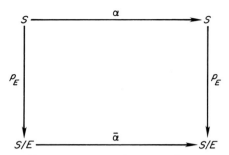

Fig. 2.18. Yes, if $s_1 \, E \, s_2 \Rightarrow \alpha(s_1) \, E \, \alpha(s_2)$.

Example 2.6-2 illustrates the basic idea that interesting developments can occur when functions mesh together well with relations. The idea is fundamental, and will appear often as the discussion progresses.

It is useful to have intuitive ways of thinking and speaking about equivalence relations. One such approach is to regard the establishment of an equivalence relation E on a set S as reducing the resolution of the set. In the present day of increasing video interface, this frequently experienced specification can be an especially useful analogy.

Exercises

2.6-1. On the subset

$$\{1, 2, 3, 4\}$$

of the integers $\mathbb{Z}$, determine all possible equivalence relations E. For each of these relations, specify the action of the projection p_E.

2.6-2. For the equivalence relation described in Example 2.6-1, define

$$T = \mathbb{Z},$$

and establish a function $f : S \to T$ which satisfies (2-120). For the function which you have chosen, determine the action of

$$\bar{f} : S/E \to T.$$

2.6-3. Let E be an equivalence relation on the set S. Further, let

$$\alpha_i : S \to S, \qquad i = 1, 2,$$

be two functions with the property

$$s_1 \, E \, s_2 \Rightarrow \alpha_i(s_1) \, E \, \alpha_i(s_2), \qquad i = 1, 2.$$

Now consider the composition

$$\alpha_2 \circ \alpha_1 : S \to S.$$

ᴊoes $s_1 E s_2$ imply

$$(\alpha_2 \circ \alpha_1)(s_1) E (\alpha_2 \circ \alpha_1)(s_2)?$$

If so, how is $\overline{\alpha_2 \circ \alpha_1}$ related to $\overline{\alpha}_2$ and $\overline{\alpha}_1$? Comment upon the possibility of applying these notions to the series connection of systems.

2.6-4.   There are many instances in the application of system theory which use complicated interconnections of the same system. One example is the integrator. With this in mind, reconsider Exercise 2.6-3, but with $\alpha_2 = \alpha_1 = \alpha$. Develop a counterpart for Fig. 2.18 when α is replaced by α^i for $i \in \mathbb{N}$.

2.6-5.   Much emphasis in algebraic system theory is placed upon concepts such as reachability and observability. These ideas are broached later in the volume. There is a fundamental counterpart of those notions which can be brought forward here. Indeed, let

$$f : S \to T$$

be any function. Then

$$f = g \circ h,$$

where

$$g : S/E(f) \to T$$

is an injection and

$$h : S \to S/E(f)$$

is a surjection. Establish this result.

2.7   SIGNAL SET EQUIVALENCES

In Section 2.4, a set-dynamical system (SDS) is denoted by

$$(J, \le ; U, X, Y; f, g). \tag{2-125}$$

Here J is the index set, equipped with a simple order relation $\le$; U, X, and Y are the input, state, and output sets, respectively; f is the local transition function; and g is the local output function.

Just as the simple order relation $\le$ is a very useful addition to the index set J, so other relations on the input, state, or output sets are possible additions to the basic SDS. Refer to the inputs, states, and outputs as *signals*, and to the corresponding sets U, X, and Y as *signal sets*.

Of special interest are equivalence relations on the product set $X \times U$, which serves as the domain for both local dynamical functions f and g.

One way to bring about such a relation is to begin by establishing relations on the individual sets X and U, respectively. Denote these in the usual manner

$$p_{E_u} : U \to U/E_u, \tag{2-126}$$

$$p_{E_x} : X \to X/E_x. \tag{2-127}$$

Then it is natural to propose the definition of an equivalence relation E on $X \times U$ by means of the statement

$$(x_1, u_1) E (x_2, u_2) \qquad \text{(2-128a)}$$

if *both*

$$x_1 E_x x_2 \qquad and \qquad u_1 E_u u_2. \qquad \text{(2-128b,c)}$$

It is a worthwhile demonstration to establish that (2-128) is indeed an equivalence relation.

Recall first that E_x and E_u must have the reflexive property in their own right. As a consequence of this,

$$x E_x x \qquad \text{(2-129a)}$$

for all x in X and

$$u E_u u \qquad \text{(2-129b)}$$

for all u in U. But (2-129) implies

$$(x, u) E (x, u) \qquad \text{(2-130)}$$

for all pairs (x, u) in $X \times U$. So E is reflexive.

Turn now to symmetry. Suppose that

$$x_1 E_x x_2 \qquad and \qquad u_1 E_u u_2. \qquad \text{(2-131a,b)}$$

Then the symmetry property of E_x and of E_u provides that

$$x_2 E_x x_1 \qquad and \qquad u_2 E_u u_1. \qquad \text{(2-132a,b)}$$

But

$$(x_1, u_1) E (x_2, u_2) \qquad \text{(2-133)}$$

would, by definition of E, imply (2-131); and (2-131) implies (2-132). Thus (2-133) must, by (2-132) and definition of E, imply

$$(x_2, u_2) E (x_1, u_1), \qquad \text{(2-134)}$$

which yields symmetry of E.

As a last step, consider transitivity. Suppose that

$$(x_1, u_1) E (x_2, u_2) \qquad and \qquad (x_2, u_2) E (x_3, u_3). \qquad \text{(2-135a,b)}$$

Now (2-135a) implies

$$x_1 E_x x_2, \qquad u_1 E_u u_2; \qquad \text{(2-136a)}$$

and (2-135b) implies

$$x_2 E_x x_3, \qquad u_2 E_u u_3. \qquad \text{(2-136b)}$$

The transitive properties of E_x and E_u, together with (2-136), then give immediately that

$$x_1 \, E_x \, x_3, \qquad u_1 \, E_u \, u_3, \tag{2-137}$$

which by definition of E leads to the desired result

$$(x_1, u_1) \, E \, (x_3, u_3), \tag{2-138}$$

as a consequence of which E is also transitive.

The proposed relation E on $X \times U$, defined by (2-128), therefore satisfies the r-s-t properties and is an equivalence relation on its own merit.

It can be very helpful to picture this construction. Begin with a product function diagram, along the lines of Fig. 2.11, but adapted here to the sets of present interest (see Fig. 2.19). Next note that E as an entity in itself induces its own quotient set $(X \times U)/E$ and projection

$$p_E: X \times U \to (X \times U)/E \tag{2-139}$$

without explicit reference to the manner in which it was constructed. Clearly, it would seem that $(X \times U)/E$ and $X/E_x \times U/E_u$ must be essentially the same set. To see this, consider Fig. 2.20. If

$$p_E(x_1, u_1) = p_E(x_2, u_2), \tag{2-140a}$$

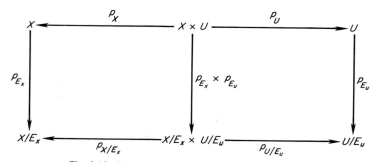

Fig. 2.19. Product function diagram for E on $X \times U$.

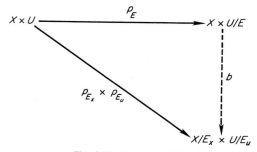

Fig. 2.20. Does b exist?

then by definition of E it must follow that

$$p_{E_x}(x_1) = p_{E_x}(x_2) \qquad \text{and} \qquad p_{E_u}(u_1) = p_{E_u}(u_2). \qquad \text{(2-140b,c)}$$

Therefore

$$(p_{E_x} \times p_{E_u})(x_1, u_1) = (p_{E_x}(x_1), p_{E_u}(u_1)) = (p_{E_x}(x_2), p_{E_u}(u_2))$$
$$= (p_{E_x} \times p_{E_u})(x_2, u_2), \qquad \text{(2-141)}$$

and it follows that b exists and is unique. From the fact that projections onto quotient sets, such as (2-126) and (2-127), are surjective, it is true that b is surjective as well, else the commutative diagram statement

$$p_{E_x} \times p_{E_u} = b \circ p_E \qquad \text{(2-142)}$$

would not be possible. Moreover, as a result of the fact that each x in X is a member of exactly one equivalence class in X/E_x, any pair

$$(x_1, x_2) \notin E_x \qquad \text{(2-143a)}$$

must have the property

$$p_{E_x}(x_1) \neq p_{E_x}(x_2). \qquad \text{(2-143b)}$$

A similar statement can be made relative to pairs

$$(u_1, u_2) \notin E_u. \qquad \text{(2-144)}$$

It is then true that

$$((x_1, u_1), (x_2, u_2)) \notin E \qquad \text{(2-145a)}$$

has to have the implication

$$(p_{E_x} \times p_{E_u})(x_1, u_1) \neq (p_{E_x} \times p_{E_u})(x_2, u_2). \qquad \text{(2-145b)}$$

From (2-145), the conclusion is that b cannot assign the same value to two different equivalence classes in $(X \times U)/E$ without violating (2-142). And so b is injective as well, thereby becoming a bijection.

In writing the basic function description, say

$$b : (X \times U)/E \rightarrow X/E_x \times U/E_u \qquad \text{(2-146)}$$

for a bijection, it is reasonably common practice to replace the arrow $\rightarrow$ by a symbol which connotes bijection on its own authority. This is often convenient in pages to follow, and the adopted notation is

$$b : (X \times U)/E \approx X/E_x \times U/E_u. \qquad \text{(2-147)}$$

If the particular form of the bijection is of no great importance, then (2-147) can be shortened to

$$(X \times U)/E \approx X/E_x \times U/E_u. \qquad \text{(2-148)}$$

This has the intuitive meaning that $(X \times U)/E$ and $X/E_x \times U/E_u$ are essentially the same sets.

It has thus been demonstrated that specifying signal equivalences E_x and E_u on the state and input sets leads naturally to an equivalence E on their product.

EXAMPLE 2.7-1

Select for X the subset $\{1, 2, 3\}$ of $\mathbb{Z}$ and for U the subset $\{1, 2\}$. Then the set $X \times U$ is comprised of

$$
\begin{array}{ll}
(1, 1), & (1, 2), \\
(2, 1), & (2, 2), \\
(3, 1), & (3, 2).
\end{array}
\qquad (2\text{-}149)
$$

For E_x, define

$$
(1, 1), \quad (2, 2), \quad (3, 3), \quad (1, 2), \quad (2, 1), \qquad (2\text{-}150\text{a})
$$

and for E_u define

$$
(1, 1), \quad (2, 2), \quad (1, 2), \quad (2, 1). \qquad (2\text{-}150\text{b})
$$

Then the resulting equivalence relation E on $X \times U$ contains

$$
\begin{array}{ll}
((1, 1), (1, 1)), & ((1, 2), (1, 2)), \\
((2, 1), (2, 1)), & ((2, 2), (2, 2)), \\
((3, 1), (3, 1)), & ((3, 2), (3, 2)), \\
((1, 1), (1, 2)), & ((1, 2), (1, 1)), \\
((1, 1), (2, 1)), & ((2, 1), (1, 1)), \\
((1, 1), (2, 2)), & ((2, 2), (1, 1)), \\
((1, 2), (2, 1)), & ((2, 1), (1, 2)), \\
((1, 2), (2, 2)), & ((2, 2), (1, 2)), \\
((2, 1), (2, 2)), & ((2, 2), (2, 1)), \\
((3, 1), (3, 2)), & ((3, 2), (3, 1)).
\end{array}
\qquad (2\text{-}151)
$$

∎

It is tempting to try to establish that the scheme (2-128) is really the only essential way to set up an E on $X \times U$. At first glance, moreover, there appears to be a natural way to bring such an argument. If E is an equivalence relation on $X \times U$, and if

$$
(x_1, u_1)\, E\, (x_2, u_2), \qquad (2\text{-}152)
$$

then it *seems* appropriate to let (2-152) infer that

$$
x_1\, E_x\, x_2 \qquad \text{and} \qquad u_1\, E_u\, u_2. \qquad (2\text{-}153\text{a,b})
$$

It is easy enough to show that the reflexive and symmetric properties of E will lead to the same properties for E_x and E_u. But the transitivity argument breaks down. Perhaps the point is best made in an example.

EXAMPLE 2.7-2

Reconsider the situation in the preceding example, but define E on $X \times U$ as follows:

$$
\begin{array}{ll}
((1, 1), (1, 1)), & ((1, 2), (1, 2)), \\
((2, 1), (2, 1)), & ((2, 2), (2, 2)), \\
((3, 1), (3, 1)), & ((3, 2), (3, 2)), \\
((1, 2), (2, 1)), & ((2, 1), (1, 2)).
\end{array}
\tag{2-154}
$$

Any attempt to apply the scheme $(2\text{-}152) \Rightarrow (2\text{-}153)$ will fail, because it will imply $(2\text{-}150)$ without generating even half of the relation $(2\text{-}151)$. ∎

This example shows that it is possible to arrange less resolution on $X \times U$ than that which would in general be set by two given equivalence relations on X and U.

In the next section, attention will be focused upon some of the dynamical constraints associated with making equivalence relations on signal sets.

Exercises

2.7-1. Show that the relations E_x and E_u proposed in $(2\text{-}153)$ are reflexive and symmetric if E in $(2\text{-}152)$ is reflexive and symmetric.

2.7-2. In Example 2.7-1, consider the binary relation defined in $(2\text{-}151)$. Now delete the last two elements in the relation, namely

$$((3, 1), (3, 2)), \quad ((3, 2), (3, 1)).$$

What properties are still true for the remaining relation?

2.7-3. In Example 2.7-2, suppose that the entries

$$((3, 1), (3, 1)), \quad ((3, 2), (3, 2))$$

were removed from $(2\text{-}154)$. What properties are still true for the remaining relation?

2.8 LOCAL DYNAMICAL EQUIVALENCES

Basically, there are four possibilities in setting up signal equivalences: on the input set U; on the state set X; on the domain $X \times U$ of the local dynamical functions; and on the output set Y. Just as in the SDS notation

(2-125), where it was useful to pair the index J with its simple order relation $\leq$ in the manner

$$(J, \leq), \tag{2-155a}$$

so it is convenient to designate the presence of various equivalence relations on the signal sets by

$$(U, E_u), \qquad (X, E_x), \qquad (X \times U, E), \qquad (Y, E_y). \quad \text{(2-155b,c,d,e)}$$

Note that four basic possibilities have the capability to generate other effects through various combinations of themselves.

Section 2.6 made the point that equivalence relations and functions may or may not go well together. In the case of an SDS, the placement of signal set equivalences must take due cognizance of the local dynamical functions f and g.

EXAMPLE 2.8

For a beginning, consider the effect of placing an equivalence relation E_u on the input set U, while making no relational changes on the other SDS sets. The effect of such a choice on the local transition function f and upon the local output function g will be largely a question of the effect of E_u upon their domain set $X \times U$. The previous section presented an approach to dealing with this issue, provided that an equivalence relation E_x on X had also been defined. Though no E_x has been given, it has been observed in Section 2.3 that the binary relation of equality, denoted $=$, is available on any set. Now, $=$ is certainly an equivalence relation; and so it is always possible to designate E_x to be $=$, without having any essential effect on the original question. Once this has been accomplished, an equivalence relation E on $X \times U$ is induced by E_x and E_u. The situation for a local transition function f is depicted, then, in Fig. 2.21. The function f may or may not be compatible with E. If it is, then a unique induced $\bar{f}$ will make the diagram

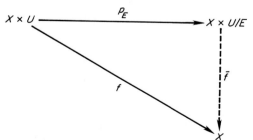

Fig. 2.21. Local transition and E.

commute. If it is not, there is no algebraic hope for properly joining the effects of E and f. The desired condition between f and E has already been brought out in Section 2.6, namely,

$$(x_1, u_1) E(x_2, u_2) \Rightarrow f(x_1, u_1) = f(x_2, u_2). \qquad (2\text{-}156)$$

For the present situation, the left member of the implication (2-156) means that

$$x_1 = x_2 \quad \text{and} \quad u_1 E_u u_2. \qquad (2\text{-}157a,b)$$

This means that (2-156) can be restated as

$$u_1 E_u u_2 \Rightarrow f(x, u_1) = f(x, u_2) \qquad (2\text{-}158)$$

for every x in X. It is not hard to exhibit such cases. Say $X = U = \mathbb{R}$, and

$$f(x, u) = x \sin u. \qquad (2\text{-}159)$$

Then an obvious E_u can be set up in terms of intervals of length 2π on the real line $\mathbb{R}$. ∎

Note that the discussion would pass along similar lines if an equivalence relation E_x had been placed instead of, or in addition to, E_u or if E had been placed directly on $X \times U$ while bypassing E_x and E_u, as in Example 2.7-2. The principle remains the same, and the fundamental implication is (2-156). Moreover, the discussion for g, which has the same domain as f, is unavoidably in the same spirit.

If the local dynamical functions f and g are compatible with given signal set equivalences on X, and U, or on $X \times U$, in the sense that induced functions $\bar{f}$ and $\bar{g}$ exist and are unique, then these signal set equivalences are said to induce *local dynamical equivalence*. The SDS implications of local dynamical equivalence are discussed in the next section.

This discussion now turns toward the issue of realizing, in a diagrammatical sense, a local dynamical equivalence. Begin with the case in which E_x and E_u are specified, and consider the local transition function. The commutative diagram of Fig. 2.22 governs this issue. From the commutativity of the upper and lower triangles, it follows that

$$b \circ p_E = p_{E_x} \times p_{E_u}, \qquad \bar{f} \circ p_E = f. \qquad (2\text{-}160a,b)$$

Inasmuch as b is a bijection, the first of these equations can be solved for p_E, which can then be eliminated from the second equation to give

$$\bar{f} \circ b^{-1} \circ (p_{E_x} \times p_{E_u}) = f. \qquad (2\text{-}161)$$

From (2-161), it is seen that a pair (x, u) in $X \times U$ can, in place of being processed through f, be preprocessed through the map

$$b^{-1} \circ (p_{E_x} \times p_{E_u}): X \times U \to (X \times U)/E \qquad (2\text{-}162)$$

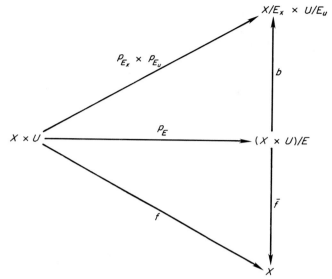

Fig. 2.22. Realizing local transition equivalence.

and then passed through the induced action of $\bar{f}$. Similar statements are made for the function g. If the equivalence E is placed directly on $X \times U$ while bypassing E_x and E_u, then the preprocessor (2-162) is just replaced by p_E, which cannot necessarily be broken down further.

An interesting question which is closely related to the definition and operation of an SDS is the way in which it is possible to picture the formation of a pair (x, u) in $X \times U$ from elements x in X and u in U. Intuitively, it should be possible simply to insert x into $X \times U$, and similarly for u.

However, the truth is more interesting than this fiction. Note that any proposed insertion function

$$i : X \to X \times U \tag{2-163}$$

would have an action of the type

$$i(x) = (x, u); \tag{2-164}$$

and the question of how to assign an element u to x is nontrivial.

There is another view of product sets which is useful in clearing up this confusion. To explicate this alternate way of thinking, consider Fig. 2.23. The set $X \cup U$ is called the *coproduct* of X and U. The functions i_X and

$$X \xrightarrow{\quad i_X \quad} X \cup U \xleftarrow{\quad i_U \quad} U$$

Fig. 2.23. The coproduct.

i_U are *injections* of X and U, respectively, into the coproduct. They are injective, and satisfy

$$(\text{Im } i_X) \cap (\text{Im } i_U) = \varnothing, \qquad (\text{Im } i_X) \cup (\text{Im } i_U) = X \cup U. \quad \text{(2-165a,b)}$$

The existence of the coproduct is easily shown. Indeed, if X and U are disjoint, the coproduct may be taken as the set theoretic union of X and U, and the injections as the insertions defined in Section 2.1. If not, then establish a bijection between one of the sets, say U, and some other set, say V, which is disjoint from X. Then the coproduct may be taken as the set theoretic union of X and V. The injection i_X is the same as before, while the injection i_U may be taken as the composition

$$i \circ b, \qquad \text{(2-166)}$$

for b the bijection $U \to V$ and i the insertion of V into $X \cup V$.

As more algebraic structure is added to the sets X and U, it will be possible to let the distinction between products and coproducts blur a bit. Without such structures, as is presently the case, the distinction is nontrivial.

As a conclusion to this section, observe that the use of an equivalence relation E_y on Y fits in with the local dynamical function g in every case. Figure 2.24 indicates why this is true. All that is required is the construction of a function

$$\tilde{g} : X \times U \to Y/E_y \qquad \text{(2-167a)}$$

in such a way that the diagram commutes. This can always be accomplished by defining

$$\tilde{g} = p_{E_y} \circ g. \qquad \text{(2-167b)}$$

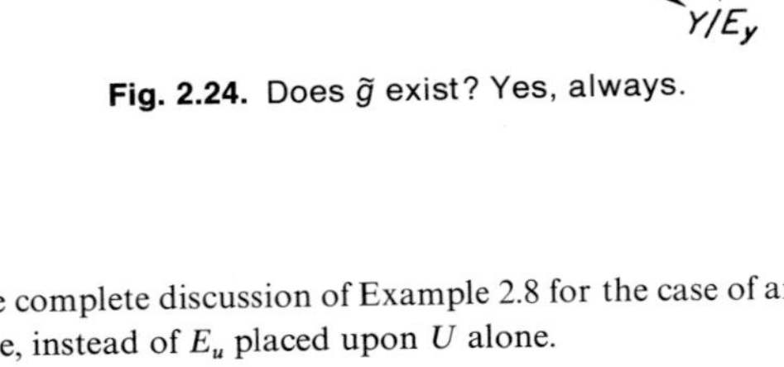

Fig. 2.24. Does $\tilde{g}$ exist? Yes, always.

Exercise

2.8. Repeat the complete discussion of Example 2.8 for the case of an equivalence relation E_x placed on X alone, instead of E_u placed upon U alone.

2.9   INDUCED SET-DYNAMICAL SYSTEMS

When signal set equivalences are emplaced in such a way as to produce local dynamical equivalence, it becomes possible to define new set-dynamical systems, which derive naturally from the quotient sets and induced local dynamical functions. Such an SDS is called, in this section, an *induced set-dynamical system*.

Many instances of the use of this concept appear in the systems literature and some of the better known applications of the idea will be described in later chapters.

In view of the discussions of earlier sections, an important case to consider is the simultaneous placing of E_u, E_x, and E_y equivalence relations.

The original SDS is described by

$$(J, \leq; U, X, Y; f, g). \tag{2-168}$$

The use of E_x and E_u immediately induces an equivalence relation E on $X \times U$. Following the guidelines of Section 2.8, place restrictions on E_x and E_u so as to insure local transition equivalence. This is accomplished by requiring that, for all

$$(x_1, x_2) \in E_x \tag{2-169a}$$

and for all

$$(u_1, u_2) \in E_u, \tag{2-169b}$$

the original local transition function must satisfy

$$f(x_1, u_1) = f(x_2, u_2). \tag{2-169c}$$

Then the condition of local dynamical equivalence is satisfied for the transition function.

Insight can be enhanced by studying the result of these assumptions on the local transition equation

$$x_{k+1} = f(x_k, u_k). \tag{2-170}$$

To proceed along these lines, apply the projection p_{E_x} to both members of (2-170), obtaining

$$\bar{x}_{k+1} = (p_{E_x} \circ f)(x_k, u_k), \tag{2-171}$$

provided the usual notational agreement

$$\bar{x}_{k+1} = p_{E_x}(x_{k+1}) \tag{2-172}$$

is honored. From (2-161), it can be written that

$$p_{E_x} \circ f = p_{E_x} \circ \bar{f} \circ b^{-1} \circ (p_{E_x} \times p_{E_u}). \tag{2-173}$$

Then (2-171) becomes

$$\bar{x}_{k+1} = \bar{f}_1(\bar{x}_k, \bar{u}_k), \tag{2-174}$$

where the *induced local transition function* $\bar{f}_1$ is defined by

$$\bar{f}_1: X/E_x \times U/E_u \rightarrow X/E_x \tag{2-175a}$$

with the action

$$\bar{f}_1 = p_{E_x} \circ \bar{f} \circ b^{-1}, \tag{2-175b}$$

and where use has been made of the straightforward construction

$$(p_{E_x} \times p_{E_u})(x, u) = (\bar{x}, \bar{u}). \tag{2-175c}$$

Equation (2-174) is the local transition equation for the induced SDS.

Next restrain E_x and E_u so as to obtain local output equivalence. Analogously to (2-169), require that, for all pairs (x_1, x_2) and (u_1, u_2) satisfying (2-169a,b), the original local output function must satisfy

$$g(x_1, u_1) = g(x_2, u_2). \tag{2-176}$$

Then operate on the local output equation

$$y_k = g(x_k, u_k) \tag{2-177}$$

with the projection p_{E_y}, which leaves

$$\bar{y}_k = (p_{E_y} \circ g)(x_k, u_k). \tag{2-178}$$

In Section 2.8, however, it was established that local output equivalence would permit a rewriting of g along the lines

$$g = \bar{g} \circ p_E = \bar{g} \circ b^{-1} \circ (p_{E_x} \times p_{E_u}); \tag{2-179}$$

and this in turn permits the form

$$\bar{y}_k = \bar{g}_1(\bar{x}_k, \bar{u}_k) \tag{2-180}$$

for (2-178), where the *induced local output function* $\bar{g}_1$ is defined by

$$\bar{g}_1: X/E_x \times U/E_u \rightarrow Y/E_y \tag{2-181a}$$

with the action

$$\bar{g}_1 = p_{E_y} \circ \bar{g} \circ b^{-1}, \tag{2-181b}$$

and with use once again of (2-175c). Equation (2-180) is the local output equation for the induced SDS.

Note the similarities present in the induced local dynamical function definitions (2-175b) and (2-181b).

The original SDS, denoted by (2-168), has now led to an induced SDS, which can be denoted by

$$(J, \leq; U, E_u; X, E_x; Y, E_y; \bar{f_I}, \bar{g_I}). \tag{2-182}$$

An induced SDS can be visualized easily in terms of diagrams. Figure 2.25 is a useful choice in this regard. In this figure, operation of an SDS is viewed according to Fig. 2.8. Establishing the signal set equivalences can be represented in such a diagram by the projections associated with each of the equivalence relations. This has been carried out in Fig. 2.26.

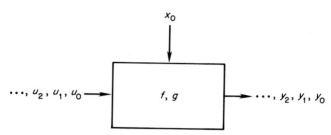

Fig. 2.25. Picture of Fig. 2.8.

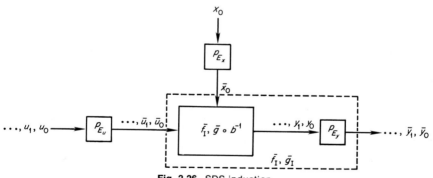

Fig. 2.26. SDS induction.

The induced SDS has been indicated in the figure by dashed lines. Its input set is U/E_u; its state set is X/E_x; and its output set is Y/E_y. The diagram suggests another, quite natural, alternate candidate for the title of induced SDS. This is based on the observation that E_y can always be added to Y without any restrictions on the nature of E_y. Therefore a reasonable case can be made for excluding the p_{E_y} box from the dashed enclosure. In that case, the induced SDS could be denoted by

$$(J, \leq; U, E_u; X, E_x; Y; \bar{f_I}, \bar{g} \circ b^{-1}). \tag{2-183}$$

Advantages may be attached to one or the other of (2-182) and (2-183) in a given application. In this work, however, an induced SDS will be interpreted according to (2-182), unless there is specific mention to the contrary.

The other possibility of interest in Section 2.8 was that in which an equivalence relation E was placed directly on $X \times U$ without any definition of an E_x and E_u. Moreover, it was shown in that section that such an E need not induce an E_x or an E_u.

For this case, all that can be explicitly done with (2-170) and (2-177) is to insert the factorizations $\bar{f} \circ p_E$ and $\bar{g} \circ p_E$ in the manner

$$x_{k+1} = (\bar{f} \circ p_E)(x_k, u_k), \qquad y_k = (\bar{g} \circ p_E)(x_k, u_k), \qquad \text{(2-184a,b)}$$

under the assumption that, for every pair

$$((x_1, u_1), (x_2, u_2)) \in E \qquad \text{(2-185a)}$$

the local dynamical functions satisfy

$$f(x_1, u_1) = f(x_2, u_2), \qquad g(x_1, u_1) = g(x_2, u_2). \qquad \text{(2-185b,c)}$$

The diagram of Fig. 2.25 does not elaborate in any significant way, and the SDS of (2-168) induces only to

$$(J, \le; U, X; Y, E_y; \bar{f} \circ p_E, p_{E_y} \circ \bar{g} \circ p_E). \qquad \text{(2-186)}$$

Now, any equivalence relation E which is not the equality relation $=$ results in a projection p_E which is *not* injective. As a consequence, none of the projections shown in Fig. 2.26 can be undone in the sense of their having a left inverse p_E^{-L}. Note that undoing p_E in the diagram would mean placement of a box *after* it; and this corresponds to left inversion rather than right inversion. If, therefore, an SDS of type (2-168) has the possibility of an induced version of type (2-183), then the resolution of the original SDS must have been greater than necessary. This is because the induced SDS of type (2-183) is producing the same outputs after a reduction of internal resolution.

An example illustrating the ideas of this section is provided in the section following.

2.10   EXAMPLE

In this section, it is desired to give a comprehensive example illustrating the ideas associated with induced set-dynamical systems.

The example has its base in Example 2.2-2. Recall the intuitive background. Widgets, Inc., is a national firm which manufactures three colors of widgets: red, white, and blue. Each color widget comes in two sizes: very large and extra large. Sealed in packages of twelve each, widgets must be ordered in lots of that size. You operate a retail store which stocks widgets, and have made a beginning in establishing an SDS to take care of widget orders. Four base sets were specified in Section 2.2. In addition to $\mathbb{N}$, there were

$$V_1 = \{\text{red, white, blue}\}, \qquad (2\text{-}187a)$$
$$V_2 = \{\text{very large, extra large}\}, \qquad (2\text{-}187b)$$
$$V_3 = \{\text{order, do not order}\}. \qquad (2\text{-}187c)$$

The input set U was chosen to be $V_1 \times V_2$; inputs are placed by customers, who select for example a (white, very large) widget. Customers wishing to purchase more than one color and size can be treated as placing successive orders. The state set X is selected to be

$$X = (V_1 \times V_2 \times \mathbb{N})^6 ; \qquad (2\text{-}188a)$$

elements in X have the form

$$x = (x_1, x_2, \dots, x_6) \qquad (2\text{-}188b)$$

for each x_i having the representation

$$x_i = (v_1, v_2, n). \qquad (2\text{-}188c)$$

It is sometimes quite convenient to regard an element x_i, as in (2-188c), from the slightly different point of view

$$x_i = ((v_1, v_2), n). \qquad (2\text{-}189a)$$

This amounts to the same thing as the use of

$$((V_1 \times V_2) \times \mathbb{N})^6 \qquad (2\text{-}189b)$$

in place of (2-188a). Inasmuch as there is a straightforward bijection

$$(V_1 \times V_2 \times \mathbb{N})^6 \approx ((V_1 \times V_2) \times \mathbb{N})^6, \qquad (2\text{-}190)$$

there is little need to distinguish (2-189b) from X symbolically. Similarly, the output set Y is selected to be

$$Y = (V_1 \times V_2 \times V_3)^6 ; \qquad (2\text{-}191a)$$

elements in Y have the form

$$y = (y_1, y_2, \dots, y_6) \qquad (2\text{-}191b)$$

for each y_i of the type

$$y_i = (v_1, v_2, v_3). \qquad (2\text{-}191c)$$

Again, it is often convenient to write

$$y_i = ((v_1, v_2), v_3) \qquad (2\text{-}192)$$

in place of (2-191c).

Now the local output equation for this SDS ordering system must take the general form

$$y_k = g(x_k, u_k), \tag{2-193}$$

so that notations of the form

$$(y_1, y_2, \ldots, y_6)_k \quad \text{and} \quad (x_1, x_2, \ldots, x_6)_k \tag{2-194a,b}$$

arise. In place of the potentially confusing $(x_i)_k$ and $(y_i)_k$ possibilities for indicating the ith component at index k, it is perhaps conceptually more direct to permit the temporary reversion used in Section 2.5, namely, $x_i(k)$ and $y_i(k)$. Then denote

$$y_i(k) = g_i(x_k, u_k). \tag{2-195a}$$

It is of interest to point out again that there is no genuine need to write (2-195a) in the manner

$$y_i(k) = g_i(x(k), u(k)), \tag{2-195b}$$

though some readers may prefer to do so. Each function g_i has the general character

$$X \times U \to (V_1 \times V_2) \times V_3, \tag{2-196}$$

though the action depends on i. It would be possible to redefine the right member of (2-195a) in a style such as

$$g_i(x_k, u_k) = h(x_k, u_k, i); \tag{2-197}$$

however the notation (2-195a) is well accepted and compact.

The action of each g_i can be expressed by

$$g_i(x, u) = (u, \text{order}) \tag{2-198a}$$

if both

$$p_U(x_i) = u \quad \text{and} \quad p_N(x_i) \leq 12; \tag{2-198b,c}$$

otherwise the action of g_i is

$$g_i(x, u) = ((v_1, v_2), \text{do not order}). \tag{2-198d}$$

This completes the definition of g, because

$$g(x, u) = (g_1(x, u), g_2(x, u), \ldots, g_6(x, u)). \tag{2-199}$$

Next consider the local transition equation, which must take the form

$$x_{k+1} = f(x_k, u_k). \tag{2-200}$$

Analogously to (2-195a), define a collection of six functions

$$f_i: X \times U \to (V_1 \times V_2) \times \mathbb{N}, \qquad (2\text{-}201)$$

which can then be used to define the local transition function f by

$$f(x, u) = (f_1(x, u), f_2(x, u), \ldots, f_6(x, u)). \qquad (2\text{-}202)$$

Suppose that x_i has the general form (2-189a), and let $\tilde{n}$ be the largest natural number in $\mathbb{N}$ which satisfies $\tilde{n} < n$. Then the action of f_i can be given by

$$f_i(x, u) = (u, \tilde{n}) \qquad \text{if} \qquad p_U(x_i) = u; \qquad (2\text{-}203\text{a,b})$$

otherwise the action of f_i is

$$f_i(x, u) = x_i. \qquad (2\text{-}203\text{c})$$

It is assumed that widget deliveries are relatively rapid in comparison with customer purchases, so that inventory does not make any attempt to fall below zero.

The SDS scheme above for reordering at the retail level appears overly complicated from the wholesaler's point of view, however. The wholesaler, therefore, sets up an equivalence relation E_n on $\mathbb{N}$, according to the rule

$$n_1 E_n n_2, \qquad (2\text{-}204\text{a})$$

if both n_1 and n_2 are no greater than 12 or if both n_1 and n_2 are greater than 12. $\mathbb{N}/E_n$ then contains exactly two equivalence classes, one containing the natural numbers

$$\{n \,|\, n \in \mathbb{N} \text{ and } n \le 12\} \qquad (2\text{-}204\text{b})$$

and the other containing the natural numbers

$$\{n \,|\, n \in \mathbb{N} \text{ and } n > 12\}. \qquad (2\text{-}204\text{c})$$

In order to complete his definition of an equivalence relation on X, the wholesaler selects E_u to be equality, and permits the pair E_u and E_n to induce an equivalence relation on $(V_1 \times V_2) \times \mathbb{N}$, along the lines of Section 2.8. This new equivalence relation, call it E_1, can then be paired with itself to produce an equivalence relation E_2 on $((V_1 \times V_2) \times \mathbb{N})^2$, and E_2 can be paired with E_1 to produce an equivalence relation E_3 on $((V_1 \times V_2) \times \mathbb{N})^3$. Continuing in this way he obtains the relation E_6 on $((V_1 \times V_2) \times \mathbb{N})^6$ and denotes it by E_x. Finally, he pairs E_x and E_u to obtain the equivalence relation E on $X \times U$.

From (2-198c), it can be inferred that local output equivalence holds for the equivalence relation proposed to the retailer. On the other hand, local transition equivalence does not hold, so that the wholesaler's scheme cannot lead to an induced SDS at the retail level.

Of course, the fundamental reason for the wholesaler's failure to induce a new SDS at the retail level is that *someone* has to count the number of widgets sold.

But now suppose that, to promote business, the retailer decides to have a "two-cent sale" in which every customer purchasing a widget of one color will receive widgets of the other two colors, in the same size, for only two cents over the price of one widget. Because of the inflated price of widgets, it is safe to assume that every customer will take advantage of the sale.

The wholesaler, rebuffed in his earlier effort to bring about streamlined accounting at the retail outlets, returns with another proposal: to establish an equivalence relation E_{v_1} on V_1, in such a way that

$$\text{red } E_{v_1} \text{ white,} \qquad \text{white } E_{v_1} \text{ blue;} \qquad (2\text{-}205a,b)$$

the result of this is that V_1/E_{v_1} contains just one element. With rather obvious minor modifications in U, X, Y, f, and g, designed so that three widgets can be sold at one time, it is an easy step to establish that local dynamical equivalence holds, and that the wholesaler's last initiative really does succeed in inducing an SDS. Details are left to the exercises.

N.B.:   The reader may wish to complete the list of statements in (2-205) in order to obtain an equivalence relation.

Exercises

2.10-1. What is the nature of the bijection asserted in (2-190)?

2.10-2. Reconsider the development of this section's example if the option (2-197) were chosen.

2.10-3. At the conclusion of the section, the retailer decides to have a two-cent sale. Make the suggested modifications in U, X, Y, f, and g so that three widgets can be sold at one time. Then determine whether local dynamical equivalence holds, so that the retailer has succeeded in inducing a new SDS.

2.11   DISCUSSION

In the study of dynamical systems, a great deal of algebraic insight can be lost through the seemingly harmless process of endowing the various domains and codomains with so rich an algebraic structure that it is hard to determine which parts of the structure are necessary for the study and which parts are only convenient luxuries. For example, the use of vector spaces is very common; and vector spaces represent a very rich algebraic structure. How many of the common results for dynamical systems over vector spaces could be achieved with less structure? The answer to this question may not always be revealed by the "usual" analyses, which have sometimes dipped freely into the "luxury" part of structure, even when such a dip was not necessary. In terms of the metaphor used at the beginning of this chapter, a size 48 suit may fit many wearers—in the sense that they can get inside—but it may reveal very little about the features of the wearer, once within.

The spirit of this philosophy has led to the present chapter, the goal of which has been to do some nontrivial things with dynamical systems while holding the algebraic structure on sets to a minimum. Four basic sets have been introduced: the input set U; the state set X; the output set Y; and an index set J.

Upon these four sets only one type of algebraic construction has been permitted—the binary relation. Actually a subset of the *product* of two sets S and T, binary relations B on S to T are typically referred to as being *on S* if T is the same set as S. Four types of binary relations have been discussed. The equality relation is available on any set. Simple order relations are convenient to associate with the index set J of a set-dynamical system. Partial order relations have not been used in this chapter but will be used in the next. Equivalence relations were probably the most useful idea in the chapter.

The equivalence relation can be viewed as a generalization of the concept of equality on sets. Indeed, equality is an equivalence relation of the most prevalent and fundamental type. A reason for the great interest attached to the equivalence relation can be found in the way in which analysts like to classify elements of sets. Anytime such a classification procedure breaks a set down into disjoint subsets whose union is the entire set, then an equivalence relation is established. Of course, all that is required for this to happen is that each element of the set be classified into one and only one of the subsets.

Functions automatically establish an equivalence relation on their domain. The classification is simple: all elements to which the function assigns the same value are grouped into one subset. In this way, a close link is established between injective functions and the binary relation of equality. The fact is that every equivalence relation can be regarded, if so desired, as being the consequence of some function. However, it should be noted that more than one function can lead to the same equivalence relation.

Because of the way in which functions and equivalence relations intertwine, it is especially worthwhile to study the question whether or not a function is compatible with an equivalence relation. The essence of this notion has been covered in the key triangle diagram in which a function is called compatible with such a relation if it assigns the same value to every element in each equivalence class induced by the relation. When such is the case, it is possible to reduce the resolution of the function by redefining it on the quotient set of the relation. This is one of the most basic concepts in modern algebra. It is therefore not surprising that it has also become a workhorse idea in the beginning of algebraic system theory. The idea will be used over and over in the sequel, in a great variety of different ways. One of the principal goals of this chapter has been to set the idea in place within a context gleaned from a very economical set of assumptions. In this way, the reader can start to develop an intuition for the relation before it becomes imbedded inside more complicated structures.

The centrality of the key triangle for both the algebra and the system theory suggests the need for a thorough discussion of the way in which it interfaces with the definition of a set-dynamical system. Special attention must be paid here, then, to the compatibility of local transition functions and local output functions with various equivalence relations. As the text progresses, added set structure will permit a better and better elaboration of this interface.

Two examples of the application of the SDS concept have been given in this chapter: order reduction and induced systems. While these do serve to introduce the reader to various algebraic issues with which a familiarity is eventually desirable, they may not pack enough wallop to satisfy the more conservative. For this reason, more elaborate developments have been woven into the fabric of the following chapter.

Finally, there are many more issues which could be taken up here, were there enough space. The entire subject of automata represents an attractive digression. However, the topics selected here have been chosen to support several central themes of the volume, and these will be further developed in later chapters.

3 OBSERVERS AND REGULATION

The preceding chapter has dealt with the notion of a set-dynamical system (SDS) which has associated with it sets of inputs, states, outputs, and indices, a pair of local dynamical functions which determine transition and output, and various binary relations. Introductory manipulations based upon those definitions were provided.

By now the reader has probably begun to develop somewhat of a feel for the way in which sets and functions can be used to advantage in system theoretic problems. The present chapter begins to build upon this foundation by adding more structure, but *not* to the basic input, state, and output sets of the SDS. A query may well be anticipated as to just where these added structures are to be placed.

A brief perusal of Chapter 2 should make it clear that the only remaining places to add structure are upon the functions or upon the relations. Actually, a preliminary example of such a construction has already appeared in Section 2.1. Suppose that there are two functions

$$f : U \to V \tag{3-1}$$

and

$$g : T \to U. \tag{3-2}$$

Without any other assumptions on f, g, U, V, and T, it was established there that there was in force a *composition function*

$$\circ : V^U \times U^T \to V^T \tag{3-3}$$

which could be used to construct from f and g a new function $T \to V$ in the manner

$$\circ (f, g). \tag{3-4}$$

As in the case of the binary relations, the notation (3-4) is typically replaced by

$$f \circ g, \tag{3-5}$$

and the action of the composition function $\circ$ is expressed by

$$(f \circ g)(t) = f(g(t)). \tag{3-6}$$

It should be noted that the construction of composition takes place without any further assumptions on U, V, and T. It is in this general spirit that the chapter now proceeds.

3.1   POSETS

Recall from Section 2.3 that a binary relation B on a set S is a partial order relation if it is reflexive, antisymmetric, and transitive. The frequent symbol for a partial order relation is $\leq$.

A *poset* is just a set P equipped with a partial order relation $\leq$. The poset can be denoted $(P, \leq)$, when necessary, to bring out clearly the symbol standing for the relation.

EXAMPLE 3.1-1

The simple order relation $\leq$ on the index set J of an SDS is a partial order relation; and so $(J, \leq)$ is a poset. Of course the partial order on the index set is assumed to satisfy also a fourth property, namely that any pair (j_1, j_2) of elements in J^2 must satisfy either

$$j_1 \leq j_2 \quad \text{or} \quad j_2 \leq j_1. \tag{3-7a,b}$$

This makes $(J, \leq)$ a special type of poset, which is called a *chain*. The term simple order relation has already been introduced for a binary relation which establishes a chain. Another term in common use is complete order relation. In this sense a chain is said to be completely ordered. ∎

EXAMPLE 3.1-2

Recall the symbolism $\mathbb{P}(S)$ for the power set of S, namely, the set of all subsets of S. Then

$$(\mathbb{P}(S), \subset) \tag{3-8}$$

is a poset, with respect to the binary relation of subset inclusion. ∎

The interesting thing about a poset is that an arbitrary pair of its elements may or *may not* stand in the relation $\leq$ to each other. This is the difference between a chain and a general poset.

Sometimes, it can be a considerable help to establish a diagram showing how the elements of a poset relate to one another. The following example illustrates this point.

EXAMPLE 3.1-3

For P, select the set

$$P = \{1, 2, 4, 5, 10, 20\}, \tag{3-9}$$

which is just the subset of $\mathbb{N}$ consisting of all those elements of $\mathbb{N}$ which divide the natural number 20. It is an easy exercise to establish that P is indeed a poset under the relation

$$p_1 \leq p_2 \qquad \text{if} \qquad p_1 | p_2, \tag{3-10a,b}$$

where (3-10b) is to be read "p_1 divides p_2." Clearly, $(2, 5)$ is not a member of the relation, while $(2, 10)$ is. Figure 3.1 is an easy way to describe this. A rising line is drawn from one natural number to another if that natural number divides the other. Note, however, that no rising line is drawn from 2 to 20, even though 2 divides 20. This is the result of a covenant not to draw lines which can be easily inferred from transitivity. In this way, the diagram can be kept reasonably uncluttered. ∎

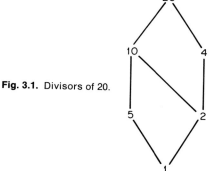

Fig. 3.1. Divisors of 20.

The use of diagrams can be helpful in achieving an understanding of some of the subtleties associated with posets.

Consider a subset S of a poset $(P, \leq)$. Suppose that there is an element s_1 in S with the property that

$$s \in S \qquad \text{and} \qquad s \leq s_1 \qquad \text{imply} \qquad s = s_1. \tag{3-11a,b,c}$$

Then s_1 is called a *minimal element* of S. In other words, there is no other element in S which stands in the relation $\leq$ to s_1. Minimality is sometimes a deceptively simple idea. Subsets may have no minimal elements, or they may have more than one.

EXAMPLE 3.1-4

Let P be $\mathbb{Z}$, and establish that $p_1 \leq p_2$ if $p_2 - p_1$ is nonnegative. Call this the *usual* partial order on $\mathbb{Z}$. Select for S the subset of negative integers. S has no minimal element. ∎

EXAMPLE 3.1-5

Let P be the set $\{a, b, c, d\}$, and let the partial order be defined by Fig. 3.2. Select the subset $\{a, b, c\}$ for S, and note that S has two minimal elements, namely, a and c. ∎

Fig. 3.2. $\{a, b, c\}$ has two minimal elements.

In like manner, a *maximal element* s_2 of a subset S is an element with the property

$$s \in S \qquad \text{and} \qquad s_2 \leq s \qquad \text{imply} \qquad s_2 = s. \qquad \text{(3-12a,b,c)}$$

Examples similar to 3.1-4 and 3.1-5 can of course be constructed for maximal elements.

Closely related, at least in appearances, to the concepts of minimal and maximal element are the ideas of lower bound and upper bound.

Consider once again a subset S of the poset $(P, \leq)$. An element p_1 in P is called a *lower bound* for S if $p_1 \leq s$ for all s in S. Similarly, an element p_2 in P is called an *upper bound* for S if $s \leq p_2$ for all s in S.

Note that the lower bound and upper bound need not reside in the set S, but can be outside the set S within the parent set P. If a bound happens to exist *within* S, then poset properties immediately imply that it is the *only* such bound within S. To see this, note that being, say, a lower bound will require of two candidate bounds s_1 and s_2 in S the demanding conditions

$$s_1 \leq s_2, \qquad s_2 \leq s_1, \qquad \text{(3-13a,b)}$$

which lead to uniqueness by antisymmetry. Outside the set S, however, no such inference has to occur. Indeed, in Fig. 3.1, let S be given by $\{10, 20\}$. Then lower bounds p_1 and p_2 can be taken to be 5 and 1, which are clearly *not* equal. When a lower bound occurs within the subset S, it is usually called a *least element*; correspondingly, an upper bound in S is called a *greatest element*.

On the lower end of the partial order relation $\leq$, there are now three terms in place: minimal element, lower bound, and least element. On the upper end, there are three more analogous terms: maximal element, upper bound, and greatest element.

For the applications which follow in this chapter, this list of terms remains, nonetheless, incomplete. This is one of the reasons that early readings in posets can become confusing.

From Chapter 1, it may be recalled that one of the main algebraic penchants is to construct higher-level sets from lower-level sets. The new idea that is needed is really just another example of this level-to-level construction.

Start with a subset S of a poset $(P, \leq)$. Next determine the subset LB(S) of P which consists of all lower bounds of the subset S. LB(S) may or may not have an element in S; if it does, then that element is unique. Any other elements in LB(S) must be outside S in the parent poset P. Now LB(S), as a subset of P in its own right, may have a greatest element. If it does, then that greatest element is called the *infimum* of S, and is denoted by inf S. Clearly,

$$p \in \text{LB}(S) \qquad \text{implies} \qquad p \leq \inf S; \qquad\qquad (3\text{-}14a,b)$$

and

$$s \in S \qquad \text{implies} \qquad \inf S \leq s. \qquad\qquad (3\text{-}14c,d)$$

If S has a least element, then that least element is its infimum. However, S may have an infimum without having a least element.

Next establish for the subset S of P the set UB(S) which consists of all its upper bounds in P. UB(S) may or may not have an element in S; if it does, that element is unique. Any other elements in UB(S) must be outside S in the parent poset P. As a subset of P, UB(S) may have a least element. If it does, then that least element is called the *supremum* of S, and is denoted by sup S. Clearly,

$$p \in \text{UB}(S) \qquad \text{implies} \qquad \sup S \leq p; \qquad\qquad (3\text{-}15a,b)$$

and

$$s \in S \qquad \text{implies} \qquad s \leq \sup S. \qquad\qquad (3\text{-}15c,d)$$

If S has a greatest element, then that greatest element is its supremum. However, S may have a supremum without having a greatest element.

On a poset $(P, \leq)$, a subset S may or may not have an infimum or a supremum. If inf S or sup S exists, there is the added question whether or not it is an element of S. Later in this chapter, occasion will arise to address both these issues in the context of an observer's construction.

Exercises

3.1-1. Establish that (3-9) is a poset under the relation (3-10).
3.1-2. Construct examples similar to Examples 3.1-4 and 3.1-5 for maximal elements.

3.2 LATTICES

The preceding section has introduced the notion of a poset, within which context the discussion of this chapter will take place. In particular, it will shortly prove convenient to establish poset structure on sets of equivalence relations. These equivalence relations may be on one of the SDS signal sets, as for example the state set X, but the poset structure is to be placed on the relations, not upon X.

Posets of interest in this chapter actually possess special structure over and above that brought about by the partial order relation $\leq$. In fact, they have two more functions associated with them. These functions are known as binary operations.

A *binary operation* on a set S is just a function

$$S \times S \to S \tag{3-16}$$

from the product of S with itself to S. Momentarily, assign the symbol b to this function. Then its action would be given by specifying

$$b(s_1, s_2) \tag{3-17}$$

for all pairs (s_1, s_2) in S^2. In the same spirit as that of binary relations, however, it is traditional to write

$$s_1 \, b \, s_2 \tag{3-18}$$

in place of the usual function notation (3-17). The reason for this will be more apparent after the following example.

EXAMPLE 3.2-1

Select S to be the set $\mathbb{R}$ of real numbers. Then b can be chosen as the usual operation of real addition:

$$r_1 + r_2. \tag{3-19a}$$

Note that (3-19a) is the ordinary convention, and not

$$+(r_1, r_2). \tag{3-19b}$$

Again, b can be chosen as the usual operation of real multiplication:

$$r_1 \cdot r_2. \tag{3-20a}$$

Once more, note that (3-20a) is the accepted way of writing such operations, and not

$$\cdot (r_1, r_2). \tag{3-20b}$$

The binary operations (3-19) and (3-20) of real addition and real multiplication are among the most familiar of such constructions.   ∎

The very familiarity of illustrations such as those in Example 3.2-1 can often make it quite difficult to imagine new types of binary operations and to put them into practice. So as not to prejudice the thinking at this stage, then, a more arbitrary symbol □ is used in place of the function symbol b. Thus (3-18) is replaced by

$$s_1 \,\square\, s_2. \tag{3-21}$$

In much the same way that binary relations are known by their properties, such as reflexivity, symmetry, antisymmetry, and transitivity, so binary operations are also measured against standard properties. If

$$s_1 \,\square\, s_2 = s_2 \,\square\, s_1 \tag{3-22a}$$

for all s_1 and s_2 in S, then the operation □ is said to be *commutative*. If

$$(s_1 \,\square\, s_2) \,\square\, s_3 = s_1 \,\square\, (s_2 \,\square\, s_3) \tag{3-22b}$$

for all s_1, s_2, and s_3 in S, then the operation □ is said to be *associative*. In fact, when □ is associative, the parentheses in (3-22b) are usually omitted, in the manner

$$s_1 \,\square\, s_2 \,\square\, s_3, \tag{3-22c}$$

inasmuch as either reasonable interpretation of (3-22c) amounts to the same thing, by (3-22b). If there exists an element e in S such that

$$s \,\square\, e = s = e \,\square\, s \tag{3-22d}$$

for all s in S, then e is said to be a *unit* for the binary operation □.

EXAMPLE 3.2-2

Suppose that the reader has determined a unit e for the binary operation □. But now another reader has found a unit $\tilde{e}$ for the same operation. What is the relationship between e and $\tilde{e}$? Since e is a unit, it must be true that

$$\tilde{e} = e \,\square\, \tilde{e};\qquad\qquad(3\text{-}23\text{a})$$

and since $\tilde{e}$ is a unit, it must be true that

$$e = e \,\square\, \tilde{e};\qquad\qquad(3\text{-}23\text{b})$$

but $=$ is an equivalence relation, so that transitivity provides

$$\tilde{e} = e\qquad\qquad(3\text{-}23\text{c})$$

by (3-22d). Thus a unit for $\square$ must be unique, if it exists. ∎

Now suppose that a binary operation $\square$ on S has a unit e. An element s in S may or may not have the property that there exists an element $\hat{s}$ in S satisfying

$$s \,\square\, \hat{s} = e = \hat{s} \,\square\, s.\qquad\qquad(3\text{-}24)$$

If it does, then $\hat{s}$ is called an *inverse* of s under $\square$.

EXAMPLE 3.2-3

A second reader, having found that his unit $\tilde{e}$ in the preceding example was no better than yours, now sees that you have found an inverse $\hat{s}$ for s under $\square$. To redeem the situation, said reader brings forward another candidate $\tilde{s}$ for inverse of s. Inasmuch as your element $\hat{s}$ is an inverse, it follows that

$$s \,\square\, \hat{s} = e;\qquad\qquad(3\text{-}25\text{a})$$

and since the other reader's $\tilde{s}$ is an inverse, it must be true that

$$s \,\square\, \tilde{s} = e;\qquad\qquad(3\text{-}25\text{b})$$

by transitivity of the relation $=$,

$$s \,\square\, \hat{s} = s \,\square\, \tilde{s}.\qquad\qquad(3\text{-}25\text{c})$$

Now apply your inverse to the left member of (3-25c) in the manner

$$\hat{s} \,\square\, (s \,\square\, \hat{s}) = (\hat{s} \,\square\, s) \,\square\, \hat{s} = e \,\square\, \hat{s} = \hat{s};\qquad\qquad(3\text{-}25\text{d})$$

a similar construction in the right member of (3-25c) yields $\tilde{s}$, so that

$$\hat{s} = \tilde{s}.\qquad\qquad(3\text{-}25\text{e})$$

Thus the other reader may possibly have an inverse distinct from yours, but only if $\square$ is not associative. ∎

If W is a subset of S, and if S is equipped with a binary operation $\square$, then W is said to be *closed under* $\square$ when w_1 and w_2 in W imply $w_1 \,\square\, w_2$

and $w_2 \ \Box \ w_1$ in W. This concept is useful repeatedly in the sequel, as for example in Section 3.7. An important feature of the closure idea is that it permits restrictions, in the sense of Section 2.1, of these functions to subsets.

Now turn to the construction of binary operations on posets. The ideas build upon Section 3.1. Let $(P, \leq)$ be a poset, and choose two elements p_1 and p_2 from P. Then

$$\{p_1, p_2\} \tag{3-26}$$

is a subset of P. The *meet* of p_1 and p_2, when it exists, is defined to be the infimum of the set (3-26). The *join* of p_1 and p_2, when it exists, is defined to be the supremum of the set (3-26).

The pieces are now available to put together the definition of a lattice. A *lattice* is a poset $(L, \leq)$ in which any two elements p_1 and p_2 have both a meet and a join. This is of course a strong restriction on a poset, in the sense that it demands the existence of certain infimums and supremums.

In a lattice, the meet and join constructions are binary operations. A notation for them is suggested by extending Example 3.1-2, which deals with the poset $(\mathbb{P}(S), \subset)$ defined by inclusion on the power set $\mathbb{P}(S)$ of all subsets of some given set S. The set theoretic intersection of two subsets of $\mathbb{P}(S)$ serves as a meet, and the set theoretic union serves as a join. The general symbols for meet and join can be regarded as parallel to those for intersection ($\cap$) and union ($\cup$). The meet symbol is chosen to be $\wedge$, and the join symbol is $\vee$.

A lattice, then, can be denoted by the list $(L, \leq, \wedge, \vee)$ as a set L with a partial order relation $\leq$, and two binary operations $\wedge$ and $\vee$.

The interplay between the partial order relation and the binary operations is sometimes intricate. Most fundamental, however, are the algebraic properties of $\wedge$ and $\vee$. In summary, these turn out as follows: (1) $\wedge$ and $\vee$ are associative; (2) $\wedge$ and $\vee$ are commutative; (3) $\wedge$ and $\vee$ are *idempotent*, which means that

$$p \wedge p = p \qquad \text{and} \qquad p \vee p = p, \tag{3-27a,b}$$

respectively; and (4) there is a property

$$p_1 \wedge (p_1 \vee p_2) = p_1 = p_1 \vee (p_1 \wedge p_2) \tag{3-28}$$

for all p_1 and p_2 in L. For ideas as to how the partial order relation interfaces with the meet and the join, the reader may wish to examine the exercises at the end of this section. Nonetheless, it turns out that these four algebraic properties associated with the two binary operations above are sufficient to define a lattice. In fact, if two binary operations which satisfy these four properties are given, then a partial order relation can be induced by defining.

$$p_1 \leq p_2 \qquad \text{if} \qquad p_1 \wedge p_2 = p_1. \tag{3-29a,b}$$

Property (4) above, which relates the two binary operations $\wedge$ and $\vee$, points out the necessity for asking questions about how various binary operations on the same set interact with each other.

The most basic of these types of questions has to do with distributivity. Let $\square_1$ and $\square_2$ be two binary operations on the same set S. Then $\square_1$ is said to *distribute* over $\square_2$ if

$$s_1 \square_1 (s_2 \square_2 s_3) = (s_1 \square_1 s_2) \square_2 (s_1 \square_1 s_3) \tag{3-30}$$

for all s_1, s_2, and s_3 in S.

In general, the meet and join in a lattice structure do not distribute over each other.

No special attempt is made here to illustrate lattices, inasmuch as some of the main issues in later sections of the chapter revolve around lattices. The principal lattice of interest will be constructed on the set of binary relations associated with a given set. Before discussing this relational lattice, however, it will be well to develop a better motivation by completing certain aspects of the discussion involved with set-dynamical systems.

Exercises

3.2-1. Verify the distributive inequalities

$$p_1 \vee (p_2 \wedge p_3) \leq (p_1 \vee p_2) \wedge (p_1 \vee p_3),$$
$$(p_1 \wedge p_2) \vee (p_1 \wedge p_3) \leq p_1 \wedge (p_2 \vee p_3).$$

3.2-2. Verify the modular inequality

$$p_1 \leq p_3 \Rightarrow p_1 \vee (p_2 \wedge p_3) \leq (p_1 \vee p_2) \wedge p_3.$$

3.3 GLOBAL DYNAMICAL FUNCTIONS

The local dynamical functions of an SDS serve primarily to indicate what transpires when a singleton input is applied to the system. To accomplish this interpretation, the SDS is assumed to be in state x; when the input u is applied, an output y is produced according to the local output function $g : X \times U \rightarrow Y$, and the state transitions to a next value $f(x, u)$ as specified by the local transition function.

In Section 2.4, the index set J was introduced in order to relate successive usages of the local dynamical functions. This goal, of relating successive local activities in the SDS, was met by placing a simple order relation $\leq$ on the index set, thus making $(J, \leq)$ a completely ordered set or chain, in the terminology of Section 3.1.

A control on which of the indices in J was to be used in a particular situation was established by defining index functions $\alpha: \mathbb{N} \to J$ which were injective and order preserving, in the sense of (2-61). Being in a state x could then be associated with a natural number n by the mechanism $x(j(n))$, and applying an input u could be described by $u(j(n))$. The state transition and output then were easily expressed by

$$x(j(n + 1)) = f(x(j(n)), u(j(n))), \tag{3-31a}$$

$$y(j(n)) = g(x(j(n)), u(j(n))). \tag{3-31b}$$

Now define a *one-sided sequence* as any element in $T^{\mathbb{N}}$ for T an arbitrary codomain. It is then suggested by (3-31) that every SDS is going to have associated with it a number of one-sided sequences. Because all sequences in this chapter are one-sided, it is convenient henceforth to refer to them simply as sequences. In fact, there is going to be a sequence of inputs, a sequence of states, and a sequence of outputs. Moreover, if an *initial state* $x(j(0))$ is provided, it is a consequence of (3-31) that an input sequence in $U^{\mathbb{N}}$ will lead naturally to a state sequence in $X^{\mathbb{N}}$ and to an output sequence in $Y^{\mathbb{N}}$. As a result, in addition to being associated with a number of sequences, an SDS is also associated with certain functions which carry sequences to sequences. It is these last functions which are the subject of the present section, namely, global dynamical functions.

Following the conventions established in (2-63)–(2-65), replace (3-31) by

$$x_{k+1} = f(x_k, u_k), \qquad y_k = g(x_k, u_k). \tag{3-32a,b}$$

The SDS then starts in an initial state x_0. An input u_0 is applied. This produces a next state x_1 and an output y_0. An input u_1 is applied. Then the next state is x_2 and the output is y_1. Continuing in this way yields a sequence

$$u_0, u_1, u_2, \ldots \tag{3-33a}$$

of inputs, a sequence

$$x_0, x_1, x_2, \ldots \tag{3-33b}$$

of states, and a sequence

$$y_0, y_1, y_2, \ldots \tag{3-33c}$$

of outputs.

A word about notation may be helpful here. A function $t: \mathbb{N} \to T$ is a one-sided sequence. It is sometimes quite suggestive to consider the function in terms of its values

$$t(0), \quad t(1), \quad t(2), \quad \ldots. \tag{3-34a}$$

If (3-34a) is regarded as the image Im t of t, then the idea of (2-32) in Section 2.1 suggests the useful notation

$$t_*(\mathbb{N}) \tag{3-34b}$$

as an alternative to (3-34a). When $\mathbb{N}$ is fixed, as it is for this dicussion, it may be suppressed. This suggests the use of

$$t_* \tag{3-34c}$$

as a symbol for a sequence in $T^{\mathbb{N}}$. This is especially useful in the SDS context, because it permits immediate access to u_*, x_*, and y_* as symbols for sequences of inputs, states, and outputs.

Depiction of sequences has a natural feel to it if it is carried out in keeping with (3-33) and (3-34a). This way of sketching the sequence can be visualized as in Fig. 3.3. On the other hand, there is a sort of opposite way to accomplish the same thing, and this has been indicated in Fig. 3.4. The opposite viewpoint appears to have a cultural edge in some types of dynamical system diagrams. Consider, for example, Fig. 3.5. Here an input sequence is portrayed as entering an SDS. The opposite sequence picturing seems to have a much wider acceptance here than the ordinary sequence picturing. Perhaps the

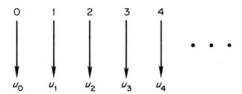

Fig. 3.3. Ordinary sequence visualization.

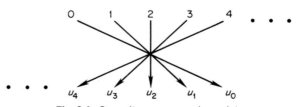

Fig. 3.4. Opposite sequence viewpoint.

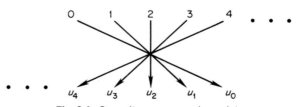

Fig. 3.5. Input sequence for SDS.

reader will disagree. However, the opposite viewpoint is used in the sequel in certain instances.

Return now to a consideration of (3-33). It is natural to define the *global output function*

$$h: X \times U^{\mathbb{N}} \to Y^{\mathbb{N}} \tag{3-35}$$

by the action

$$h(x_0, u_0, u_1, u_2, \dots) = (y_0, y_1, y_2, \dots). \tag{3-36}$$

In view of the simplified notation (3-34c), it is possible to rewrite (3-36) in the less cumbersome way

$$h(x_0, u_*) = y_*. \tag{3-37}$$

Along the same lines, it is equally inevitable to define a *global transition function*

$$\pi: X \times U^{\mathbb{N}} \to X^{\mathbb{N}} \tag{3-38}$$

by the action

$$\pi(x_0, u_0, u_1, u_2, \dots) = (x_0, x_1, x_2, \dots). \tag{3-39}$$

The simplified notation permits (3-39) to be replaced by

$$\pi(x_0, u_*) = x_*. \tag{3-40}$$

The *global dynamical functions* of an SDS are then the global output function and the global transition function.

It is useful to develop a bit more structure on SDS sequences. Basically, this structure can follow from the observation that there is a bijection

$$T^{\mathbb{N}} \approx T \times T \times T \times \cdots, \tag{3-41}$$

where the right member has a countably infinite number of terms. The essence of this bijection has already been brought out in (3-34). Of course, in the finite case, the same idea was also used in Section 1.2, in the context $\mathbb{R}^2$. As a product, the right member of (3-41) has associated with it the usual product projections, as discussed in Section 2.5. In view of the identical nature of all the sets in the product here, however, it is helpful to establish a special notation for the projection. To project out the nth set in the product, use the projection

$$p_T[n]: T \times T \times T \times \cdots \to T. \tag{3-42}$$

This is pictured in Fig. 3.6. To project out a number of consecutive factors, use the standard interval notation

$$[n_1, n_2] = \{n_1, n_1 + 1, \dots, n_2 - 1, n_2\}, \tag{3-43}$$

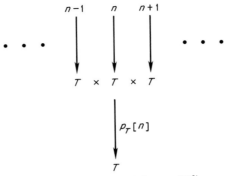

Fig. 3.6. Projecting nth factor of T^N.

and write

$$p_T[n_1, n_2]: T \times T \times T \times \cdots \to T^m \qquad (3\text{-}44a)$$

if there are m elements in (3-43). In this chapter, $n_1 \le n_2$ is assumed. Note that, strictly speaking, the subscript on p in the left member of (3-44a) ought to be T^m in place of T. The m is suppressed for reasons of simplicity. Action of the projection function (3-44a) is given by

$$p_T[n_1, n_2](t_*) = (p_T[n_1](t_*), p_T[n_1 + 1](t_*), \ldots, p_T[n_2](t_*)). \qquad (3\text{-}44b)$$

For a pictorial rendering of (3-44), the diagram of Fig. 3.7 may be useful.
Consider now the *sequence segment*

$$t(n_1), t(n_1 + 1), \ldots, t(n_2). \qquad (3\text{-}45)$$

Denote (3-45) by $t_*[n_1, n_2]$. Then

$$t_*[n_1, n_2] = p_T[n_1, n_2](t_*). \qquad (3\text{-}46)$$

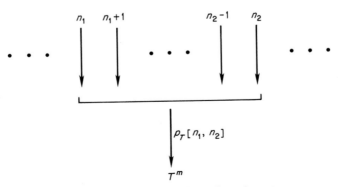

Fig. 3.7. The projection $p_T[n_1, n_2]$.

The entire sequence t_* can itself be (re)constructed in this way, for

$$t_* = t_*[0, \infty) = p_T[0, \infty)(t_*), \tag{3-47a,b}$$

where $[n_1, \infty)$ has been used as the traditional replacement for $[n_1, n_2]$ when n_2 is infinity. In this way, it is possible to generate from $T^{\mathbb{N}}$ a set of sequence segments $SS(T^{\mathbb{N}})$.

The definitions of the global dynamical functions extend directly to these new sets. The extension for the global output function is

$$h: X \times SS(U^{\mathbb{N}}) \to SS(Y^{\mathbb{N}}) \tag{3-48}$$

and can be written in the manner

$$h(x_{k_1}, u_*[k_1, k_2]) = y_*[k_1, k_2]. \tag{3-49}$$

It seems reasonable to leave the symbol h unaltered from (3-37), because (3-37) can be regarded as a special case of (3-49), with k_1 equal to zero and k_2 infinite. Similarly, the extension for the global transition function is

$$\pi: X \times SS(U^{\mathbb{N}}) \to SS(X^{\mathbb{N}}); \tag{3-50}$$

it can be written in the manner

$$\pi(x_{k_1}, u_*[k_1, k_2]) = x_*[k_1, k_2 + 1]. \tag{3-51}$$

The extension of the global transition function brings up an interesting point. What happens if no input is applied? From Section 2.1, remember that the empty set $\varnothing$ is a subset of every set. In the treatment of the set $SS(U^{\mathbb{N}})$, it is suggestive to denote the empty sequence by $\not{u}$. Then

$$\pi(x_{k_1}, \not{u}) = x_*[k_1] = x_{k_1}, \tag{3-52}$$

which states that, if no input is applied to the SDS, then the state remains unchanged. The reader should take care to observe that (3-52) does not say that $x_{k_1 + 1}$ is equal to x_{k_1}. Rather it says that no transition takes place. If the empty sequence in $SS(Y^{\mathbb{N}})$ is denoted by $\not{y}$, then

$$h(x_k, \not{u}) = \not{y}, \tag{3-53}$$

which means that the SDS produces no output whatsoever.

In concluding this section, it may be remarked that it is a common practice with some authors to use the symbol t_* for any element in $SS(T^{\mathbb{N}})$ and to suppress the natural numbers n_1 and n_2 in $t_*[n_1, n_2]$. Then (3-49) would look like (3-37). This does have some advantages, as for example when it is desired to concatenate two elements of $SS(T^{\mathbb{N}})$. The notation of this section also has some conceptual advantages, however; and so n_1 and n_2 will not generally be suppressed.

3.4 REACHABILITY

Two of the most heavily used systems concepts have been those related to how well an input sequence is able to influence the state of an SDS and to how well the state of an SDS can influence the output sequence. Both of these issues are of course an influence upon how economical a given state set X is, with regard to parameterizing input–output behavior of the SDS.

In this chapter, a bit more emphasis is placed upon the latter idea than upon the former, in the sense that ideas from lattice theory will be applied. However, both are discussed, for the sake of completeness, and this section begins with the former concept.

A state x in X is said to be *reachable from a state x_0* in X if there is an input sequence segment $u_*[0, k]$ in $SS(U^{\mathbb{N}})$ such that the resulting state sequence segment

$$x_*[0, k + 1] = \pi(x_0, u_*[0, k]) \tag{3-54a}$$

in $SS(X^{\mathbb{N}})$ satisfies

$$x_{k+1} = x. \tag{3-54b}$$

An SDS is reachable from a state x_0 in X if every state x in X is reachable from x_0.

The construction (3-54b) can be formalized, and it is sometimes quite convenient to do so. Note that the sequence segment $x_*[0, k + 1]$ in (3-54a) in essence is a member of X^{k+2}. Denote by

$$\tilde{p}_X[k + 1] : X^{k+2} \to X \tag{3-55}$$

the projection of X^{k+2} onto the last of its factors, along the lines of Fig. 3.8.

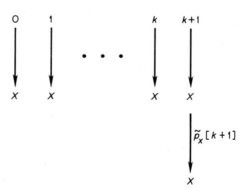

Fig. 3.8. Visualization of (3-55).

Then an action

$$((\tilde{p}_X[k+1]) \circ \pi)(x_0, u_*[0,k]) \tag{3-56}$$

is possible, and results in a function

$$\phi: X \times SS(U^{\mathbb{N}}) \to X, \tag{3-57}$$

whose action provides the state which results from the input sequence $u_*[0,k]$ applied to an SDS in initial state x_0 according to the rule

$$\phi(x_0, u_*[0,k]) = x_{k+1}. \tag{3-58}$$

It is natural to extend the idea of (3-58) so as to encompass the case

$$\phi(x_{k_1}, u_*[k_1, k_2]) = x_{k_2+1}. \tag{3-59}$$

The function ϕ will be called the *finite transition function*. In terms of ϕ, the concept of reachability can be expressed quite succinctly.

   The idea is as follows. If the argument x in (3-57) is fixed, ϕ becomes essentially a function

$$SS(U^{\mathbb{N}}) \to X, \tag{3-60}$$

which may be denoted by ϕ_x, and which has the defining action

$$\phi_x(u_*[k_1, k_2]) = \phi(x, u_*[k_1, k_2]). \tag{3-61}$$

Then an SDS is reachable from a state x if and only if ϕ_x is surjective.

EXAMPLE 3.4-1

   The concepts of reachability, as introduced above, can be used to establish an interesting relation B on X. To do this, simply agree that

$$x_1 B x_2 \tag{3-62}$$

if x_2 is reachable from x_1. It is of interest to check some of the properties of this relation. B is reflexive, because

$$\pi(x, \psi) = x, \tag{3-63}$$

which means that x is reachable from x for every x in X. Moreover, B is transitive. Suppose that

$$x_1 B x_2 \quad \text{and} \quad x_2 B x_3; \tag{3-64a,b}$$

then it must be established that (x_1, x_3) is also in B. Let $u_*[0, k_{12}]$ be the input sequence segment which satisfies

$$\phi(x_1, u_*[0, k_{12}]) = x_2, \tag{3-65a}$$

and let $u_*[0, k_{23}]$ be the corresponding segment satisfying

$$\phi(x_2, u_*[0, k_{23}]) = x_3. \tag{3-65b}$$

A composition activity can be carried out in (3-65) to give

$$\phi(\phi(x_1, u_*[0, k_{12}]), u_*[0, k_{23}]) = x_3. \tag{3-66}$$

But this suggests the specification of another input sequence segment $u_*[0, k_{13}]$, where k_{13} is one less than the sum of the number of natural numbers in $[0, k_{12}]$ and in $[0, k_{23}]$, namely, $k_{12} + k_{23} + 1$. A straightforward way to construct $u_*[0, k_{13}]$ is to note that

$$u_*[0, k_{12}] \in U^{k_{12}+1}, \qquad u_*[0, k_{23}] \in U^{k_{23}+1}. \tag{3-67a,b}$$

Then form

$$U^{k_{13}+1} = U^{k_{12}+1} \times U^{k_{23}+1}, \tag{3-68}$$

and define

$$u_*[0, k_{13}] \in U^{k_{13}+1} \tag{3-69a}$$

by

$$u_*[0, k_{13}] = (u_*[0, k_{12}], u_*[0, k_{23}]). \tag{3-69b}$$

It is then an easy exercise to establish that

$$\phi(x_1, u_*[0, k_{13}]) = x_3. \tag{3-70}$$

This completes the argument that the binary relation B is transitive. B is not, however, symmetric, so that no equivalence classes are established on X. A binary relation B which is reflexive and transitive is sometimes called a *quasi-ordering*. For readers having some familiarity with topology, two topological spaces can be quasi-ordered by the relation that one is homeomorphic with a subset of the other. The algebraic counterpart of this is the binary relation of inclusion in a set theoretic sense on a power set. ∎

The idea of pairing two sequence segments, as in (3-69), can be generalized. Consider $SS(T^{\mathbb{N}})$, and let $t_*[n_1, n_2]$ and $t_*[n_3, n_4]$ be any two sequence segments in that set. Now regard

$$t_*[n_1, n_2] \in T^{m_{12}}, \qquad t_*[n_3, n_4] \in T^{m_{34}}, \tag{3-71a,b}$$

where m_{12} is the number of entries in the interval $[n_1, n_2]$ and m_{34} is the corresponding number for $[n_3, n_4]$. By forming

$$T^{m_{12}} \times T^{m_{34}}, \tag{3-72}$$

the segments can be paired in the manner

$$(t_*[n_1, n_2], t_*[n_3, n_4]), \tag{3-73}$$

which is then bijectively related to some element

$$t_*[n_5, n_6] \in T^{m_{56}}, \tag{3-74}$$

where m_{56} is the sum of m_{12} and m_{34}. The nature of such a bijection has been discussed already in a prior section, namely, Section 2.5. There is a minor difficulty with (3-74), however, because there is no natural way to assign the natural numbers n_5 and n_6 in terms of n_1 through n_4.

The key to resolving this difficulty has been discussed at the conclusion of the preceding section, and is well exemplified by the transitivity discussion in Example 3.4-1, especially in (3-69b). Essentially, the notation $t_*[n_1, n_2]$ is merely a conceptual convenience. Almost all of the things which need to be said in the current context can be adequately phrased in the form $t_*[0, n]$. In other words, for an SDS as defined in Section 2.4, in which $\mathbb{N}$ is not a factor of the domain product for the local dynamical functions, the beginning point and the end point of a sequence segment is rarely of as much interest as its length. Moreover, for much work, even the length can be suppressed. Thus, for the present type of discussion, it makes good sense to notate an arbitrary element of $SS(T^{\mathbb{N}})$ by t_*. If t_*^1 and t_*^2 are two such elements, then there is a binary operation

$$\circ : SS(T^{\mathbb{N}}) \times SS(T^{\mathbb{N}}) \to SS(T^{\mathbb{N}}) \tag{3-75a}$$

whose action is given by

$$t_*^1 \circ t_*^2 = (t_*^1, t_*^2). \tag{3-75b}$$

It should be noted that this operation is associative and has a unit, which may be taken as t.

As a consequence, an SDS which permits only binary relations on its sets does in fact induce without much further ado a binary operation on $SS(U^{\mathbb{N}})$, which is part of the domain product of its global dynamical functions. This is in much the same spirit as the composition discussion of the introductory paragraphs to this chapter.

EXAMPLE 3.4-2

An interesting illustration of the use of this notation is available in connection with the finite transition function ϕ on $X \times SS(U^{\mathbb{N}})$ to X. If the argument x is fixed, it has been seen that a function $\phi_x : SS(U^{\mathbb{N}}) \to X$ results. Along the lines of Example 2.3-1, each such function ϕ_x induces an equivalence relation on $SS(U^{\mathbb{N}})$. The statement

$$u_*^1 E u_*^2 (\phi_x) \tag{3-76}$$

then means that u_*^1 and u_*^2 carry an initial state x into the same resulting state

$$\phi_x(u_*^1) = \phi_x(u_*^2). \tag{3-77}$$

As pointed out in Section 2.6, the usage $E(\phi_x)$ can be helpful. Here, the quotient set

$$SS(U^{\mathbb{N}})/E(\phi_x) \tag{3-78}$$

is an illustration of such a usage. The equivalence classes (3-78) are of interest in the study of reachability. ∎

Exercises

3.4-1. Establish (3-70).
3.4-2. In Example 3.4-1, a binary relation B is established in (3-62). Show that B is not symmetric.

3.5 OBSERVABILITY

Turning now to the companion concept of reachability, consider the way in which the state of an SDS can influence the output sequence.

At the heart of this issue is the global output function

$$h: X \times U^{\mathbb{N}} \to Y^{\mathbb{N}} \tag{3-79a}$$

of the SDS, with action specified by

$$h(x_0, u_0, u_1, u_2, \dots) = (y_0, y_1, y_2, \dots). \tag{3-79b}$$

For (3-79b), the shorter version

$$h(x_0, u_*) = y_* \tag{3-79c}$$

can be used, in the manner of Section 3.3. The reader will note that (3-79) is not the extended version of the global output function, which has $X \times SS(U^{\mathbb{N}})$ for a domain and $SS(Y^{\mathbb{N}})$ for a codomain. Though the extended version is quite useful, especially for the transition functions and for discussions of reachability, the simpler form (3-79) is quite adequate for the purposes of this section.

Several types of questions can be asked about the global output function h. In one way or another, most of these questions center upon its nature relative to injection. The best of all possible worlds would have h injective, and this would mean that knowledge of an output sequence y_* would be completely adequate to determine both the initial state x_0 and the input sequence u_*. Such a situation is not impossible, but it is hardly the everyday case.

If h itself is not to be injective, then there are two significant ways to alter h in order to form new functions which may be injective. One of these ways is to regard the initial state x_0 as fixed at some specified x in X. Under this assumption, h becomes in essence a function

$$U^{\mathbb{N}} \to Y^{\mathbb{N}}, \tag{3-80}$$

which can be denoted by h_x. The reader can compare this situation with the corresponding statements for the function ϕ_x in the section preceding. When h_x is injective, it means that, for the SDS in state x, it is always possible in principle to invert the SDS, in the sense that the output sequence y_* can be used to construct the input sequence u_*. The subject of invertibility of systems will be treated in some detail later in the volume, after more algebraic structure has been placed on the SDS input, state, and output sets. The notion of invertibility is fundamental to many applications of system theory in communications and control. At this junction, however, the more interesting argument to consider fixed is the input sequence u_*.

Fixing the input sequence u_* in effect makes h into a function

$$X \to Y^{\mathbb{N}}, \tag{3-81}$$

which is denoted by h_{u_*}. If h_{u_*} is injective, then the SDS is *observable relative to the input sequence* u_*.

It sometimes happens that an SDS definition makes no provision for inputs. Such an SDS might be described by a local transition function

$$f : X \to X \tag{3-82}$$

and a local output function

$$g : X \to Y. \tag{3-83}$$

Then the list

$$(J, \leq ; X, Y; f, g) \tag{3-84}$$

would serve to specify it. There is a difficulty, however, in the fact that the SDS discussion of Section 2.4 really makes no provision for the SDS (3-84) to operate, for operation has been explained as the result of applying an input.

This disparity is more apparent than actual. In the case of a proposed SDS such as (3-84), simply define a singleton set

$$U = \{1\} \tag{3-85}$$

of inputs. Because the set of inputs has but one element, it is suggestive to denote that element by 1. No connection is intended with any other 1s with which the reader might be familiar. Now define an SDS local transition function

$$\tilde{f} : X \times U \to X \tag{3-86}$$

by the action

$$\tilde{f}(x, u) = f(x), \tag{3-87}$$

where the right member is computed according to (3-82). But u can take only the one value 1, so that, for all x in X

$$f(x) = \tilde{f}(x, 1) = \tilde{f}_1(x), \qquad (3\text{-}88)$$

and thus one may as well identify f with $\tilde{f}_1 : X \rightarrow X$. In a similar way, define a local output function

$$\tilde{g} : X \times U \rightarrow Y \qquad (3\text{-}89)$$

by

$$\tilde{g}(x, u) = g(x), \qquad (3\text{-}90)$$

with the right member computed by (3-83). Again,

$$g(x) = \tilde{g}(x, 1) = \tilde{g}_1(x) \qquad (3\text{-}91)$$

for all x in X, and g may as well be identified with $\tilde{g}_1 : X \rightarrow Y$.

In this way, (3-84) may be explained as an SDS

$$(J, \leq ; \{1\}, X, Y; \tilde{f}, \tilde{g}) \qquad (3\text{-}92)$$

in which

$$f = \tilde{f}_1, \qquad g = \tilde{g}_1, \qquad (3\text{-}93\text{a,b})$$

and the input set $\{1\}$ has been suppressed. An SDS of the type (3-84) is called *autonomous*.

It should be noted that an autonomous SDS cannot be represented as an SDS driven by the empty sequence μ in $SS(U^{\mathbb{N}})$. The reason for this is the fact that an autonomous SDS is operating, whereas an SDS driven by μ is not operating.

With the help of the scheme (3-92) for visualizing an autonomous SDS, global transition and global output functions can be inferred easily for an autonomous SDS. Only the output function is of immediate interest here, and so the transition function is left as an exercise.

The global output function for an SDS of type (3-92) has the action (3-79b) in which

$$(u_0, u_1, u_2, \dots) = (1, 1, 1, \dots). \qquad (3\text{-}94)$$

It is appropriate, perhaps, to designate the fixed sequence (3-94) by 1_*. Then the action

$$h(x, 1_*) \qquad (3\text{-}95\text{a})$$

on $X \times \{1\}^{\mathbb{N}}$ to $Y^{\mathbb{N}}$ can be replaced effectively by an action

$$h_{1_*}(x) \qquad (3\text{-}95\text{b})$$

on X to $Y^{\mathbb{N}}$. In fact, since 1_* is the only element in the set $\{1\}^{\mathbb{N}}$, the subscript 1_* in (3-95b) may as well be suppressed, giving the global output function action

$$h(x) \qquad (3\text{-}95c)$$

on X to $Y^{\mathbb{N}}$.

In a similar way, there is a global transition function

$$\pi: X \to X^{\mathbb{N}} \qquad (3\text{-}96)$$

for an autonomous SDS.

EXAMPLE 3.5-1

One of the ways in which an autonomous system arises is by feedback. Let the SDS local transition and output equations be given by

$$x_{k+1} = f(x_k, u_k), \qquad y_k = g(x_k, u_k). \qquad (3\text{-}97a,b)$$

Let V be a set of exogenous inputs, which may represent both disturbances, which are unmeasurable phenomena, and commands. The SDS (3-97) can be regarded as an object whose input is to be controlled, in a way based upon its state and upon the exogenous input. Define a feedback function

$$fb: X \times V \to U \qquad (3\text{-}98)$$

with an action

$$fb(x_k, v_k) = u_k. \qquad (3\text{-}99)$$

Then (3-97) becomes

$$x_{k+1} = f(x_k, fb(x_k, v_k)), \qquad y_k = g(x_k, fb(x_k, v_k)), \qquad (3\text{-}100a,b)$$

which can then be simplified to

$$x_{k+1} = \alpha(x_k, v_k), \qquad y_k = \beta(x_k, v_k). \qquad (3\text{-}101a,b)$$

Under feedback control, the SDS is of type

$$(J, \le; V, X, Y; \alpha, \beta). \qquad (3\text{-}102)$$

Autonomous operation occurs if V is a singleton set $\{1\}$, in which case the SDS

$$(J, \le; X, Y; \alpha_1, \beta_1) \qquad (3\text{-}103)$$

is obtained. Usually, the subscript 1 in (3-103) would be suppressed, giving finally

$$(J, \le; X, Y; \alpha, \beta). \qquad (3\text{-}104)$$

EXAMPLE 3.5-2

A variant on this example is to consider output feedback

$$\text{fb}: Y \times V \to U, \qquad\qquad (3\text{-}105)$$

with the action

$$\text{fb}(y_k, v_k) = u_k. \qquad\qquad (3\text{-}106)$$

In this case, (3-97) becomes

$$x_{k+1} = f(x_k, \text{fb}(y_k, v_k)), \qquad y_k = g(x_k, \text{fb}(y_k, v_k)). \qquad (3\text{-}107\text{a,b})$$

A controlled SDS description can be obtained if (3-107b) can be solved for y_k as a function of x_k and v_k, in other words, if there exists a solution

$$y_k = \theta(x_k, v_k) \qquad \text{for} \qquad \theta: X \times V \to Y \qquad (3\text{-}108\text{a,b})$$

of the equation (3-107b). Whether or not this is true depends, of course, on the specific nature of the actions of g, fb. ∎

EXAMPLE 3.5-3

Recall Example 2.3-1, which showed that every function induces an equivalence relation. Note that

$$h_{u_*}: X \to Y^{\mathbb{N}}, \qquad\qquad (3\text{-}109)$$

as a function, induces an equivalence relation on X. If (3-109) is extended to $SS(Y^{\mathbb{N}})$, then an entire family of equivalence relations can be generated, depending upon how much of the output sequence y_* is available. Sets of equivalence relations have lattice properties, as discussed in the following sections. ∎

Exercise

3.5. Consider the scheme (3-92) for visualizing an autonomous SDS. The text gives a treatment of global output functions for such a system. In an analogous way, determine the global transition function for an autonomous SDS. Refer to (3-96).

3.6   THE RELATIONAL LATTICE

Example 3.5-3 raised the question of families of relations on the same set, in that case the state set X. In this section, some of the properties of such families are explored, prior to their application later in the chapter.

For a beginning, denote by $B(S)$ the set of all binary relations on a set S. It turns out that there is a natural way to place a partial order relation $\leq$ on

$B(S)$ so as to make it into a poset. To accomplish this, simply agree that two elements B_1 and B_2 in $B(S)$ satisfy

$$B_1 \leq B_2 \quad \text{if} \quad (s_1, s_2) \in B_1 \quad \text{implies} \quad (s_1, s_2) \in B_2. \quad \text{(3-110a,b,c)}$$

Then $(B(S), \leq)$ is a poset. With reference to Section 3.1, this poset has a greatest element, which is the relation consisting of the entire S^2. It also has a least element, which is the empty set.

EXAMPLE 3.6-1

Let $S = \mathbb{Z}$, and consider Fig. 3.9. All the binary relations are subsets of $\mathbb{Z}^2$, according to the definition (2-48). Define

$$B_1 = \{(-3, -1), (0, 0)\}, \quad \text{(3-111a)}$$
$$B_2 = \{(-1, 2), (0, 3)\}, \quad \text{(3-111b)}$$
$$B_3 = \{(0, 0), (1, 1), (3, 2), (4, 0)\}, \quad \text{(3-111c)}$$
$$B_4 = \{(1, -2)\}, \quad \text{(3-111d)}$$
$$B_5 = \{(1, 1), (3, 2)\}. \quad \text{(3-111e)}$$

Then $B_5 \leq B_3$; however, no others among (3-111) stand in the relation $\leq$ to each other. ∎

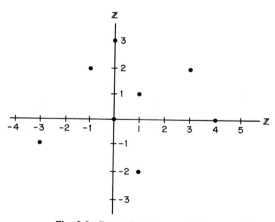

Fig. 3.9. Example binary relations.

A lattice structure can be developed on the poset $(B(S), \leq)$. To do this, two binary operations—of meet and of join—have to be developed.
The meet

$$\wedge : B(S) \times B(S) \rightarrow B(S) \quad \text{(3-112)}$$

can be put into place as follows. If B_1 and B_2 are elements of $B(S)$, define the action of $\wedge$ by

$$B_1 \wedge B_2 = B_1 \cap B_2, \tag{3-113}$$

where the right member is set theoretic intersection, understood in S^2. As a subset of S^2, (3-113) surely defines a binary relation on S. In fact, a brief reexamination of (3-110) shows that $B_1 \leq B_2$ is a statement of set theoretic inclusion in S^2. The join

$$\vee : B(S) \times B(S) \to B(S) \tag{3-114}$$

in a similar manner can be defined to have action, for B_1 and B_2 in $B(S)$, of

$$B_1 \vee B_2 = B_1 \cup B_2, \tag{3-115}$$

where the right member is set theoretic union, again understood in S^2.

It should be noted that the meet of two relations in S^2 can very well be the empty relation, which is the least element of $B(S)$.

The lattice $(B(S), \leq, \wedge, \vee)$ arises quite easily, then, when it is remembered that a binary relation on S is a subset of S^2. There is then a bijection

$$B(S) \approx \mathbb{P}(S^2) \tag{3-116}$$

from the set of all binary relations on S to the power set of S^2. The lattice structure $(\mathbb{P}(S^2), \subset, \cap, \cup)$ discussed in Section 3.2 can then be identified with that of $(B(S), \leq, \wedge, \vee)$. For sake of viewpoint however, the latter is called a *relational lattice*.

EXAMPLE 3.6-2

With reference to the first example of this section,

$$B_1 \wedge B_3 = \{(0,0)\}, \tag{3-117a}$$
$$B_3 \wedge B_5 = B_5, \tag{3-117b}$$
$$B_2 \vee B_4 = \{(-1,2),(0,3),(1,-2)\}. \tag{3-117c}$$

∎

Now the reader will recall from Section 3.2 that a lattice is a poset in which any two elements have both an infimum and a supremum. It is next desirable to extend this notion to arbitrary subsets.

A lattice $(L, \leq, \wedge, \vee)$ in which *every* subset has both an infimum and a supremum is said to be *complete*. The extension here is that the subsets may now have more than two elements. In fact, the number of elements in a subset need not be finite.

EXAMPLE 3.6-3

Consider any complete lattice $(L, \leq, \wedge, \vee)$. Inasmuch as *any* subset of L must have an infimum, select L itself as a subset. Then L must have an infimum, which must satisfy

$$\inf L \leq v \qquad (3\text{-}118)$$

for every v in L; moreover, $\inf L$ is an element of L, because the lattice is complete. As a lower bound (3-118) of L and as an element of L, $\inf L$ must then be the unique least element of L. Likewise, L must have a greatest element $\sup L$.

∎

An important feature of a bounded poset $(P, \leq)$ is this: if each of the nonempty subsets of P has an infimum, then each of the nonempty subsets will have a supremum also. This means that the infimum property for arbitrary nonvoid sets leads to the complete lattice structure. It is relatively easy to see how this feature occurs. Let W be an arbitrary nonempty subset of P; $\inf W$ is available by assumption. Form the set $\mathrm{UB}(W)$ of upper bounds of W in P. $\mathrm{UB}(W)$ cannot be empty, because the poset $(P, \leq)$ is bounded. Then

$$\inf(\mathrm{UB}(W)) \qquad (3\text{-}119)$$

exists by assumption as well. As an infimum of upper bounds, (3-119) must itself be an upper bound for any lower bound to $\mathrm{UB}(W)$. But any element w in W is a lower bound for $\mathrm{UB}(W)$, and so

$$\inf(\mathrm{UB}(W)) \in \mathrm{UB}(W). \qquad (3\text{-}120)$$

According to Section 3.1, then, $\inf(\mathrm{UB}(W))$ must be the unique least element of $\mathrm{UB}(W)$. It then follows that

$$\inf(\mathrm{UB}(W)) = \sup W, \qquad (3\text{-}121)$$

as desired.

This idea can be applied immediately to the lattice $(\mathbb{P}(S), \subset, \cap, \cup)$ associated with any arbitrary set S.

EXAMPLE 3.6-4

Let S be an arbitrary set, and let $\mathbb{P}(S)$ be its power set according to Section 2.1. As in Section 3.2, then, $(\mathbb{P}(S), \subset, \cap, \cup)$ is a lattice. Moreover, it is bounded below by the empty set and above by S itself. Let W be any nonempty subset of $\mathbb{P}(S)$, indexed by $v \in V$ in the manner

$$W = \{S_v | S_v \subset S \text{ and } v \in V\}. \qquad (3\text{-}122)$$

Then

$$\inf W = \bigcap_{v \in V} S_v \tag{3-123}$$

makes $(\mathbb{P}(S), \subset, \cap, \cup)$ into a complete lattice. ∎

As a consequence of this example and of (3-116), it follows that the relational lattice $(B(S), \le, \wedge, \vee)$ is complete as well. All that is necessary in Example 3.6-4 is to replace the arbitrary set S by S^2.

Now Section 2.3 has brought out the fact that binary relations are often classified according to their properties. For example, such a relation may be reflexive, or symmetric, or transitive.

It is fundamental to ask the question whether such properties are preserved under the lattice operations of meet and join. For example, is the meet of two reflexive relations a reflexive relation? The reader has probably already answered in the affirmative. In the next section, this issue is discussed, in the context of what is called the equivalence lattice.

Exercises

3.6-1. At the outset of discussion in this section, a proposed poset structure

$$(B(S), \le)$$

was laid out on $B(S)$, the set of all binary relations on the set S, with (3-110) outlining the suggested scheme for partial ordering. Verify that (3-110) does indeed establish a partial order relation.

3.6-2. Explain how S^2 is a greatest element for the poset of Exercise 3.6-1. How is the empty set a least element?

3.7 THE EQUIVALENCE LATTICE

The section preceding this one has raised the question whether or not the standard properties of binary relations, such as reflexivity, symmetry, and so forth, are preserved under the lattice operations of meet and join in $(B(S), \le, \wedge, \vee)$.

EXAMPLE 3.7-1

Consider two binary relations B_1 and B_2 in $(B(S), \le, \wedge, \vee)$, and suppose that they are both reflexive, that is,

$$(s, s) \in B_1, \qquad (s, s) \in B_2, \tag{3-124a,b}$$

for all s in S. Then (s, s) must clearly be in the set theoretic intersection of B_1 and B_2, considered as subsets of S^2, for all s in S. Thus

$$(s, s) \in B_1 \wedge B_2. \tag{3-124c}$$

Further, (s, s) must be in the set theoretic union of B_1 and B_2, considered as subsets of S^2, for all s in S. Thus

$$(s, s) \in B_1 \vee B_2, \tag{3-124d}$$

as well.

∎

This example permits illustration of another idea about lattices. Let W be any subset of a lattice $(L, \leq, \wedge, \vee)$. If W is closed under the binary operations $\wedge$ and $\vee$, in the sense of Section 3.2, then $(W, \leq, \wedge, \vee)$ is said to be a *sublattice* of $(L, \leq, \wedge, \vee)$. Some interpretation is necessary. The partial order relation in the sublattice is to be regarded as a subset of $W \times W$, itself a subset of $L \times L$, and the $\wedge$ and $\vee$ operations in the sublattice are restrictions

$$W \times W \rightarrow W \tag{3-125}$$

of the lattice operations

$$L \times L \rightarrow L. \tag{3-126}$$

Example 3.7-1 shows that the subset of binary relations which are reflexive is actually a sublattice. This means, of course, that it is a lattice in its own right, but contained within the greater lattice of all binary relations.

The fact that reflexive relations are a sublattice suggests a similar question for symmetric relations.

EXAMPLE 3.7-2

Two binary relations B_1 and B_2 in $(B(S), \leq, \wedge, \vee)$ are symmetric, which is to say that

$$(s_1, s_2) \in B_i \qquad \text{implies} \qquad (s_2, s_1) \in B_i, \tag{3-127a,b}$$

for i equal to 1 and 2. If, therefore, (s_1, s_2) is an element of both B_1 and B_2, then so is (s_2, s_1). Moreover, if (s_1, s_2) is an element of either B_1 or B_2, then so is (s_2, s_1). Thus

$$(s_1, s_2) \in B_1 \wedge B_2 \qquad \text{implies} \qquad (s_2, s_1) \in B_2 \wedge B_1, \tag{3-128a,b}$$

and

$$(s_1, s_2) \in B_1 \vee B_2 \qquad \text{implies} \qquad (s_2, s_1) \in B_1 \vee B_2. \qquad \text{(3-129a,b)}$$

Closed under meet and join, the symmetric relations are also a sublattice.

∎

In passing, note that the r-s relations, which are both reflexive and symmetric, are also a sublattice, as a consequence of these two examples.

Unfortunately, the transitive property is not compatible with being a sublattice of the relational lattice.

EXAMPLE 3.7-3

Once again, examine B_1 and B_2 in $(B(S), \leq, \wedge, \vee)$, and assume that they are transitive relations. If (s_1, s_2) is an element of both B_1 and B_2, and if (s_2, s_3) is an element of both B_1 and B_2, then it follows from the transitive property of B_1 that (s_1, s_3) is an element of B_1 and from the transitive property of B_2 that (s_1, s_3) is an element of B_2. Thus the transitive relations are closed under meet $\wedge$. However, the transitive relations are not closed under the join of the relational lattice. Let S be $\mathbb{Z}$, and define B_1 and B_2 by

$$B_1 = \{(0,0), (1,1), (0,1), (1,0)\}, \qquad \text{(3-130a)}$$
$$B_2 = \{(0,0), (2,2), (1,2), (2,1), (1,1)\}. \qquad \text{(3-130b)}$$

The relational lattice join is just

$$B_1 \vee B_2 = \{(0,0), (1,1), (2,2), (0,1), (1,0), (1,2), (2,1)\}, \qquad \text{(3-131)}$$

which, for example, does not contain $(0, 2)$ even though it does contain $(0,1)$ and $(1,2)$.

∎

In the study of observers in the section following, it is helpful to place a lattice structure on the set of equivalence relations associated with a given set S. Denote this set of equivalence relations by $E(S)$. It is clear that

$$E(S) \subset B(S), \qquad \text{(3-132)}$$

in a set theoretic sense. Moreover, the greatest element of the relational lattice, which corresponds to the binary relation consisting of all of S^2, is an element of $E(S)$ as well. Indeed, define the equivalence relation E^1 by

$$s_1 \, E^1 \, s_2 \qquad \text{if} \qquad (s_1, s_2) \in S^2. \qquad \text{(3-133a,b)}$$

Then $E^1 \in B(S)$ is the desired greatest element. The equivalence subset $E(S)$ inherits the partial order relation $\leq$ from the relational lattice as well. Thus it is a poset bounded above by E^1. It is also bounded below, but not by the least element of the relational lattice. The reader will recall from the previous section that that least element was the empty subset in S^2. The least element in $E(S)$ is the equivalence relation E^0 defined by

$$s_1 E^0 s_2 \qquad \text{if} \qquad s_1 = s_2. \qquad\qquad \text{(3-134a,b)}$$

It is clear that E^0 is the equality relation $=$, as discussed in Section 2.3. Thus $(E(S), \leq)$ is a poset which is bounded above, by E^1, and below, by E^0. From Section 3.6, then, if an infimum operation can be successfully introduced on this poset, it will become a complete lattice in its own right.

The introduction of an infimum operation onto the bounded poset $(E(S), \leq)$ can be aided by the observation that the meet operation on the relational lattice preserved the r-s-t properties, as shown in Examples 3.7-1–3.7-3. This very basic observation means that the relational lattice meet can be restricted to the poset $(E(S), \leq)$. So $(E(S), \leq, \wedge)$, as a sub-algebraic system of $(B(S), \leq, \wedge)$, has an infimum available on its subsets. This is the same type of infimum construction which was made in the relational lattice, namely,

$$E_1 \wedge E_2 = E_1 \cap E_2, \qquad\qquad \text{(3-135)}$$

where the right member is set theoretic intersection, understood on the set S^2. It is very useful in the remaining sections of this chapter to keep this fact in mind.

Recall now from Section 3.6 the feature of bounded posets, namely, that the presence of an infimum on nonempty subsets implies the existence of a supremum on nonempty subsets. In effect, that supremum was defined to be the infimum of the set of upper bounds for a given nonempty subset, with at least one upper bound being assured from the boundedness of the poset itself.

But $(E(S), \leq, \wedge)$ is a bounded poset, with an infimum induced according to (3-135) by that on the relational lattice. Accordingly, there is a join operation available on equivalence relations as well, denoted by $\vee_E$ to distinguish it from $\vee$ in the relational lattice. With this join, $(E(S), \leq, \wedge, \vee_E)$ becomes a complete lattice, called the *equivalence lattice* of S. It is a subset of the relational lattice, but not a sublattice of it. It has in essence the same meet, but of necessity displays a different join.

However, the distinction between these two joins is perhaps not a major point, because both joins can be regarded as coming from the same type of infimum construction. In other words, both joins are consequences of applying the same infimum idea to subsets of upper bounds. It happens, as this is carried out, that the equivalence lattice supremum cannot be understood as a union, whereas the relational lattice supremum can be so understood.

EXAMPLE 3.7-4

To see some of these issues in pictorial form, adapt Example 2.6-1. The set S is taken to be a subset $\{1, 2, 3\}$ of the integers $\mathbb{Z}$. There are the following possibilities for equivalence relations on this S:

$$E^0 = \{(1,1),(2,2),(3,3)\}, \tag{3-136a}$$

$$E_1 = \{(1,1),(2,2),(3,3),(1,2),(2,1)\}, \tag{3-136b}$$

$$E_2 = \{(1,1),(2,2),(3,3),(1,3),(3,1)\}, \tag{3-136c}$$

$$E_3 = \{(1,1),(2,2),(3,3),(2,3),(3,2)\}, \tag{3-136d}$$

$$E^1 = \{(1,1),(2,2),(3,3),(1,2),(2,1),(1,3),(3,1),(2,3),(3,2)\}. \tag{3-136e}$$

These have been pictured in Fig. 3.10, where the letters on the plot indicate which of the equations (3-136) is being shown. Consider the construction

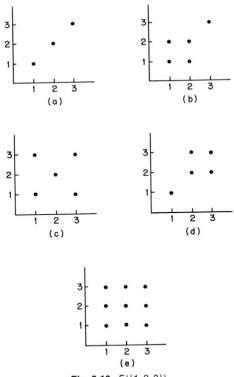

Fig. 3.10. $E(\{1,2,3\})$.

$$E_1 \vee E_2, \tag{3-137}$$

where the join from the relational lattice has been employed. This construction does not give an element of $E(S)$. However, $E_1 \vee E_2$ does have an upper bound in $E(S)$, namely, E^1. Because E^1 is the only upper bound, it is also the infimum of upper bounds for $E_1 \vee E_2$. Thus

$$E_1 \vee_E E_2 = E^1. \tag{3-138}$$

Similarly,

$$E_2 \vee_E E_3 = E^1 \tag{3-139}$$

and

$$E_1 \vee_E E_3 = E^1 ; \tag{3-140}$$

but

$$E^0 \vee E_1 = E^0 \vee_E E_1 = E_1. \tag{3-141}$$

The whole partial order scheme is easily visualized in Fig. 3.11. ∎

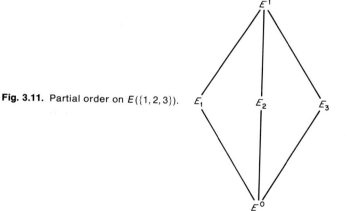

Fig. 3.11. Partial order on $E(\{1, 2, 3\})$.

In concluding this discussion of the equivalence lattice $(E(S), \leq, \wedge, \vee_E)$, it should be noted that the meet of two equivalence relations is never empty, as it can be for two arbitrary binary relations. This is a consequence of the least element E^0 for $E(S)$ being the equality relation. The bottom line of all meets in an equivalence lattice is thus the equality relation. Achieving this bottom line is a goal of the next section.

Exercises

3.7-1. Show that the binary relations in

$$(B(S), \leq, \wedge, \vee),$$

which are both reflexive and symmetric, constitute a sublattice.

3.7-2. Show that the equivalence relation E^0 defined in (3-134) is a least element in $E(S)$.

3.8 OBSERVER MODELS

In Section 3.5, it was stated that an SDS

$$(J, \leq; U, X, Y; f, g) \tag{3-142}$$

was observable relative to the input sequence u_* if the function

$$h_{u_*}: X \to Y^{\mathbb{N}} \tag{3-143}$$

was injective. If the SDS input set U was just the singleton set $\{1\}$, it was seen that the sequence notation

$$u_* = 1_* = (1, 1, 1, \ldots) \tag{3-144}$$

might be suppressed, with the result that (3-143) takes on the appearance

$$h: X \to Y^{\mathbb{N}}. \tag{3-145}$$

With only one possible input sequence 1_*, the terminology "relative to the input sequence 1_*" is not needed, and the SDS is simply said to be *observable* if the function h of (3-145) is injective. Note that h is being considered as a special type of global output function.

One way in which such a situation often arises has been seen in the autonomous SDS

$$(J, \leq; X, Y; \alpha, \beta), \tag{3-146}$$

where the local transition equation is of the form

$$x_{k+1} = \alpha(x_k) \tag{3-147a}$$

and the local output equation has the appearance

$$y_k = \beta(x_k). \tag{3-147b}$$

For an illustration of how such an autonomous SDS (3-146) could arise from an SDS (3-142) under feedback control, the reader is invited to reconsider Example 3.5-1.

This section begins with an examination of what it means for the global output function arising from (3-146) to be injective. This study can be carried out with the aid of the equivalence lattice. It is easy to see that the action of the global output function is given by

$$h(x_0) = (\beta(x_0), (\beta \circ \alpha)(x_0), (\beta \circ \alpha^2)(x_0), \ldots), \tag{3-148}$$

where α^2 is a notation for $\alpha \circ \alpha$. For h to be injective, therefore, it must be true that knowledge of the action of the functions

$$\beta : X \to Y, \qquad \beta \circ \alpha : X \to Y, \qquad \beta \circ \alpha^2 : X \to Y, \qquad \ldots \tag{3-149a,b,c}$$

on a state x_0 in X must be enough to determine that state uniquely. But each of the functions in (3-149) determines x_0 only up to its membership in an equivalence class.

Consider the set of all equivalence relations $E(X)$ on the state set X. The last section has established that $E(X)$ admits the structure of a complete lattice $(E(X), \leq, \wedge, \vee_E)$. In the notation of Section 2.6, (3-149) leads immediately to the equivalence relations

$$E(\beta), \quad E(\beta \circ \alpha), \quad E(\beta \circ \alpha^2), \ldots \tag{3-150}$$

on X. Suppose now for a moment that there is some other state $\tilde{x}_0$ such that

$$(x_0, \tilde{x}_0) \in E(\beta \circ \alpha^i), \qquad i \in \mathbb{N}, \tag{3-151}$$

where $\alpha^0 : X \to X$ is to be understood as 1_X. This is the sort of thing which would occur when the SDS is not observable. In view of the intersection theoretic nature of the relational lattice meet, which restricts to the equivalence lattice meet, it is clear that (3-151) is to be interpreted as

$$(x_0, \tilde{x}_0) \in \inf\{E(\beta \circ \alpha^i), i \in \mathbb{N}\}. \tag{3-152}$$

For observability of the SDS, $(x_0, \tilde{x}_0)$ must be a member of the equality relation E^0. Thus the SDS (3-146) is observable if and only if

$$\inf\{E(\beta \circ \alpha^i), i \in \mathbb{N}\} = E^0. \tag{3-153}$$

In this case, the pair (β, α) is said to be an *observable pair*.

Of course, it may happen that a pair (β, α) fails to be observable. In such a case, the most knowledge that can be obtained about the value of the initial state x_0 is in which of the equivalence classes associated with

$$E(\beta, \alpha) = \inf\{E(\beta \circ \alpha^i), i \in \mathbb{N}\} \tag{3-154}$$

it has its membership. The equivalence relation $E(\beta, \alpha)$ defined by (3-154) will be used later to construct a state set for an SDS to "observe" the SDS (3-146). Some of the properties of $E(\beta, \alpha)$ are brought out in the following examples.

EXAMPLE 3.8-1

From the definition of $E(\beta, \alpha)$, as an infimum for the equivalence relations $E(\beta \circ \alpha^i)$, $i \in \mathbb{N}$, it is clear that

$$E(\beta, \alpha) \le E(\beta). \tag{3-155}$$

This modest observation has an interesting diagrammatical consequence. Consider Fig. 3.12. If

$$x_1 \, E(\beta, \alpha) \, x_2, \tag{3-156a}$$

then by definition of the partial order in (3-155),

$$x_1 \equiv x_2(\beta), \tag{3-156b}$$

which means that

$$\beta(x_1) = \beta(x_2). \tag{3-156c}$$

Thus there is a unique $\bar{\beta} : X/E(\beta, \alpha) \to Y$, as shown in the figure.   ∎

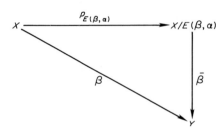

Fig. 3.12. Induced local output map.

This example shows that there is local dynamical equivalence insofar as the local output function β is concerned. It is next appropriate to consider local transition equivalence.

EXAMPLE 3.8-2

Continuing with the argument, note that

$$E(\beta, \alpha) \le E(\beta \circ \alpha^i), \qquad i \in \mathbb{N}. \tag{3-157}$$

This means that

$$x_1 \, E(\beta, \alpha) \, x_2 \qquad \text{implies} \qquad x_1 \equiv x_2(\beta \circ \alpha^i), \qquad i \in \mathbb{N}, \tag{3-158a,b}$$

which in turn gives

$$(\beta \circ \alpha^{i+1})(x_1) = (\beta \circ \alpha^{i+1})(x_2), \qquad i \in \mathbb{N}. \tag{3-158c}$$

Rewrite this last equation and obtain

$$(\beta \circ \alpha^i)(\alpha(x_1)) = (\beta \circ \alpha^i)(\alpha(x_2)), \qquad i \in \mathbb{N}, \tag{3-158d}$$

which means that

$$\alpha(x_1) \, E(\beta, \alpha) \, \alpha(x_2). \tag{3-158e}$$

But then, by Example 2.6-2, there exists a unique

$$\bar{\alpha}: X/E(\beta, \alpha) \to X/E(\beta, \alpha) \tag{3-159}$$

in such a way as to make the diagram of Fig. 3.13 commute. ∎

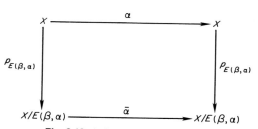

Fig. 3.13. Induced local transition.

The inference of this example is that local transition equivalence holds as well. Thus, local dynamical equivalence conditions are met by placing the equivalence relation $E(\beta, \alpha)$ on X. These remarks are minor extensions of the ideas of Section 2.8, which can be accommodated by the same process used to handle autonomous systems—namely, by regarding Section 2.8 in the context of a singleton input set $\{1\}$ for U.

Now that local dynamical equivalence has been established, it is possible to construct an SDS which observes the SDS (3-146). This is accomplished by applying the theory of Section 2.9 on induced set-dynamical systems.

Begin with the local transition equation (3-147a), and apply the projection

$$p_{E(\beta, \alpha)}: X \to X/E(\beta, \alpha) \tag{3-160}$$

to obtain

$$\bar{x}_{k+1} = (p_{E(\beta, \alpha)} \circ \alpha)(x_k) \tag{3-161a}$$

$$= (\bar{\alpha} \circ p_{E(\beta, \alpha)})(x_k) \tag{3-161b}$$

$$= \bar{\alpha}(\bar{x}_k). \tag{3-161c}$$

Note that (3-161b) follows from (3-161a) because the diagram of Fig. 3.13 commutes. Next, the local output equation becomes

$$y_k = \beta(x_k) \tag{3-162a}$$

$$= (\bar{\beta} \circ p_{E(\beta, \alpha)})(x_k) \tag{3-162b}$$

$$= \bar{\beta}(\bar{x}_k). \tag{3-162c}$$

Again, (3-162b) follows from (3-162a) because the diagram of Fig. 3.12 commutes. From (3-161) and (3-162) emerges the structure of an induced SDS

$$(J, \leq; X, E(\beta, \alpha); Y; \bar{\alpha}, \bar{\beta}), \tag{3-163}$$

which is called an *observer model* for the SDS (3-146). Of course, the pair $(\bar{\beta}, \bar{\alpha})$ is observable, by construction; this means that

$$E(\bar{\beta}, \bar{\alpha}) = E^0. \tag{3-164}$$

If loaded with the correct initial state

$$p_{E(\beta, \alpha)}(x_0) = \bar{x}_0, \tag{3-165}$$

the observer model produces exactly the same output sequence y_* as the autonomous SDS itself. Denote by

$$\bar{h}: X/E(\beta, \alpha) \to Y^{\mathbb{N}} \tag{3-166a}$$

the global output function of the observer model. Then

$$h = \bar{h} \circ p_{E(\beta, \alpha)}, \tag{3-166b}$$

which is a common type of factoring procedure in algebra. Note that $\bar{h}$ is injective, while $p_{E(\beta, \alpha)}$ is surjective. It turns out that every function $f: S \to T$ can be represented by the composition of an injective function with a surjective function. For example, the surjective function might be defined from $S \to \operatorname{Im} f$, while the injective function could be taken as an insertion $\operatorname{Im} f \to T$. The reader will note, however, that this is not the way in which (3-166b) has been accomplished. In this case, the surjective function has been chosen as $p_{E(f)}: S \to S/E(f)$, and the injective function has been taken as $\bar{f}: S/E(f) \to T$.

Once $\bar{x}_0$ has been determined in the observer model, its progress as a function of k in $\mathbb{N}$ can be determined by the transition functions of the observer model.

As a final point, it is a frequent event in the literature to see Figs. 3.12 and 3.13 combined. This has been done in Fig. 3.14.

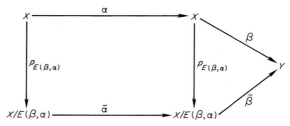

Fig. 3.14. Induced observer model.

Exercise

3.8.   Refer to (3-163). Establish in detail that the pair $(\bar{\beta}, \bar{\alpha})$ is observable.

3.9   REGULATOR PROBLEM FORMULATION

The regulator problem is a mainstay in the theory of feedback control systems. Typical statements of this problem in the applied mathematical literature have reached a level at which it is not always apparent to the casual reader just what is the intuitive nature of regulator problems in general. Therefore, this section discusses various aspects of such formulations, within the SDS context.

It is convenient in this portion of the narrative to suppress the poset $(J, \leq)$ in the notation for an SDS. $(J, \leq)$ is to be regarded as a constant for the duration of the chapter.

3.9.1   Plant

In every regulator problem, there is an object whose activities are to be influenced. For cultural reasons, this object is called a plant. The plant is a physical system. Its input set is ordinarily factored according to

$$A \times D, \tag{3-167}$$

where A is a set of *actuation requests* and D is a set of *disturbances*. In SDS symbolism, then, a plant is described by the list

$$(A \times D, X_p, Y; f_p, g_p), \tag{3-168}$$

where the subscript is a help to remember the plant in later discussions involving several different set-dynamical systems.

Generally, a regulator accepts requests for certain types of outputs from the plant and achieves them in the face of the unwanted disturbances. To help in carrying out its task, the regulator is permitted to make *measurements* on the plant, with the aid of a *local measurement function*

$$m_p : X_p \times A \times D \to Y_m \tag{3-169}$$

expressed by the action

$$y_m(k) = m_p(x_p(k), a(k), d(k)). \tag{3-170}$$

Again, the custom of writing $y_m(k)$ in place of $(y_m)_k$ has been continued. Equipped with the measuring apparatus (3-169), the plant (3-168) is given a more complete description

$$(A \times D, X_p, Y, Y_m; f_p, g_p, m_p). \tag{3-171}$$

Consider next the requests for certain types of outputs. A set of *output sequence requests* Y_r can be expressed as a subset

$$Y_r \in \mathbb{P}(Y^{\mathbb{N}}) \tag{3-172}$$

of the set of output sequences $Y^{\mathbb{N}}$. When elements of Y_r are supplied to the regulator as desirable, the regulator is required to emulate these desirable sequences in a way satisfactory to the designer. While it is attempting to do so, the regulator has to act in the face of elements from a set of disturbance sequences

$$D_r \in \mathbb{P}(D^{\mathbb{N}}). \tag{3-173}$$

For symmetry in (3-172) and (3-173), the subscript r has been used twice. In (3-172), it clearly stands for request. In (3-173), it may be interpreted as reject.

3.9.2   Exosystem

Considered as impinging from the outside, the output requests and disturbances are usually lumped together and described as *exogenous inputs* to the regulator.

It is possible to regard exogenous inputs to the regulator as the outputs of another SDS, which is then aptly called an *exosystem*. The exosystem is specified by the designer in such a way as to characterize the requests and disturbances. Such an exosystem would be an SDS of the form

$$(V, X_e, D, Y; f_e, g_e, m_e). \tag{3-174}$$

In (3-174), the output set D is listed before the measurement set Y, because the disturbances cannot be measured whereas the output requests are available to the regulator. Thus the exosystem has a disturbance output

$$d(k) = g_e(x_e(k), v(k)) \tag{3-175}$$

which couples to the measured plant (3-171) as the second entry in its input pair; it has a request output

$$y(k) = m_e(x_e(k), v(k)) \tag{3-176}$$

which is in essence a measurement available to the regulator. This accounts for the use of the function symbol m_e.

The exosystem (3-174) can be considered as a generator, through its associated global output function and global measurement function, of the sets Y_r and D_r in (3-172) and (3-173), respectively.

3.9.3 Extended Plant

In much of the regulator literature, the plant and exosystem are combined into an *extended plant*. The adjective extended is suppressed more often than not, so that the reader is more or less required to understand how this is done. This subsection explains the procedure.

As a combined entity, the plant and exosystem will have an input set $A \times V$ consisting of actuation requests that can be made by the regulator and the inputs for the exosystem which cannot be affected by the regulator. For the state set, select the natural candidate $X_e \times X_p$. For the output set, select $Y \times Y$. This is an important subtlety. The first component in the output set $Y \times Y$ contains the output generated by the function m_e in (3-176); the second component contains the plant output produced by the function g_p in (3-171). For the output measurement set, select $Y \times Y_m$. Here the first component stands again for the measurement brought about by (3-176); the second component contains the plant output measurement created by m_p in (3-171). Note that (3-176) is being used in two ways—as an output, where it is to be compared in some way with the plant output; and as a measured output, where it is used to excite the regulator.

The extended plant can then be listed as

$$(A \times V, X_e \times X_p, Y^2, Y \times Y_m; f_{ep}, g_{ep}, m_{ep}). \qquad (3\text{-}177)$$

Actions of the local dynamic and measurement functions are

$$f_{ep}(x_e, x_p, a, v) = (f_e(x_e, v), f_p(x_p, a, g_e(x_e, v))), \qquad (3\text{-}178)$$

$$g_{ep}(x_e, x_p, a, v) = (m_e(x_e, v), g_p(x_p, a, g_e(x_e, v))), \qquad (3\text{-}179)$$

$$m_{ep}(x_e, x_p, a, v) = (m_e(x_e, v), m_p(x_p, a, g_e(x_e, v))). \qquad (3\text{-}180)$$

As mentioned when the extended plant idea was brought forward, much of the literature starts with (3-177), but without any of the sets factored.

3.9.4 Regulation Conditions

The codomain of the local output function for the extended plant is $Y \times Y$. The first factor represents requests by the exosystem, and the second factor represents attainments by the plant. It is desired to relate these two factors.

One way to relate them is to use the equality relation on Y. For an output from the extended plant to be an element in this equivalence relation, the exosystem request and plant output must be identical. This is a very stringent requirement. A more general way to accomplish the same thing is to specify a binary relation B on Y and to require that the request and the output stand in the relation B to one another. This can be expressed by stating that

$$g_{ep}(x_e, x_p, a, v) \in B \subset Y^2. \tag{3-181}$$

If (3-181) is the case, then the extended plant is said to satisfy the *regulation condition* expressed by the binary relation *B*.

3.9.5 Controller

To bring an extended plant into a desired regulation condition means to construct a regulator. The central link in such a construction is the *controller*.

A controller has as its input set $Y \times Y_m$, the measurement set of the extended plant. Its output set is *A*, the set of actuation requests. And a descriptive list for it is written

$$(Y \times Y_m, X_c, A; f_c, g_c). \tag{3-182}$$

The local output equation

$$a(k) = g_c(x_c(k), m_{ep}(x_e(k), x_p(k), a(k), v(k))) \tag{3-183}$$

establishes a feedback link around the extended plant, from its output measurements to its actuation requests. Note that (3-183) can be written with its indices suppressed in the manner

$$a = g_c(x_c, m_{ep}(x_e, x_p, a, v)). \tag{3-184}$$

From (3-184), it is evident that the presence of the actuation request set *A* as a factor in the domain of m_{ep} means that there is a possibility that (3-184) must be solved for a function

$$\theta: X_c \times X_e \times X_p \times V \to A \tag{3-185}$$

which has the action

$$a = \theta(x_c, x_e, x_p, v). \tag{3-186}$$

Controllers having a local output function g_c for which a unique θ of type (3-185) exists will be called *admissible*.

3.9.6 Regulators

An admissible controller combines with the extended plant to form a *regulator*, if the regulation condition (3-181) is satisfied.

SDS description of the regulator is a necessary prerequisite for the next section. The input set is clearly *V*; a state set $X_c \times X_e \times X_p$ is natural; the output set remains Y^2. Thus,

$$(V, X_c \times X_e \times X_p, Y^2; f, g) \tag{3-187}$$

is a suitable description, provided that the actions are

$$f(x_c, x_e, x_p, v) = (f_c(x_c, m_{ep}(x_e, x_p, \theta(x_c, x_e, x_p, v), v), v)),$$
$$f_{ep}(x_e, x_p, \theta(x_c, x_e, x_p, v), v)), \tag{3-188}$$
$$g(x_c, x_e, x_p, v) = g_{ep}(x_e, x_p, \theta(x_c, x_e, x_p, v), v). \tag{3-189}$$

It is interesting to observe that g and g_{ep} are *not* the same function, because they have differing domains.

If V is a singleton, as is often the case, the regulator is autonomous, and may be described by

$$(X_c \times X_e \times X_p, Y^2; \alpha, \beta). \tag{3-190}$$

The regulation condition then looks like

$$\beta(x) \in B \subset Y^2, \tag{3-191}$$

where

$$x = (x_c, x_e, x_p) \quad \text{and} \quad X = X_c \times X_e \times X_p \tag{3-192a,b}$$

is a straightforward abbreviation. Note that the regulation condition (3-191) can be expressed in X if the notation of Section 2.1 is employed. Then (3-191) becomes

$$x \in \beta^*(B). \tag{3-193}$$

Condition (3-193) should make it apparent that regulation can be associated with being in specified subsets of the state set X.

In summary, an autonomous regulator can be described by

$$(X, Y^2; \alpha, \beta), \tag{3-194}$$

and the regulation condition can be stated in terms of the regulator state x satisfying a membership condition

$$x \in X_d \subset X \tag{3-195}$$

in some desired subset X_d of X. In turn, X_d can be understood as the inverse image under the local output function β of a binary relation B on Y.

3.10   THE INTERNAL MODEL PRINCIPLE

One of the most fascinating and quite general properties of regulators is their tendency to include as part of the dynamical action of the controller a copy, in an appropriate sense, of the dynamical action of the exosystem. This copy is often called an internal model of the exosystem.

This section assumes that the local output function of the plant takes the form

$$g_p : X_p \times D \to Y \tag{3-196}$$

and that the local measurement function of the plant does likewise, that is,

$$m_p : X_p \times D \to Y_m. \tag{3-197}$$

Then the extended plant local output and local measurement functions have the domains and codomains

$$g_{ep} : X_e \times X_p \times V \to Y^2, \qquad m_{ep} : X_e \times X_p \times V \to Y \times Y_m, \tag{3-198a,b}$$

with the actions

$$g_{ep}(x_e, x_p, v) = (m_e(x_e, v), g_p(x_p, g_e(x_e, v))), \tag{3-199a}$$
$$m_{ep}(x_e, x_p, v) = (m_e(x_e, v), m_p(x_p, g_e(x_e, v))). \tag{3-199b}$$

Two interesting consequences follow from these assumptions. First, (3-181) is replaced by

$$g_{ep}(x_e, x_p, v) \in B \subset Y^2; \tag{3-200}$$

second, (3-184) becomes

$$a = g_c(x_c, m_{ep}(x_e, x_p, v)). \tag{3-201}$$

The latter condition assures that every controller is admissible; the former condition implies that satisfaction of the regulation condition brought about by B on Y implies nothing concerning the controller state x_c.

Further, as discussed in Section 3.9.6, let the regulator be autonomous, with V being a singleton. This means that V is suppressed as a factor in domain products and that v is omitted from the argument list in action expressions. The regulator can then be described by

$$(X, Y^2; \alpha, \beta) \tag{3-202}$$

with X expressed in (3-192).

Because the controller has no access to the exosystem, save for its output request, it is to be expected that the exosystem state evolves according to the local transition function f_e. In fact, this can easily be checked. Equation (3-188) represents the action of the local transition function f for the regulator. A study of (3-188) shows that f produces the next extended plant state with f_{ep}. The action of f_{ep} is specified in (3-178), where it is seen that the next exosystem state is calculated by f_e, as suggested above. These remarks can be summarized in the commutative diagram of Fig. 3.15.

In talking of how the exosystem works as a part of the regulator, it would be nice to be able to restrict the action of f to X_e. For, in that case,

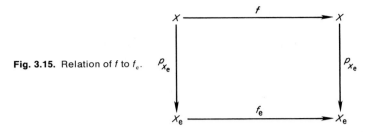

Fig. 3.15. Relation of f to f_e.

the restriction would be a good model upon which to base this section. Unfortunately, such a restriction is not possible, because

$$X_e \not\subset X. \qquad (3\text{-}203)$$

If this statement surprises the reader, then a review of the coproduct treatment in Section 2.8 may be helpful. As an immediate result of (3-203), it is necessary to assume the existence of an X subset in the same spirit.

Exosubset Assumption (ESA).   Corresponding to the action of the local transition function f_e on X_e, it is assumed that there exists a subset

$$E \subset X \qquad (3\text{-}204\text{a})$$

with the property that

$$f_*(E) \subset E. \qquad (3\text{-}204\text{b})$$

Under the exosubset assumption (ESA), f has a restriction

$$f \,|\, E : E \to E, \qquad (3\text{-}205)$$

which can be related to f as in Fig. 3.16. In this figure, the vertical arrows are insertions of the subset E into X.

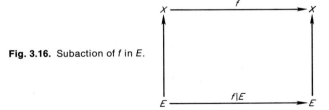

Fig. 3.16. Subaction of f in E.

Unlike the exosystem, which has the intuitive need to be inserted into X for the present discussion, the controller needs only to be constructed from X. This is handled easily by the product projection

$$p_{X_c} : X \to X_c \qquad (3\text{-}206)$$

having the familiar action

$$p_{X_c}(x_c, x_e, x_p) = x_c. \tag{3-207}$$

Because of the frequency of usage of this projection, it is convenient to simplify the notation by writing

$$p = p_{X_c}. \tag{3-208}$$

Of course, as an equality of functions, (3-208) implies equality of domains, codomains, and action.

What is meant by an internal model of the exosystem in the controller? Define the restriction

$$p|E : E \rightarrow X_c. \tag{3-209}$$

Then the controller is said to contain an *internal model* of the exosystem if there exists a function

$$\bar{f} : X_c \rightarrow X_c \tag{3-210}$$

such that

$$\bar{f} \circ p|E = (p|E) \circ f|E \tag{3-211}$$

and if

$$\text{Im}\, p|E \approx E. \tag{3-212}$$

The latter condition insures that all of the exosystem is modeled in the controller. The former condition is expressed by Fig. 3.17. With three more assumptions, the existence of such an internal model can be demonstrated. The first of these three assumptions permits construction of an $\bar{f}$, but in a different diagram, that of Fig. 3.18. X_d is the desired subset (3-195) of X, and the restricted functions are defined in the usual way.

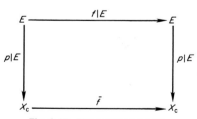

Fig. 3.17. The internal model.

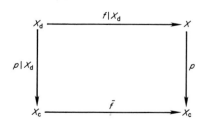

Fig. 3.18. A candidate for $\bar{f}$.

Feedback Assumption (FBA).   Let $E(p|X_d)$ be the equivalence relation induced by $p|X_d$ on X_d, and let $E(p \circ f|X_d)$ be that induced by $p \circ f|X_d$ on X_d. Then

$$E(p|X_d) \leq E(p \circ f|X_d). \tag{3-213}$$

The feedback assumption FBA has the intuitive meaning that, if the regulator state satisfies the regulation condition, the next state of the controller depends only upon the present state of the controller. This has the pleasing connotation that the feedback loop is inoperative when the regulation condition is satisfied.

Although FBA does permit the construction of an $\bar{f}$ in Fig. 3.18, it remains to show that $\bar{f}$ is uniquely defined. To accomplish this, recall from the beginning of the section that $x \in X_d$ places no constraint on x_c. Accordingly,

$$p_*(X_d) = X_c, \tag{3-214}$$

and so $p \mid X_d$ is surjective. This means that $\bar{f}$ must be defined for all x_c in X_c by

$$\bar{f} = p \circ f \mid X_d \circ (p \mid X_d)^{-R} \tag{3-215}$$

for any right inverse

$$(p \mid X_d)^{-R} : X_c \to X_d \tag{3-216}$$

of $p \mid X_d$, and FBA assures that the action (3-215) is not dependent upon the particular right inverse chosen.

A second additional assumption will be enough to meet the commutative requirement (3-211).

Exoregulation Assumption (ERA).   The exosubset E, whose existence is postulated by ESA, must satisfy

$$E \subset X_d, \tag{3-217}$$

where X_d is the subset of regulator states which satisfy the regulation condition.

ERA is a reasonable assumption. Recall that X_d is generated by a binary relation

$$B = \{(y_r, y_p) \mid y_r \in Y \text{ and } y_p \in Y\} \subset Y^2. \tag{3-218}$$

The requests y_r arise from the exosystem alone, whereas y_p is constructed by the entire regulator. ERA simply means that every exosystem state is effective in producing output requests. Under ESA, this must be expressed in terms of E, as has been done in (3-217).

With ERA, functions defined on X_d can be restricted to E. Begin with the commutative statement

$$\bar{f} \circ p \mid X_d = p \circ f \mid X_d \tag{3-219}$$

of Fig. 3.18. Simply restrict both sides of (3-219) to E. This gives

$$(\bar{f} \circ p \mid X_d) \mid E = (p \circ f \mid X_d) \mid E. \tag{3-220}$$

In the left member,

$$(\bar{f} \circ p | X_d) | E = \bar{f} \circ p | E; \qquad (3\text{-}221)$$

in the right member,

$$(p \circ f | X_d) | E = p \circ (f | X_d) | E. \qquad (3\text{-}222)$$

Notice in Fig. 3.18 that $f | X_d$ is defined with codomain X, while in Fig. 3.17 the function $f | E$ is defined to have codomain E. Let

$$i: E \to X \qquad (3\text{-}223)$$

be the insertion. Then

$$(f | X_d) | E = i \circ f | E. \qquad (3\text{-}224)$$

It is easy to see that

$$p \circ i = p | E; \qquad (3\text{-}225a)$$

and so the right member becomes

$$(p | E) \circ f | E. \qquad (3\text{-}225b)$$

Condition (3-211) is thus established.

It remains to arrange things in such a way that condition (3-212) is met. For this end, introduce the third additional assumption.

Exo-Observability Assumption (EOA).   As a consequence of ESA, it is possible to define an autonomous SDS

$$(E, X_c; f | E, p | E). \qquad (3\text{-}226)$$

It is assumed that $(p | E, f | E)$ is an observable pair, in the terminology of Section 3.8.

Again, EOA has a pleasant intuitive meaning, which suggests that the controller be able to observe the states of the exosystem, through the overall dynamics of the regulator.

With EOA, $p | E$ can be established as injective. According to (3-153), the observability of the pair $(p | E, f | E)$ means that

$$\inf\{E(p | E \circ (f | E)^i), i \in \mathbb{N}\} = E^0, \qquad (3\text{-}227)$$

where the reader will recall that E^0 denotes the equality relation, least element of the equivalence lattice on the exosubset E. The fact that $p | E$ is injective is established by showing

$$E(p | E) = E^0; \qquad (3\text{-}228)$$

(3-228) is achieved by showing that

$$E(p|E) = \inf\{E(p|E \circ (f|E)^i), i \in \mathbb{N}\}. \qquad (3\text{-}229)$$

Because of the antisymmetry of partial order relations, it is sufficient to show

$$\inf\{E(p|E \circ (f|E)^i), i \in \mathbb{N}\} \le E(p|E) \qquad (3\text{-}230a)$$

and

$$E(p|E) \le \inf\{E(p|E \circ (f|E)^i), i \in \mathbb{N}\}. \qquad (3\text{-}230b)$$

The first step is easy, by considering $i = 0$ in (3-230a). For the second step, suppose that

$$x_1 \equiv x_2(p|E). \qquad (3\text{-}231)$$

In view of the fact that (3-211) has been satisfied, it follows from

$$(\bar{f} \circ p|E)(x_1) = (\bar{f} \circ p|E)(x_2) \qquad (3\text{-}232)$$

that, by commutativity of the internal model diagram in Fig. 3.17,

$$(p|E \circ f|E)(x_1) = (p|E \circ f|E)(x_2), \qquad (3\text{-}233)$$

which in turn implies that

$$(f|E)(x_1) \equiv (f|E)(x_2)(p|E). \qquad (3\text{-}234)$$

Repetition of the steps (3-231) to (3-234) evidently gives

$$(f|E)^i(x_1) \equiv (f|E)^i(x_2)(p|E), \qquad i \in \mathbb{N}. \qquad (3\text{-}235)$$

But (3-231) and (3-235) mean that $E(p|E)$ is a lower bound for $\{E(p|E \circ (f|E)^i), i \in \mathbb{N}\}$. With the infimum being the greatest element of the set of lower bounds, (3-230b) follows.

The assumptions of this section have included ESA, FBA, ERA, and EOA. When all these assumptions are in force, it has been shown that an internal model of the exosystem is present in the controller of a regulator. This result is known as the *internal model principle* (IMP).

In many ways, the exosubset assumption ESA is the most intriguing. It, in turn, is a consequence of the different roles played in the last two chapters by products and coproducts.

Exercises

3.10-1.   As the discussion of FBA proceeds from (3-214), it is asserted that $\bar{f}$ must be defined for all x_c in X_c by (3-215) for any right inverse (3-216) of $p|X_d$. Why is this the case?

3.10-2.   Why is it true that the fact that $p|E$ is injective can be established by showing (3-228)?

3.11 DISCUSSION

This chapter has investigated the way in which relatively common concepts, such as observability and regulation, can be studied within the spirit of Chapter 2. According to those conventions, the input, state, and output sets of a set-dynamical system are permitted only the structure of binary relations.

Surprisingly, the set of binary relations $B(S)$ on a set S turns out to have a rather rich structure. Because of this natural structure, the ideas of poset and lattice have been introduced. The poset is just a set equipped with a binary relation satisfying the r-a-t properties of reflexivity, antisymmetry, and transitivity. The lattice also has two binary operations, called meet and join. Close ties exist between the partial order relation and the pair of binary operations. In rather general circumstances, the pair of binary operations can be used to induce a partial order, and conversely.

Understanding the application of lattice ideas to observers and to the internal model principle requires special study of the relational lattice and the equivalence lattice. As motivation for those concepts, the notions of global dynamical functions, reachability, and observability lay the groundwork for the later, major sections on observer models, regulator problems, and internal models.

More emphasis has been given in this chapter to observability than to reachability. In the following chapter, the situation achieves better balance.

A great deal of Chapters 2 and 3 could be elaborated more fully if the product of two sets could be equipped with a natural insertion from the factors, or if the coproduct of two sets could be equipped with a natural projection onto its subsets. The basic point of interest here is the lack of any binary operation structure on the sets of the SDS. Without such operations, there can be no units for them. The following chapter permits such structure, and the reader will then have an opportunity to coalesce the product and coproduct ideas into a biproduct which has both insertions and projections.

Along these lines, the discussion of Section 3.2 on binary operations, their terminology and their notations, is a useful review before proceeding to Chapter 4.

4 GROUP MORPHIC SYSTEMS

Up to this point, the narrative has permitted only binary relations on the sets U, X, and Y associated with an SDS. Further structure was allowed to develop on the set of binary relations, however.

In this section, the sets U, X, and Y can each have their own binary operations. This means that there can be two levels of discussion insofar as binary operations are concerned: the level of U, X, and Y, and the level of $B(U)$, $B(X)$, and $B(Y)$. A major reason for the preliminaries of Chapters 2 and 3 has been to assist the reader in developing an intuition for the latter level of discourse, without any encumbrance from preconceived operations on the former level.

Consider the state set X. A binary operation, such as the meet $\wedge$, on the lattice $(B(X), \leq, \wedge, \vee)$ has been defined by

$$B_1 \wedge B_2 = B_1 \cap B_2, \qquad (4\text{-}1)$$

in (3-113), where B_1 and B_2 are elements of $B(X)$—which means that they are subsets of X^2, as explained in Section 2.3—and the right member stipulates set theoretic intersection in X^2. On the level of $B(X)$, then, the binary operations deal with subsets of X^2. Notice, however, that binary operations on X itself will involve individual elements of X^2. So it is generally easy to keep the levels well distinguished.

An exception to this ease of interpretation may occur for some readers in the case of $(E(X), \leq, \wedge, \vee_E)$. This exception is usually the consequence of the feature

$$p_E: X \to X/E \qquad (4\text{-}2)$$

for each E in $E(X)$, which divides X into equivalence classes, such that each element of X is a member of one and only one class. Recall Section 2.6. It will be seen in this chapter that a binary operation on X can induce a binary operation on X/E. This construction is distinct from the meet $\wedge$ and the join $\vee_E$ of the equivalence lattice.

Even this potential point of confusion can be avoided with little effort if the reader keeps in mind that a function is specified by three items: domain, codomain, and action. The binary operations on the equivalence lattice of X are

$$E(X) \times E(X) \to E(X), \tag{4-3}$$

while the induced operations described above will be

$$X/E \times X/E \to X/E. \tag{4-4}$$

4.1 GROUPS AND MORPHISMS

A *group* is a set G, equipped with an associative binary operation

$$\square : G \times G \to G \tag{4-5}$$

which has a unit e and under which each element g in G has an inverse $\hat{g}$ in G. The notation $(G, \square, e)$ for a group is frequently put to use. The group is *commutative* if its binary operation is commutative.

EXAMPLE 4.1-1

The omnipresent real numbers $\mathbb{R}$ provide a transparent example $(\mathbb{R}, +, 0)$ where $+$ is the usual addition of real numbers with unit the real number zero. Traditionally, the inverse $\hat{r}$ of a real number r under $+$ is written $-r$. ∎

EXAMPLE 4.1-2

Consider the four-sided polygon of Fig. 4.1. In part (a) of the figure, vertices have been numerically designated for reference purposes. If the polygon is rotated clockwise through an angle of 90°, the resulting vertex configuration is sketched in part (b). Denote this quarter-turn construction by g_1, which is one of the elements in a well-known group. Another element in this group is g_2, a reflection in the axis passing from the upper left vertex to the lower right vertex. The result of g_2 applied to the polygon in part (a)

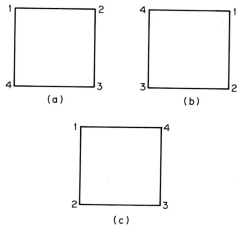

Fig. 4.1. Development of a group.

is part (c). Needless to say, g_1 and g_2 can be applied one after the other to the polygon of part (a). If g_1 is applied prior to g_2, the resulting vertex arrangement is

$$\begin{matrix} 4 & 3 \\ 1 & 2 \end{matrix};$$

(4-6)

if g_2 precedes g_1,

$$\begin{matrix} 2 & 1 \\ 3 & 4 \end{matrix}$$

(4-7)

is obtained. The construction "apply g_1, then apply g_2" can be understood as yet a third construction

$$g_2 \,\square\, g_1$$

(4-8)

in G; similarly, the construction leading to (4-7) would be written

$$g_1 \,\square\, g_2.$$

(4-9)

Adjoin the construction "leave the polygon as is," and denote it by e. Then $(G, \square, e)$ is a group, with

$$G = \{e, g_1, g_1 \,\square\, g_1, g_1 \,\square\, g_1 \,\square\, g_1,$$
$$g_2, g_1 \,\square\, g_2, g_1 \,\square\, g_1 \,\square\, g_2, g_1 \,\square\, g_1 \,\square\, g_1 \,\square\, g_2\}.$$

(4-10)

In checking this assertion, the reader may find it helpful to use the observations that

$$g_1 \, \square \, g_1 \, \square \, g_1 \, \square \, g_1 = e, \tag{4-11a}$$

$$g_2 \, \square \, g_2 = e, \tag{4-11b}$$

$$g_1 \, \square \, g_1 \, \square \, g_1 \, \square \, g_2 = g_2 \, \square \, g_1. \tag{4-11c}$$

$(G, \square, e)$ is a *dihedral group*. ∎

Suppose now that two groups $(S, \square, e_S)$ and $(T, *, e_T)$ are available. Notice how the group notation allows an easy specification of the set symbol, the binary operation symbol, and the unit symbol. Pass to the next level of construction and establish a function

$$f : S \to T. \tag{4-12}$$

Does the fact that S and T now have group structure pertain to the function f? The answer is yes. One basic effect is apparent from the next example.

EXAMPLE 4.1-3

The binary operation $*$, on $(T, *, e_T)$ induces a similar operation on T^S. To see this, let

$$f_i \in T^S, \qquad i = 1, 2. \tag{4-13}$$

Define a binary operation

$$* : T^S \times T^S \to T^S \tag{4-14}$$

by the action

$$(f_1 * f_2)(s) = f_1(s) * f_2(s). \tag{4-15}$$

The binary operation in the left member is on T^S, whereas in the right member it is on T. It is customary to use the same symbol for both, because they are distinguished by their domains and codomains. The binary operation (4-14) is associative because $(T, *, e_T)$ is a group, which means that

$$* : T \times T \to T \tag{4-16}$$

is associative. When S and T are $\mathbb{R}$, for instance, the reader has encountered such constructions frequently, whenever two functions were added in the manner

$$\sin(x) + \cos(x). \tag{4-17}$$

Section 1.3 used this idea as part of its discussion on connections. ∎

It is of interest to remark that Example 4.1-3 requires structure only on T, not on S. Actually, the example can be strengthened. Define a function

$$e_{TS}: S \to T \tag{4-18}$$

by the action

$$e_{TS}(s) = e_T \tag{4-19}$$

for all s in S. Then e_{TS} is a unit for the binary operation (4-14). Next define a function

$$\hat{f}: S \to T \quad \text{by} \quad \hat{f}(s) = \widehat{f(s)}, \tag{4-20a,b}$$

for f as in (4-12). Then a brief reference to (4-15) establishes that

$$(f * \hat{f})(s) = f(s) * \hat{f}(s) = f(s) * \widehat{f(s)}$$
$$= e_T = e_{TS}(s), \tag{4-21}$$

and similarly for $(\hat{f} * f)(s)$. So

$$f * \hat{f} = e_{TS} = \hat{f} * f, \tag{4-22}$$

making $\hat{f}$ an inverse for f under (4-14). And $(T^S, *, e_{TS})$ becomes a group in its own right. This new group is quite useful in the next chapter.

Possibly the most widely studied class of functions consists of those which are linear. It turns out that the real power behind the concept of linearity has to do with the way in which functions interact with the binary operations on their domains and codomains.

Relative to the function f of (4-12) whose domain is a group $(S, \square, e_S)$ and whose codomain is a group $(T, *, e_T)$, the notion of linearity is generalized in the following way. f is a *morphism of groups* if, for all s_1 and s_2 in S,

$$f(s_1 \square s_2) = f(s_1) * f(s_2). \tag{4-23}$$

The statement "f is a morphism" replaces the statement "f is linear." Intuitively, it could be said that a morphism is compatible with the binary operation structures on its domain and codomain.

EXAMPLE 4.1-4

Let T be $(\mathbb{R}, +, 0)$ of Example 4.1-1, and let S be $(\mathbb{R}^{\mathbb{R}}, +, 0)$. Notice that the 0 in T is the real number zero, while the 0 in S is the function which assigns the real number zero to every element in its domain. Denote by $\tilde{S} \subset S$ the subset of S which is Riemann integrable. It is easy to check that $\tilde{S}$ is a group $(\tilde{S}, +, 0)$, under the restriction of $+ : S \times S \to S$ to $\tilde{S} \times \tilde{S}$. Then the Riemann integral is a morphism. ∎

EXAMPLE 4.1-5

Let T be $(\bar{\mathbb{R}}, \cdot, 1)$, where $\bar{\mathbb{R}} \subset \mathbb{R}$ contains the nonzero real numbers, $\cdot$ is real multiplication, and 1 is the real number one. $(\bar{\mathbb{R}}, \cdot, 1)$ is a group, where inversion under $\cdot$ is associated with division. For S choose the set of 2×2 matrices with elements in $\mathbb{R}$ and having determinants in $\bar{\mathbb{R}}$. For $\square$, select ordinary matrix multiplication, with unit

$$\begin{bmatrix} 1 & 0 \\ 0 & 1 \end{bmatrix}. \tag{4-24}$$

If s is in S, its inverse $\hat{s}$ is calculated by the usual matrix inversion. Let the action of $f : S \to T$ be determined by

$$f(s) = \text{determinant of } s. \tag{4-25}$$

Then

$$f(s_1 \,\square\, s_2) = f(s_1) \cdot f(s_2) \tag{4-26}$$

for all s_1 and s_2 in S. ∎

Morphisms have certain basic properties. Consider the statement

$$e_S \,\square\, e_S = e_S, \tag{4-27}$$

and apply f to both members to obtain

$$f(e_S \,\square\, e_S) = f(e_S). \tag{4-28}$$

Because f is a morphism, (4-28) advances to

$$f(e_S) * f(e_S) = f(e_S), \tag{4-29}$$

to which $\widehat{f(e_S)}$ can be adjoined in the manner

$$\widehat{f(e_S)} * f(e_S) * f(e_S) = \widehat{f(e_S)} * f(e_S). \tag{4-30}$$

Applying associativity of $*$ in the left member and the unit property in both members, (4-30) reduces to

$$f(e_S) = e_T. \tag{4-31}$$

Thus, a morphism of groups carries the unit of its domain to the unit of its codomain. Sometimes, this is expressed by saying that a morphism of groups is also a morphism of units. A further property follows by rewriting (4-31) as

$$e_T = f(e_S) = f(s \,\square\, \hat{s}) = f(s) * f(\hat{s}) \tag{4-32}$$

for an arbitrary s in S and its inverse $\hat{s}$ under $\square$. From the uniqueness of the inverse $\widehat{f(s)}$ when the binary operation is associative (see Example

3.2-3), it follows that

$$\widehat{f(s)} = f(\hat{s}) \tag{4-33}$$

for all s in S.

A nonvoid subset S contained in a group $(G, \square, e)$ is a *subgroup* if the two conditions

$$\text{(i)} \quad s \in S \Rightarrow \hat{s} \in S, \tag{4-34a}$$

$$\text{(ii)} \quad (s_1, s_2) \in S^2 \Rightarrow s_1 \square s_2 \in S \tag{4-34b}$$

are satisfied. If the nonvoid qualifier is dropped, then a third condition $e \in S$ should be added. The condition pair (4-34) is often replaced by the equivalent single condition

$$(s_1, s_2) \in S^2 \Rightarrow s_1 \square \hat{s}_2 \in S. \tag{4-35}$$

It is straightforward to show that (4-34) implies (4-35). For the opposite implication, let $s \in S$; then (4-35) implies $(s, s) \in S^2$, which implies

$$e = s \square \hat{s} \in S. \tag{4-36}$$

Now consider $(e, s) \in S^2$; then (4-35) gives

$$e \square \hat{s} = \hat{s} \in S, \tag{4-37}$$

which establishes (4-34a). Next if (s_1, s_2) is in S^2, it follows by (4-34a) that $(s_1, \hat{s}_2)$ is in S^2. Then (4-35) provides

$$s_1 \square \hat{\hat{s}}_2 \in S, \tag{4-38}$$

which proves (4-34b), since

$$\hat{\hat{g}} = g \tag{4-39}$$

for g an element of any group $(G, \square, e)$.

Note that a subgroup S is a group on its own merits, when the binary operation $\square$ is restricted to $S \times S$. It is conventional to write $(S, \square, e)$ for this group.

If $f: S \to T$ is a morphism of groups, then Im f is a subgroup of T. The *kernel* of f, denoted Ker f, is the subset

$$\text{Ker } f = \{s \,|\, s \in S \text{ and } f(s) = e_T\}. \tag{4-40}$$

The kernel of f is a subgroup of S. If the morphism f is surjective, then f is an *epimorphism* or is *epic*; if f is injective, then f is a *monomorphism* or is *monic*. It is a reasonable exercise to show that f is monic if and only if

$$\text{Ker } f = e_S. \tag{4-41}$$

Finally, if f is a bijection, then it is termed an *isomorphism*. When domain and codomain are the same group, sometimes a morphism is called an *endomorphism*. In that case, the word *automorphism* may replace the word *isomorphism*.

Kernels play an important part in what follows, and they are defined in terms of the unit of a binary operation.

Exercises

4.1-1. Show that e_{Ts} as defined in (4-19) is a unit for the binary operation (4-14).

4.1-2. Reconsider Example 4.1-4. There is no difficulty in making a restriction of

$$+ : S \times S \to S$$

in the manner

$$\tilde{+} : \tilde{S} \times \tilde{S} \to S.$$

Indeed, show that this would be possible for *any* $\tilde{S} \subset S$. The interesting step is to show that

$$\text{Im } \tilde{+} \subset \tilde{S}.$$

Do this for the $\tilde{S}$ of the example. Comment on the relevance of this step in regard to making $(\tilde{S}, +, 0)$ a group.

4.1-3. Explain in detail how the passage from (4-30) to (4-31) is to be accomplished.

4.1-4. Equation (4-39) asserts that

$$\hat{\hat{g}} = g$$

for any element g in a group $(G, \Box, e)$. Why is this the case?

4.1-5. Let $f : S \to T$ be a morphism of groups. Show that Im f is a subgroup of T.

4.1-6. For the morphism of Exercise 4.1-5, show that Ker f is a subgroup of S.

4.1-7. For the morphism of Exercise 4.1-5, show that f is monic if and only if

$$\text{Ker } f = e_S.$$

4.1-8. Let $(G, \Box, e_G)$ be a group. A subgroup S contained in G is *normal* if, for each s in S and for each g in G,

$$g \Box s \Box \hat{g} \in S.$$

Is Ker f in Exercise 4.1-6 a normal subgroup?

4.1-9. Verify that (4-10) is in fact a group.

4.1-10. Explain the way in which (4-17) relates to the discussion of Section 1.3.

4.2   THE SDS WITH GROUP STRUCTURE

Recall the set-dynamical system

$$(J, \leq ; U, X, Y; f, g) \tag{4-42}$$

of Chapter 2. Suppress the index chain $(J, \leq)$, and endow each of the SDS sets with group structure, indicated by

$$(U, \triangle, e_U), \qquad (X, \square, e_X), \qquad (Y, *, e_Y). \qquad \text{(4-43a,b,c)}$$

Then (4-42) can be expressed as

$$(U, \triangle, e_U; X, \square, e_X; Y, *, e_Y; f, g). \qquad \text{(4-44)}$$

Notice that the notations for the units e_S for groups developed on a set S have been standardized. They may also be suppressed, then, without confusion. It is useful, however, to retain the binary operation symbols in the manner

$$(U, \triangle; X, \square; Y, *; f, g). \qquad \text{(4-45)}$$

It was seen in Example 4.1-3 that the mere addition of group structure on a codomain offered new constructive possibilities for functions having that codomain. In this section, special attention is focused upon what happens when the local dynamical functions $f: X \times U \to X$ and $g: X \times U \to Y$ are required to be morphisms.

Fundamental to such a study is the question of how (4-43a) and (4-43b) lead to a group structure on the product $X \times U$. To see how this works, define a binary operation

$$**: (X \times U) \times (X \times U) \to (X \times U) \qquad \text{(4-46a)}$$

with the action

$$(x_1, u_1) ** (x_2, u_2) = (x_1 \square x_2, u_1 \triangle u_2). \qquad \text{(4-46b)}$$

The operation $**$ is associative because the operations $\square$ and $\triangle$ are associative. A unit can be given for $**$, namely,

$$(e_X, e_U), \qquad \text{(4-47a)}$$

which is checked by the calculations

$$(x, u) ** (e_X, e_U) = (x \square e_X, u \triangle e_U) = (x, u) \qquad \text{(4-47b)}$$

and

$$(e_X, e_U) ** (x, u) = (e_X \square x, e_U \triangle u) = (x, u). \qquad \text{(4-47c)}$$

Finally, for any element x in the group $(X, \square, e_X)$, there is a unique inverse $\hat{x}$ under $\square$. Similarly, for any element u in the group $(U, \triangle, e_U)$, there is a unique $\hat{u}$. Then for any (x, u) in $X \times U$, there is a unique inverse

$$(\widehat{x, u}) = (\hat{x}, \hat{u}) \qquad \text{(4-48a)}$$

under $**$, as seen by the calculations

$$(x, u) ** (\hat{x}, \hat{u}) = (x \square \hat{x}, u \triangle \hat{u}) = (e_X, e_U), \qquad \text{(4-48b)}$$
$$(\hat{x}, \hat{u}) ** (x, u) = (\hat{x} \square x, \hat{u} \triangle u) = (e_X, e_U). \qquad \text{(4-48c)}$$

So

$$(X \times U, **, (e_X, e_U)) \tag{4-49}$$

is a group, called the *product group* of the groups (4-43a) and (4-43b).

It is now possible to talk about the local dynamical functions being morphisms of groups. Consider the local transition function. The morphism condition is

$$f((x_1, u_1) ** (x_2, u_2)) = f(x_1, u_1) \,\square\, f(x_2, u_2) \tag{4-50}$$

for all pairs (x_1, u_1) and (x_2, u_2) in $X \times U$. By definition of the binary operation ** in (4-46), the condition (4-50) is replaced by

$$f(x_1 \,\square\, x_2, u_1 \,\triangle\, u_2) = f(x_1, u_1) \,\square\, f(x_2, u_2). \tag{4-51}$$

Let the local transition function f be a morphism of groups, that is, let it satisfy (4-50) or (4-51) for all pairs (x_1, u_1) and (x_2, u_2) in $X \times U$. Then observe that

$$
\begin{aligned}
f(x, u) &= f((x, u) ** (e_X, e_U)) && \text{(4-52a)} \\
&= f(x \,\square\, e_X, u \,\triangle\, e_U) && \text{(4-52b)} \\
&= f(x \,\square\, e_X, e_U \,\triangle\, u) && \text{(4-52c)} \\
&= f((x, e_U) ** (e_X, u)) && \text{(4-52d)} \\
&= f(x, e_U) \,\square\, f(e_X, u). && \text{(4-52e)}
\end{aligned}
$$

Now define functions

$$f_1 : X \to X \qquad \text{and} \qquad f_2 : U \to X \tag{4-53a,b}$$

with actions

$$f_1(x) = f(x, e_U) \qquad \text{and} \qquad f_2(u) = f(e_X, u). \tag{4-53c,d}$$

The functions f_1 and f_2 are morphisms. For example, consider f_1. The calculation is

$$
\begin{aligned}
f_1(x_1 \,\square\, x_2) &= f(x_1 \,\square\, x_2, e_U) && \text{(4-54a)} \\
&= f(x_1 \,\square\, x_2, e_U \,\triangle\, e_U) && \text{(4-54b)} \\
&= f(x_1, e_U) \,\square\, f(x_2, e_U) && \text{(4-54c)} \\
&= f_1(x_1) \,\square\, f_1(x_2). && \text{(4-54d)}
\end{aligned}
$$

A similar calculation works for f_2.

If the local output function g is also a morphism of groups, then the same argument establishes the existence of two morphisms

$$g_1 : X \to Y \qquad \text{and} \qquad g_2 : U \to Y \tag{4-55a,b}$$

with actions

$$g_1(x) = g(x, e_U) \quad \text{and} \quad g_2(u) = g(e_X, u). \tag{4-55c,d}$$

Overall, then, it is seen that

$$f(x, u) = f_1(x) \,\square\, f_2(u), \qquad g(x, u) = g_1(x) * g_2(u) \tag{4-56a,b}$$

when f and g are morphisms. It has been customary to label the four functions in right members of (4-56) with different symbols. Therefore, establish the following function equalities,

$$a = f_1, \qquad b = f_2, \qquad c = g_1, \qquad d = g_2, \tag{4-57a,b,c,d}$$

bearing in mind that function equality implies the same domain, the same codomain, and the same action.

In view of (4-56) and (4-57), the description (4-45) can be specialized to

$$(U, \triangle; X, \square; Y, *; a, b, c, d). \tag{4-58}$$

When an SDS is equipped with group structures (4-43), and when its local dynamical actions can be accomplished by morphisms (a, b, c, d), then it will be called a *group morphic system* (GRMPS). GRMPSs are sometimes described simply by (a, b, c, d), and their local dynamical equations are given by

$$x_{k+1} = ax_k \,\square\, bu_k, \qquad y_k = cx_k * du_k. \tag{4-59a,b}$$

Later in this section, it will be shown that (4-59) is actually a somewhat more general structure than is indicated by (4-57).

The reader should take note of the fact that the separation of f into a and b, as well as g into c and d, made use of the group units and the morphism properties of local transition and local output functions. It made no use of inverses under the binary operations, or even of associativity. This means that separation in an SDS could be achieved under much weaker assumptions than are made in this chapter. For example, a *monoid* is a set M, equipped with an associative binary operation

$$\square : M \times M \to M \tag{4-60}$$

which has unit e. A function

$$f : M \to N, \tag{4-61a}$$

where $(M, \square, e_M)$ and $(N, \triangle, e_N)$ are monoids, is a *morphism of monoids* if $f(e_M)$ is e_N and if

$$f(m_1 \,\square\, m_2) = f(m_1) \,\triangle\, f(m_2) \tag{4-61b}$$

for all m_1 and m_2 in M. Product monoids are easily constructed. So, the local transition and local output functions of an SDS will separate if they are morphisms of monoids.

EXAMPLE 4.2-1

Let

$$(U, \triangle, e_U) = (\mathbb{Z}_2, +, 0), \tag{4-62}$$

where addition in $\mathbb{Z}_2$ satisfies

$$\begin{array}{c|cc} + & 0 & 1 \\ \hline 0 & 0 & 1 \\ 1 & 1 & 0 \end{array}; \tag{4-63}$$

likewise, define

$$(Y, *, e_Y) = (\mathbb{Z}_2, +, 0). \tag{4-64}$$

For $(X, \square, e_X)$, select the dihedral group of Example 4.1-2. Define

$$a(g_1) = e, \qquad a(g_2) = g_1 \square g_2, \tag{4-65a,b}$$

$$b(1) = g_2, \tag{4-66}$$

$$c(g_1) = 0, \qquad c(g_2) = 1, \tag{4-67a,b}$$

$$d(1) = 0. \tag{4-68}$$

Of course, the fact that each of these is a morphism demands

$$\begin{aligned} a(e) &= e, & b(0) &= e, \\ c(e) &= 0, & d(0) &= 0. \end{aligned} \tag{4-69}$$

Completion of the definition of actions of a and c is made using the morphism condition. For example,

$$a(g_1 \square g_2) = a(g_1) \square a(g_2) = e \square g_1 \square g_2 = g_1 \square g_2 \tag{4-70}$$

and

$$c(g_1 \square g_2) = c(g_1) * c(g_2) = 0 + 1 = 1. \tag{4-71}$$

Demonstration that a, b, c, and d turn out to be morphisms is left as an exercise. ∎

EXAMPLE 4.2-2

The construction (4-52) has an interesting technicality. Suppose that the passage from (4-52b) to (4-52c) had been accomplished in the alternative way

$$f(x \square e_X, u \triangle e_U) = f(e_X \square x, u \triangle e_U) = f((e_X, u) ** (x, e_U))$$
$$= f(e_X, u) \square f(x, e_U). \tag{4-72}$$

Then (4-56a) would have been

$$f(x, u) = f_2(u) \,\Box\, f_1(x). \tag{4-73}$$

This appears to imply that $(X, \Box, e_X)$ must be commutative. On the other hand, the preceding example is certainly not commutative. The answer to this dilemma is contained in the observation that each construction

$$a(x) \,\Box\, b(u) \tag{4-74}$$

induces a function $f: X \times U \to X$; but this function is not necessarily a morphism. Thus every GRMPS brings about an SDS structure, but not every GRMPS corresponds to an SDS in which local dynamics have become morphic.

∎

The fascinating point of this last example can again be traced to the difference between products and coproducts. In this case, however, the issue does not center around whether or not the product group has insertions as well as projections. Further discussions take place in a later section of the chapter.

Exercises

4.2-1. Refer to Example 4.2-1. Establish in detail that a, b, c, and d as therein described are morphisms.

4.2-2. The binary operation ∗∗ introduced in (4-46) has been asserted to be associative because the operations $\Box$ and $\triangle$ are associative. Carry out the details in showing that this is the case.

4.2-3. Establish that the function (4-53b) with action (4-53d) is indeed a morphism.

4.2-4. In (4-61), the concept of a morphism of monoids was brought forward. Establish monoid structure on the state and input sets X and U, respectively. Develop then a product monoid structure on $X \times U$. Having made these steps, examine the possibility of the local transition function of an SDS separating if it is a morphism of monoids.

4.2-5. Explain why (4-74) need not be a morphism.

4.3 INTERCONNECTED GRMPSs: SERIES

The theory of systems tends to become a bit less than exciting when no provision is made for connecting systems together. In this section, some of the basic issues involved in series connections are clarified. It will be seen that an interconnection of two GRMPSs may not be GRMP, without appropriate restrictions on certain of the input, state, and output groups. GRMPSs are symbolized here by the 4-tuple (a, b, c, d) of (4-59), where

$$a: X \to X, \qquad b: U \to X, \qquad c: X \to Y, \qquad d: U \to Y, \tag{4-75a,b,c,d}$$

are morphisms of the groups (4-43).

Visualization of interconnections can be greatly aided by pictures. To this end, adopt the ideas of Chapter 1, especially Section 1.5, and represent GRMPSs as in Fig. 4.2. Recall that the binary operation symbol is used at junctions to emphasize the fact that it is these operations which permit such a concept. Notice that the identity function $1_U : U \to U$ becomes a morphism when $(U, \triangle, e_U)$ is a group, because

$$1_U(u_1 \triangle u_2) = u_1 \triangle u_2 = 1_U(u_1) \triangle 1_U(u_2) \qquad (4\text{-}76)$$

for all u_1 and u_2 in U. The same is true for $1_Y : Y \to Y$.

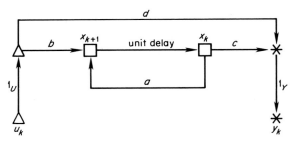

Fig. 4.2. The GRMPS.

In the spirit of Section 1.2, suppose that two GRMPSs are available,

$$(a_i, b_i, c_i, d_i), \qquad i = 1, 2. \qquad (4\text{-}77)$$

This involves a total of six groups,

$$(U_i, \triangle_i, e_{U_i}), \qquad i = 1, 2, \qquad (4\text{-}78\text{a})$$
$$(X_i, \square_i, e_{X_i}), \qquad i = 1, 2, \qquad (4\text{-}78\text{b})$$
$$(Y_i, *_i, e_{Y_i}), \qquad i = 1, 2. \qquad (4\text{-}78\text{c})$$

Then they can be combined.

By a series connection is meant the general scheme of Fig. 4.3. The basic constraint placed by this connection is that the input group of GRMPS 1 must be the output group of GRMPS 2. This can be expressed by

$$(U_1, \triangle_1, e_{U_1}) = (Y_2, *_2, e_{Y_2}). \qquad (4\text{-}79)$$

It is clear that the input group for the connection will be $(U_2, \triangle_2, e_{U_2})$, and that the output group for the connection will be $(Y_1, *_1, e_{Y_1})$. For the state

Fig. 4.3. Series connection of GRMPSs.

group, select the product group developed on $X_1 \times X_2$ according to the procedures of Section 4.2. Then a reasonable goal is to seek a **GRMPS** of type

$$(U_2, \triangle_2; X_1 \times X_2, **; Y_1, *_1; a, b, c, d) \tag{4-80}$$

and to determine the nature of the morphisms a, b, c, and d.

Begin with the local equations for **GRMPS** 2, which is uninfluenced by the connection. These are

$$x_2(k+1) = a_2 x_2(k) \,\square_2\, b_2 u_2(k), \qquad y_2(k) = c_2 x_2(k) *_2 d_2 u_2(k). \tag{4-81a,b}$$

Next write the local equations for **GRMPS** 1, taking the connection into account, and obtain

$$x_1(k+1) = a_1 x_1(k) \,\square_1\, b_1 y_2(k), \qquad y_1(k) = c_1 x_1(k) *_1 d_1 y_2(k). \tag{4-82a,b}$$

Substitution of (4-81b) into (4-82b) gives

$$y_1 = c_1 x_1 *_1 d_1(c_2 x_2 *_2 d_2 u_2), \tag{4-83a}$$

where the indices have been suppressed. Because d_1 is a morphism, (4-83a) develops to

$$y_1 = c_1 x_1 *_1 d_1(c_2 x_2) *_1 d_1(d_2 u_2) \tag{4-83b}$$
$$= c_1 x_1 *_1 (d_1 \circ c_2)(x_2) *_1 (d_1 \circ d_2)(u_2). \tag{4-83c}$$

To determine the possibility of a single **GRMPS** to describe the connection, it is necessary to investigate whether (4-83c) can be put into the form

$$y_1 = c(x_1, x_2) *_1 d u_2 \tag{4-84}$$

for appropriate morphisms $c: X_1 \times X_2 \to Y_1$ and $d: U_2 \to Y_1$.

Two side issues must be addressed. First, is $d_1 \circ c_2: X_2 \to Y_1$ a morphism? The answer is yes, by the calculation

$$(d_1 \circ c_2)(x_2 \,\square_2\, \tilde{x}_2) = d_1(c_2(x_2 \,\square_2\, \tilde{x}_2)) = d_1(c_2 x_2 *_2 c_2 \tilde{x}_2)$$
$$= d_1(c_2 x_2) *_1 d_1(c_2 \tilde{x}_2)$$
$$= (d_1 \circ c_2)(x_2) *_1 (d_1 \circ c_2)(\tilde{x}_2). \tag{4-85}$$

Second, in the group $(T^S, *, e_{TS})$ of Section 4.1, suppose that f_1 and f_2 are morphisms. Then is $f_1 * f_2$ a morphism? The answer is *maybe*, by

$$(f_1 * f_2)(s_1 \,\square\, s_2) = f_1(s_1 \,\square\, s_2) * f_2(s_1 \,\square\, s_2)$$
$$\overset{?}{=} f_1(s_1) * f_1(s_2) * f_2(s_1) * f_2(s_2)$$
$$\overset{?}{=} f_1(s_1) * f_2(s_1) * f_1(s_2) * f_2(s_2)$$
$$= (f_1 * f_2)(s_1) * (f_1 * f_2)(s_2), \tag{4-86}$$

where the question mark can be removed if $(T, *, e_T)$ is a commutative group.

Return now to (4-83c). Define

$$d = d_1 \circ d_2; \tag{4-87}$$

this is a morphism. Next, define the action of c by

$$
\begin{aligned}
c(x_1, x_2) &= c_1 x_1 *_1 (d_1 \circ c_2) x_2 \\
&= c_1(p_{X_1}(x_1, x_2)) *_1 (d_1 \circ c_2)(p_{X_2}(x_1, x_2)) \\
&= (c_1 \circ p_{X_1})(x_1, x_2) *_1 (d_1 \circ c_2 \circ p_{X_2})(x_1, x_2) \\
&= [(c_1 \circ p_{X_1}) *_1 (d_1 \circ c_2 \circ p_{X_2})](x_1, x_2),
\end{aligned}
\tag{4-88}
$$

where p_{X_i}, $i = 1, 2$, are the product projections, which are trivially morphisms. With this action, c is a morphism if $(Y_1, *_1, e_{Y_1})$ is a commutative group.

The next step is to substitute (4-81b) into (4-82a), from which follows

$$x_1(k + 1) = a_1 x_1(k) \,\square_1\, b_1(c_2 x_2(k) *_2 d_2 u_2(k)) \tag{4-89a}$$
$$= a_1 x_1(k) \,\square_1\, (b_1 \circ c_2) x_2(k) \,\square_1\, (b_1 \circ d_2) u_2(k) \tag{4-89b}$$
$$= [(a_1 \circ p_{X_1}) \,\square_1\, (b_1 \circ c_2 \circ p_{X_2})](x_1(k), x_2(k))$$
$$\square_1\, (b_1 \circ d_2) u_2(k). \tag{4-89c}$$

Notice also that (4-81a) itself can be rewritten as

$$x_2(k + 1) = (a_2 \circ p_{X_2})(x_1(k), x_2(k)) \,\square_2\, b_2 u_2(k). \tag{4-90}$$

Then

$$(x_1(k + 1), x_2(k + 1)) = ([(a_1 \circ p_{X_1}) \,\square_1\, (b_1 \circ c_2 \circ p_{X_2})](x_1, x_2) \,\square_1\, (b_1 \circ d_2) u_2,$$
$$(a_2 \circ p_{X_2})(x_1, x_2) \,\square_2\, b_2 u_2), \tag{4-91}$$

in which right member indices have been suppressed. By definition of the binary operation $**$ on the product group $X_1 \times X_2$, (4-91) is equal to

$$([(a_1 \circ p_{X_1}) \,\square_1\, (b_1 \circ c_2 \circ p_{X_2})](x_1, x_2), (a_2 \circ p_{X_2})(x_1, x_2))$$
$$** ((b_1 \circ d_2) u_2, b_2 u_2). \tag{4-92}$$

To establish a single GRMPS to describe the connection, it is necessary to put (4-92) into the form

$$a(x_1, x_2) ** b u_2 \tag{4-93}$$

for morphisms $a: X_1 \times X_2 \to X_1 \times X_2$ and $b: U_2 \to X_1 \times X_2$. The actions of a and b are defined in (4-92). Notice that there is one side issue.

The side issue can be addressed as follows. Suppose that $f_1: S \to T_1$ and $f_2: S \to T_2$ are morphisms of groups $(S, \square, e_S)$ and $(T_i, *_i, e_{T_i})$, $i = 1, 2$. Define a function

$$f: S \to T_1 \times T_2 \qquad \text{by} \qquad f(s) = (f_1(s), f_2(s)) \tag{4-94a,b}$$

for all s in S. Is f a morphism on S to the product group $T_1 \times T_2$? The answer is yes, by the calculation

$$
\begin{aligned}
f(s_1 \square s_2) &= (f_1(s_1 \square s_2), f_2(s_1 \square s_2)) \\
&= (f_1(s_1) *_1 f_1(s_2), f_2(s_1) *_2 f_2(s_2)) \\
&= (f_1(s_1), f_2(s_1)) ** (f_1(s_2), f_2(s_2)) \\
&= f(s_1) ** f(s_2).
\end{aligned}
\tag{4-95}
$$

By these presentations, the function b of (4-93), with action defined by

$$
bu_2 = ((b_1 \circ d_2)u_2, b_2 u_2),
\tag{4-96}
$$

is a morphism. Moreover, if $(X_1, \square_1, e_{X_1})$ is a commutative group, the function a of (4-93), with action defined by

$$
a(x_1, x_2) = ([(a_1 \circ p_{X_1}) \square_1 (b_1 \circ c_2 \circ p_{X_2})](x_1, x_2), (a_2 \circ p_{X_2})(x_1, x_2)),
\tag{4-97}
$$

is also a morphism.

Under assumptions, then, that the GRMPS 1 has commutative state and output groups, it has been shown that the series connection of Fig. 4.3 also admits a GRMP representation.

It is perhaps not surprising that the restrictions have to be placed on the GRMPS which is being driven. This is entirely in keeping with the discussion growing out of Example 4.1-3.

Exercises

4.3-1. Show that the product projection

$$
p_U : X \times U \to U
$$

is a morphism of groups if U and X are groups and if $X \times U$ has the corresponding product group structure.

4.3-2. Explain in detail why the function c with action given by (4-88) is a morphism if

$$
(Y_1, *_1, e_{Y_1})
$$

is a commutative group.

4.3-3. Explain in detail why the functions b and a expressed in (4-96) and (4-97) are morphisms.

4.3-4. Extend the discussion of this section to the case of *three* GRMPSs in series.

4.4 INTERCONNECTED GRMPSs: PARALLEL

The section preceding was directed to the issue of whether or not two GRMPSs connected in series could be described as an equivalent GRMPS. This section again considers the pair (4-77) of GRMPSs with groups (4-78), insofar as what happens when they are connected in parallel.

As discussed in Section 1.3, this notation is available because of the binary operation on the GRMPS output sets. Here it is assumed that these are identical groups, denoted by

$$(Y_1, *_1, e_{Y_1}) = (Y_2, *_2, e_{Y_2});$$ (4-98)

the designation $(Y, *, e_Y)$, with subscripts dropped completely, is therefore appropriate. The essence of this parallel connection is then indicated in Fig. 4.4. Notice also that the input groups should satisfy

$$(U_1, \triangle_1, e_{U_1}) = (U_2, \triangle_2, e_{U_2}),$$ (4-99)

which is designated $(U, \triangle, e_U)$.

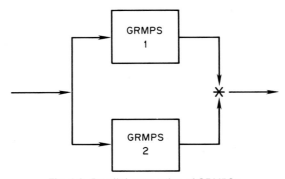

Fig. 4.4. Parallel connection of GRMPSs.

It is then desired to describe a GRMP system

$$(U, \triangle; X_1 \times X_2, **; Y, *; a, b, c, d)$$ (4-100)

which acts as the parallel connection.

The best way to begin is with the local dynamical equations

$$x_1(k + 1) = a_1 x_1(k) \square_1 b_1 u(k),$$ (4-101a)
$$x_2(k + 1) = a_2 x_2(k) \square_2 b_2 u(k),$$ (4-101b)
$$y_1(k) = c_1 x_1(k) * d_1 u(k),$$ (4-101c)
$$y_2(k) = c_2 x_2(k) * d_2 u(k).$$ (4-101d)

The overall output equation for the parallel connection is

$$y(k) = y_1(k) * y_2(k).$$ (4-102)

Suppress the indices and expand in terms of (4-101c) and (4-101d) to obtain

$$y = c_1 x_1 * d_1 u * c_2 x_2 * d_2 u.$$ (4-103)

It is necessary to put (4-103) into the form

$$y = c(x_1, x_2) * du \tag{4-104}$$

in order to fit the (a, b, c, d) idea of the GRMPS. The issues involved with (4-103) appear on the surface to be different from those associated with the corresponding series equation (4-83). In (4-103), the input terms and state terms are not yet grouped together, as will be necessary for the GRMPS description. However, the cure for this, namely, to assume the output group commutative, is the same as that used in the series case. Make this assumption; then (4-103) becomes

$$y = c_1 x_1 * c_2 x_2 * d_1 u * d_2 u. \tag{4-105}$$

Now define

$$d: U \to Y \quad \text{by} \quad d(u) = d_1 u * d_2 u; \tag{4-106a,b}$$

d is a morphism by the calculation

$$
\begin{aligned}
d(u_1 \triangle u_2) &= d_1(u_1 \triangle u_2) * d_2(u_1 \triangle u_2) \\
&= d_1 u_1 * d_1 u_2 * d_2 u_1 * d_2 u_2 \\
&= d_1 u_1 * d_2 u_1 * d_1 u_2 * d_2 u_2 \\
&= d(u_1) * d(u_2), \tag{4-107}
\end{aligned}
$$

which also uses the fact that $*$ is commutative. In a similar way, define

$$c: X_1 \times X_2 \to Y \quad \text{by} \quad c(x_1, x_2) = c_1 x_1 * c_2 x_2; \tag{4-108a,b}$$

c is a morphism by the calculation

$$
\begin{aligned}
c((x_1, x_2) ** (\tilde{x}_1, \tilde{x}_2)) &= c(x_1 \square_1 \tilde{x}_1, x_2 \square_2 \tilde{x}_2) \\
&= c_1(x_1 \square_1 \tilde{x}_1) * c_2(x_2 \square_2 \tilde{x}_2) \\
&= c_1 x_1 * c_1 \tilde{x}_1 * c_2 x_2 * c_2 \tilde{x}_2 \\
&= c_1 x_1 * c_2 x_2 * c_1 \tilde{x}_1 * c_2 \tilde{x}_2 \\
&= c(x_1, x_2) * c(\tilde{x}_1, \tilde{x}_2), \tag{4-109}
\end{aligned}
$$

which also makes use of $*$ commutativity.

It remains to combine (4-101a) and (4-101b). As in Section 4.3, the indices will be suppressed in the right members of these equations, while the calculations are carried out. These are

$$
\begin{aligned}
(x_1(k + 1), x_2(k + 1)) &= (a_1 x_1 \square_1 b_1 u, a_2 x_2 \square_2 b_2 u) \\
&= (a_1 x_1, a_2 x_2) ** (b_1 u, b_2 u) \\
&= (a_1 \times a_2)(x_1, x_2) ** (b_1 u, b_2 u). \tag{4-110}
\end{aligned}
$$

Define

$$b : U \rightarrow X_1 \times X_2 \qquad \text{by} \qquad bu = (b_1 u, b_2 u); \qquad (4\text{-}111\text{a,b})$$

b is a morphism by the calculation

$$
\begin{aligned}
b(u_1 \triangle u_2) &= (b_1(u_1 \triangle u_2), b_2(u_1 \triangle u_2)) \\
&= (b_1 u_1 \,\square_1\, b_1 u_2, b_2 u_1 \,\square_2\, b_2 u_2) \\
&= (b_1 u_1, b_2 u_1) \ ** \ (b_1 u_2, b_2 u_2) \\
&= bu_1 \ ** \ bu_2. \qquad\qquad (4\text{-}112)
\end{aligned}
$$

Finally, define

$$a = a_1 \times a_2, \qquad (4\text{-}113)$$

where $a_1 \times a_2$ is the product function defined in Section 2.5. It is not diffi-cult to show that this product function becomes a morphism when a_1 and a_2 are morphisms. Calculate

$$
\begin{aligned}
a((x_1, x_2) \ ** \ (\tilde{x}_1, \tilde{x}_2)) &= a(x_1 \,\square_1\, \tilde{x}_1, x_2 \,\square_2\, \tilde{x}_2) \\
&= (a_1(x_1 \,\square_1\, \tilde{x}_1), a_2(x_2 \,\square_2\, \tilde{x}_2)) \\
&= (a_1 x_1 \,\square_1\, a_1 \tilde{x}_1, a_2 x_2 \,\square_2\, a_2 \tilde{x}_2) \\
&= (a_1 x_1, a_2 x_2) \ ** \ (a_1 \tilde{x}_1, a_2 \tilde{x}_2) \\
&= a(x_1, x_2) \ ** \ a(\tilde{x}_1, \tilde{x}_2). \qquad (4\text{-}114)
\end{aligned}
$$

With the morphisms of (4-106), (4-108), (4-111), and (4-113), the symbolism (4-100) is established as a GRMPS, the parallel combination of two GRMPSs. Again the focus has been on the right side, in that it is the output group which makes the idea go, by being commutative.

EXAMPLE 4.4

The ideas of this section and the foregoing section can of course be used together. Consider Fig. 4.5. GRMPSs 1 and 2 are connected in series, and the resulting combination is connected in parallel with GRMPS 3. The discussion of Section 4.3 stipulates that the state group of GRMPS 1 must be commutative; and the determinations of this section state that the output groups of GRMPSs 1 and 3 must be identical and commutative. ∎

The basic idea of this section has the pleasing intuitive verbalization that the use of the abstract junction, denoted by $*$ in Figs. 4.4 and 4.5, has the best meaning when the order in which signals are combined at the junction is immaterial.

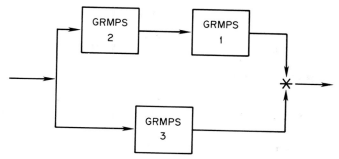

Fig. 4.5. More general connections.

Exercises

4.4-1. Reconsider Example 4.4. Find

$$(a, b, c, d)$$

for the interconnected system shown in Fig. 4.5. Define carefully their domains, codomains, and actions.

4.4-2. Extend the discussion of this section to the case of *three* GRMPSs in parallel.

4.5   REFINING THE TRIANGLE

The key triangle of Section 2.6 has played a central role in many of the discussions up to this point. Essentially, that discussion can be pictured in Fig. 4.6. Given is a function f with domain S and codomain T. Given also is an equivalence relation E on S. E breaks S up into a set of disjoint equivalence classes S/E, which have the property that each element of S belongs to one and only one equivalence class. The function p_E with domain S and codomain S/E assigns each element of S to its equivalence class. The key question growing out of these ideas is this: under what assumptions does there exist a unique function $\bar{f}: S/E \rightarrow T$ which makes the triangle commute?

Fig. 4.6. Existence of $\bar{f}$.

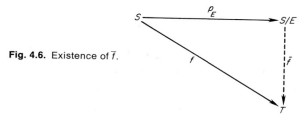

It has been shown that the necessary and sufficient condition is

$$p_E(s_1) = p_E(s_2) \Rightarrow f(s_1) = f(s_2), \tag{4-115}$$

which has been colloquially called the compatibility of f with E. In poset terms, (4-115) has the simple statement

$$E \le E(f). \tag{4-116}$$

Issues of this type have been fundamental to the procedures of Chapters 2 and 3.

This section takes up the issue of what happens when S is permitted to become a group, and when $f:S \to T$ is a morphism of groups.

Begin with the function

$$p_E:S \to S/E, \tag{4-117}$$

understood in terms of sets, as in Chapter 2. This function is pictured in Fig. 4.7. Suppose that the set S/E of equivalence classes admits group structure, say $(S/E, \square_E, e_{S/E})$, that S has group structure $(S, \square, e_S)$, and that p_E is a morphism of groups. What does this imply about E? Well, in view of the group structure on its codomain, the projection p_E satisfies

$$p_E(s_1) \square_E \widehat{p_E(s_2)} = e_{S/E} \tag{4-118}$$

for every pair $(s_1, s_2) \in E$. Recall Eq. (4-33). The fact that p_E is a morphism then implies

$$p_E(s_1) \square_E p_E(\hat{s}_2) = e_{S/E} \qquad \text{or} \qquad p_E(s_1 \square \hat{s}_2) = e_{S/E}. \tag{4-119a,b}$$

But this means that

$$s_1 \square \hat{s}_2 \in \text{Ker } p_E. \tag{4-120}$$

For $(s_1, s_2) \in E$, then, it is necessary that

$$s_1 = x_{12} \square s_2 \tag{4-121}$$

for some x_{12} in $\text{Ker } p_E$.

Fig. 4.7. Quotient set projection. $S \xrightarrow{\quad p_E \quad} S/E$

So the refining of the key triangle is unalterably involved with kernels of morphisms.

Section 4.1 stated that the kernel of a morphism $f:S \to T$ of groups is a subgroup. This is easy to show. Notice that $\text{Ker } f$ is not empty, inasmuch as it has to contain e_S. Then let s_1 and s_2 be elements in $\text{Ker } f$, that is,

$$f(s_i) = e_T, \qquad i = 1, 2, \tag{4-122}$$

and calculate

$$f(\hat{s}_i) = \widehat{f(s_i)} = \hat{e}_T = e_T \tag{4-123}$$

and

$$f(s_1 \,\square\, s_2) = f(s_1) * f(s_2) = e_T * e_T = e_T. \tag{4-124}$$

By (4-34), $\operatorname{Ker} f$ is a subgroup. Actually, $\operatorname{Ker} f$ is a special type of subgroup.
Let $(S, \square, e_S)$ be a group. A subgroup W contained in S is a *normal subgroup* of S if, for each w in W, and for each s in S,

$$s \,\square\, w \,\square\, \hat{s} \in W. \tag{4-125}$$

It follows that kernels of morphisms of groups are normal, by the calculation

$$\begin{aligned}
f(s \,\square\, w \,\square\, \hat{s}) &= f(s) * f(w) * f(\hat{s}) \\
&= f(s) * e_T * \widehat{f(s)} \\
&= f(s) * \widehat{f(s)} = e_T.
\end{aligned} \tag{4-126}$$

Not surprisingly, it turns out that normal subgroups are just right to use in
setting up equivalence relations on groups. Notice that, in a commutative
group, every subgroup is normal.

Let W be a normal subgroup of S. Set up an equivalence relation E on S
by defining

$$s_1 \, E \, s_2 \qquad \text{if} \qquad s_1 = w \,\square\, s_2 \tag{4-127a,b}$$

for some w in W. E is reflexive, because the unit $e_S \in W$. E is symmetric
because, when w is in W, so is $\hat{w}$. Finally, E is transitive due to the closure of
w under $\square$.

Consider the choice for the binary operation $\square_E$ on S/E. It needs to make
S/E into a group and to make p_E into a morphism. These constraints require

$$p_E(s_1) \,\square_E\, p_E(s_2) = p_E(s_1 \,\square\, s_2). \tag{4-128}$$

Is this a valid definition of a binary operation? Consider Fig. 4.8. A similar

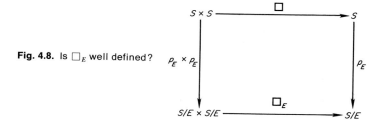

Fig. 4.8. Is $\square_E$ well defined?

question was raised in Section 3.10, Fig. 3.18. Using the fact that p_E is surjective, write

$$\Box_E = p_E \circ \Box \circ (p_E \times p_E)^{-R} \qquad (4\text{-}129\text{a})$$

for some right inverse

$$(p_E \times p_E)^{-R}: S/E \times S/E \to S \times S. \qquad (4\text{-}129\text{b})$$

In fact, the action of $\Box_E$ does not depend on which right inverse is chosen. Suppose that

$$(p_E \times p_E)^{-R}(\bar{s}_1, \bar{s}_2) = (s_1, s_2); \qquad (4\text{-}130)$$

then any other right inverse must give

$$(w_1 \,\Box\, s_1, w_2 \,\Box\, s_2) \qquad (4\text{-}131)$$

for $w_i \in W$, $i = 1, 2$. But then

$$(p_E \circ \Box)(w_1 \,\Box\, s_1, w_2 \,\Box\, s_2) = p_E(w_1 \,\Box\, s_1 \,\Box\, w_2 \,\Box\, s_2) \qquad (4\text{-}132)$$

and

$$w_1 \,\Box\, s_1 \,\Box\, w_2 \,\Box\, s_2 = w_1 \,\Box\, s_1 \,\Box\, w_2 \,\Box\, \hat{s}_1 \,\Box\, s_1 \,\Box\, s_2$$
$$= w_1 \,\Box\, \tilde{w} \,\Box\, s_1 \,\Box\, s_2 = \tilde{\tilde{w}} \,\Box\, s_1 \,\Box\, s_2, \qquad (4\text{-}133\text{a})$$

where

$$\tilde{w} = s_1 \,\Box\, w_2 \,\Box\, \hat{s}_1 \in W \qquad (4\text{-}133\text{b})$$

because W is normal, and

$$\tilde{\tilde{w}} = w_1 \,\Box\, \tilde{w} \in W \qquad (4\text{-}133\text{c})$$

because W is a subgroup. Thus

$$p_E(w_1 \,\Box\, s_1 \,\Box\, w_2 \,\Box\, s_2) = p_E(s_1 \,\Box\, s_2) \qquad (4\text{-}134)$$

as desired, and $\Box_E$ is well defined. Normality played a crucial role in (4-133a), as the reader has noted.

With $\Box_E$ established as a binary operation on S/E, and with p_E established as satisfying (4-128), it is only necessary to complete the group structure on S/E. First $\Box_E$ is associative, because $\Box$ is associative. It has a unit, namely, $p_E(e_S)$. If $p_E(s)$ is in S/E, then $p_E(\hat{s})$ is an inverse for $p_E(s)$ under $\Box_E$. These last statements have minor technicalities involved with them, such as showing that, respectively,

$$w_1 \,\Box\, s \,\Box\, w_2 \,\Box\, e_S \, E \, s \,\Box\, e_S, \qquad w_1 \,\Box\, s \,\Box\, w_2 \,\Box\, \hat{s} \, E \, s \,\Box\, \hat{s}, \qquad (4\text{-}135\text{a,b})$$

which can be established with the aid of a construction along the lines of (4-133).

For Fig. 4.7 to represent a morphism of groups, it has now been shown that the equivalence relation has to be set up in the manner (4-127) for W a normal subgroup. The reader may verify that (4-127b) can be replaced by

$$s_1 = s_2 \,\square\, w. \tag{4-136}$$

If the group is commutative, (4-127b) and (4-136) are essentially the same.

The final step is to construct a morphism $\bar{f} : S/E \to T$ of groups in Fig. 4.6. It is assumed that $f : S \to T$ is a morphism of groups and that the function $\bar{f}$ exists in the sense of Section 2.6. Then all that remains is to show that $\bar{f}$ satisfies the morphism condition. To do this, follow the calculation

$$\bar{f}(\bar{s}_1 \,\square_E\, \bar{s}_2) = (\bar{f} \circ p_E)(s_1 \,\square\, s_2) \tag{4-137}$$

for any s_i satisfying

$$p_E(s_i) = \bar{s}_i, \qquad i = 1, 2, \tag{4-138}$$

which continues to yield

$$\bar{f}(\bar{s}_1 \,\square_E\, \bar{s}_2) = f(s_1 \,\square\, s_2) \tag{4-139a}$$

$$= f(s_1) * f(s_2) \tag{4-139b}$$

$$= (\bar{f} \circ p_E)(s_1) * (\bar{f} \circ p_E)(s_2) \tag{4-139c}$$

$$= \bar{f}(\bar{s}_1) * \bar{f}(\bar{s}_2). \tag{4-139d}$$

Notice that the transitions from (4-137) to (4-139a) and from (4-139b) to (4-139c) are accomplished with the aid of the identity

$$f = \bar{f} \circ p_E, \tag{4-140}$$

established in Chapter 2. On the other hand, the step from (4-139a) to (4-139b) uses the fact that f is a morphism.

In summary, for the key diagram of Section 2.6 to refine to the GRMP case, the equivalence relation must be brought about by a normal subgroup, as in (4-127).

EXAMPLE 4.5-1

For $f : S \to T$ a morphism of groups, it has been shown that Ker f is a normal subgroup. What of Im f? Example 4.2-1 is helpful here, for

$$\text{Im } a = \{e, g_1 \,\square\, g_2\}. \tag{4-141}$$

Now select from the dihedral group of Example 4.1-2 the element

$$g_1 \,\square\, g_1 \,\square\, g_1 \tag{4-142}$$

and its inverse g_1. Then

$$(g_1 \,\square\, g_1 \,\square\, g_1) \,\square\, (g_1 \,\square\, g_2) \,\square\, g_1 = g_2 \,\square\, g_1 \notin \operatorname{Im} a. \qquad (4\text{-}143)$$

Thus $\operatorname{Im} f$ need not be a normal subgroup. ∎

EXAMPLE 4.5-2

Normal subgroups can always be established as the kernel of some morphism. In fact, select the projection morphism p_E, where E has been defined in the manner (4-127). Then

$$p_E(e_S) = p_E(w \,\square\, e_S) \qquad (4\text{-}144)$$

for every w in W. This shows that

$$W \subset \operatorname{Ker} p_E. \qquad (4\text{-}145)$$

Conversely, suppose that

$$p_E(s) = p_E(e_S); \qquad (4\text{-}146a)$$

then

$$s = w \,\square\, e_S = w, \qquad (4\text{-}146b)$$

so that

$$\operatorname{Ker} p_E \subset W. \qquad (4\text{-}147)$$

∎

EXAMPLE 4.5-3

The basic requirement for the existence of the unique function $\bar{f} : S/E \to T$ in Fig. 4.6 is (4-115). In raising that diagram up to the level of groups and morphisms, (4-127) had to be used, where W is a normal subgroup. The left member of the implication (4-115) means

$$s_2 = w \,\square\, s_1, \qquad (4\text{-}148)$$

which then makes the right member

$$f(s_1) = f(w \,\square\, s_1) = f(w) * f(s_1) \qquad (4\text{-}149a,b)$$

by reason of the fact that f is a morphism. But then

$$f(s_1) * \widehat{f(s_1)} = f(w) * f(s_1) * \widehat{f(s_1)}, \qquad (4\text{-}149c)$$

so that

$$e_T = f(w), \qquad (4\text{-}149d)$$

which means that (4-115) implies

$$W \subset \operatorname{Ker} f. \qquad (4\text{-}150)$$

Moreover, it is straightforward to show that (4-150) implies (4-115). ∎

In view of the role played by the normal subgroup W in these deliberations, it is usual to replace the notation S/E by S/W, read S modulo W or S mod W. According to the same reasoning, the subscripts E on p_E and on $\square_E$ will be replaced by W. The final triangle is drawn in Fig. 4.9.

Equipped with its binary operation $\square_W$ and its unit $p_W(e_S)$, $(S/W, \square_W, p_W(e_S))$ is then a group in its own right, called the *quotient group* of S by the normal subgroup W. For simplicity, the quotient group is usually denoted by $(S/W, \square, e_{S/W})$.

In the quotient group case, the equivalence class associated with an element s is called the *coset* of s.

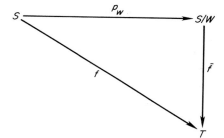

Fig. 4.9. $\bar{f}$ exists when $W \subset \operatorname{Ker} f$.

Exercises

4.5-1. In a commutative group, demonstrate that all subgroups are normal.

4.5-2. Complete the details involved in showing that (4-127) is an equivalence relation. Hints of the ideas involved may be found in the remarks after (4-127b).

4.5-3. Show that the binary operation $\square_E$ is associative under the assumption that the binary operation $\square$ is associative.

4.5-4. Show that $p_E(e_S)$ is a unit for $\square_E$. Also establish that $p_E(\hat{s})$ is an inverse for $p_E(s)$ under $\square_E$. In making these arguments, it will be helpful to consider (4-135).

4.5-5. Reconsider the developments of this section if (4-127b) is replaced by (4-136).

4.5-6. Show that (4-150) implies (4-115).

4.6 GRMPSs: FINITE TRANSITION

Section 3.4 introduced the notion of a finite transition function for a set-dynamical system. This was a function

$$\phi: X \times \mathrm{SS}(U^{\mathbb{N}}) \to X \qquad (4\text{-}151\mathrm{a})$$

whose action is given by

$$\phi(x_{k_1}, u_*[k_1, k_2]) = x_{k_2+1}. \qquad (4\text{-}151\text{b})$$

When J is not a factor of the domain product for a local transition function, it has been pointed out that often one may as well select k_1 to be zero. Then the actions of interest are of the form

$$\phi(x_0, u_*[0, k]) = x_{k+1}, \qquad (4\text{-}151\text{c})$$

for k in $\mathbb{N}$. Recall that the sequence segment notation $u_*[0, k]$ stands for the inputs

$$u_0, u_1, u_2, \ldots, u_k. \qquad (4\text{-}152)$$

This section gives the details involved with computing the action represented in (4-151c) for GRMPSs.

Start with a GRMPS

$$(U, \triangle; X, \square; Y, *; a, b, c, d). \qquad (4\text{-}153)$$

Given the initial state x_0, and an input u_0, the first state is

$$x_1 = ax_0 \,\square\, bu_0. \qquad (4\text{-}154)$$

When the next input u_1 is applied, the result is a second state given by

$$\begin{aligned} x_2 = ax_1 \,\square\, bu_1 &= a(ax_0 \,\square\, bu_0) \,\square\, bu_1 \\ &= (a \circ a)(x_0) \,\square\, (a \circ b)(u_0) \,\square\, bu_1. \end{aligned} \qquad (4\text{-}155)$$

Here it is most helpful to suppress the notation for composition. Thus,

$$a \circ a = a^2, \qquad a \circ b = ab. \qquad (4\text{-}156)$$

Suppose now that the ith state is given by

$$x_i = a^i x_0 \,\square\, a^{i-1} bu_0 \,\square\, a^{i-2} bu_1 \,\square \cdots \square\, bu_{i-1}. \qquad (4\text{-}157)$$

The application of an ith input u_i leads to

$$\begin{aligned} x_{i+1} = ax_i \,\square\, bu_i \\ &= a(a^i x_0 \,\square\, a^{i-1} bu_0 \,\square \cdots \square\, bu_{i-1}) \,\square\, bu_i \\ &= a^{i+1} x_0 \,\square\, a^i bu_0 \,\square \cdots \square\, abu_{i-1} \,\square\, bu_i. \end{aligned} \qquad (4\text{-}158)$$

An inductive argument, then, provides that (4-158) calculates the action of the finite transition function, that is,

$$\phi(x_0, u_*[0, k]) = a^{k+1} x_0 \,\square\, a^k bu_0 \,\square \cdots \square\, bu_k. \qquad (4\text{-}159)$$

Denote the expression

$$a^i bu_0 \;\square\; a^{i-1} bu_1 \;\square \cdots \square\; abu_{i-1} \;\square\; bu_i \tag{4-160}$$

by

$$\underset{j=0}{\overset{i}{\square}} \, a^{i-j} bu_j, \tag{4-161}$$

where a^0 is again understood as the morphism $1_X : X \to X$. Then the finite transition action of (4-159) has compact representation

$$\phi(x_0, u_*[0,k]) = a^{k+1} x_0 \;\square\; \left(\underset{j=0}{\overset{k}{\square}} \, a^{k-j} bu_j \right). \tag{4-162}$$

While (4-162) is technically the thing to write, the presence of two binary operation symbols tends to be awkward. When no confusion will result, then (4-162) is itself elided to

$$\phi(x_0, u_*[0,k]) = a^{k+1} x_0 \underset{j=0}{\overset{k}{\square}} \, a^{k-j} bu_j. \tag{4-163}$$

The reader should take note of the fact that the order of terms shown in (4-160) as a definition for (4-161) is important.

Constructions of the type (4-161) play a very transparent role in group theoretic systems studies. Consider this expression as defining a function

$$\underset{j=0}{\overset{i}{\square}} \, a^{i-j} b : \tilde{U}_{i+1} \to X, \tag{4-164a}$$

where $\tilde{U}_{i+1}$ is the subset

$$\{ u_*[0,i] \,|\, u_*[0,i] \in \mathrm{SS}(U^{\mathbb{N}}) \} \tag{4-164b}$$

of $\mathrm{SS}(U^{\mathbb{N}})$ containing input sequence segments beginning at index 0 and ending at index i. The set of states which can be produced by this function can be defined

$$X_{i+1} = \left(\underset{j=0}{\overset{i}{\square}} \, a^{i-j} b \right)_* (\tilde{U}_{i+1}) \tag{4-165}$$

as the image of (4-164) contained in X. Of interest in reachability determinations is the sequential growth of X_i. Helpful in the treatment of X_i is the notion of a lattice of subgroups.

Let $(G, \square, e_G)$ be a group, and consider the set $\mathbb{S}(G)$ of all subgroups of G. $\mathbb{S}(G)$ is partially ordered by set theoretic inclusion. It has a greatest element, which is the original group $(G, \square, e_G)$; and it has a least element, which is the group $(\{e_G\}, \square, e_G)$ containing just the unit e_G. It is easy to see that the poset $(\mathbb{S}(G), \subset)$ can be equipped with an infimum, for the set theoretic intersection of two elements S_1 and S_2 in $\mathbb{S}(G)$ is a subgroup. So the meet can be defined as intersection. Moreover, because $(\mathbb{S}(G), \subset)$ is a bounded poset with an infimum defined, there is a supremum defined as well. Use this supremum to define a binary join operation $\vee_{\mathbb{S}(G)}$ to make

$$(\mathbb{S}(G), \subset, \cap, \vee_{\mathbb{S}(G)}) \tag{4-166}$$

into a lattice. Unfortunately,

$$S_1 \vee_{\mathbb{S}(G)} S_2 \tag{4-167a}$$

is not necessarily the same as

$$S_1 \square S_2 = \{s_1 \square s_2 \,|\, s_i \in S_i, \, i = 1, 2\}, \tag{4-167b}$$

though in general (4-167a) contains (4-167b). The reason behind this is lack of commutativity on the part of operation $\square$, which prevents the step

$$s_1 \square s_2 \square \tilde{s}_1 \square \tilde{s}_2 = s_1 \square \tilde{s}_1 \square s_2 \square \tilde{s}_2. \tag{4-168}$$

Thus (4-167b) need not be a subgroup; then (4-167a) is the smallest subgroup which contains the subgroups S_1 and S_2. Notice that the definition (4-167b) really defines a binary operation

$$\mathbb{P}(G) \times \mathbb{P}(G) \to \mathbb{P}(G); \tag{4-169}$$

the statement that $S_1 \square S_2$ may not be in $\mathbb{S}(G)$, even though S_1 and S_2 are, merely means that $\square$ does not restrict to a binary operation on $\mathbb{S}(G)$. These distinctions will be helpful in the examples.

EXAMPLE 4.6-1

In the construction of any X_i, each element of the input group is passed through the morphism b. Let

$$b_j \in \operatorname{Im} b, \qquad j = 0, \ldots, i; \tag{4-170a}$$

then each element in X_i has the representation

$$\prod_{j=0}^{i-1} a^{i-1-j} b_j. \tag{4-170b}$$

The calculation

$$a\left(\prod_{j=0}^{i-1} a^{i-1-j} b_j\right) \square b_i = \prod_{j=0}^{i} a^{i-j} b_j \tag{4-171}$$

shows that

$$a_* X_i \square \operatorname{Im} b \subset X_{i+1}. \tag{4-172}$$

Conversely, if

$$\bigsqcup_{j=0}^{i} a^{i-j} b_j \in X_{i+1},$$
(4-173a)

then write

$$\bigsqcup_{j=0}^{i} a^{i-j} b_j = \left(\bigsqcup_{j=0}^{i-1} a^{i-j} b_j \right) \square\, b_i = a \left(\bigsqcup_{j=0}^{i-1} a^{i-1-j} b_j \right) \square\, b_i,$$
(4-173b,c)

from which

$$X_{i+1} \subset a_* X_i \,\square\, \operatorname{Im} b.$$
(4-174)

Thus

$$X_{i+1} = a_* X_i \,\square\, \operatorname{Im} b.$$
(4-175)

If $\square$ is commutative, (4-175) combines with the subgroup lattice ideas to establish that all the X_i will be subgroups. This is a consequence of the fact that $\operatorname{Im} b$ is a subgroup and that X_i a subgroup implies $a_*(X_i)$ is a subgroup, a technicality left to the reader as an exercise. ∎

EXAMPLE 4.6-2

A different kind of result may be obtained if (4-173a) is rewritten in the alternate way

$$\bigsqcup_{j=0}^{i} a^{i-j} b_j = a^i b_0 \bigsqcup_{j=1}^{i} a^{i-j} b_j.$$
(4-176)

Make the change of variable

$$k = j - 1$$
(4-177)

in the right member of (4-176); then

$$\bigsqcup_{j=0}^{i} a^{i-j} b_j = a^i b_0 \bigsqcup_{k=0}^{i-1} a^{i-1-k} b_{k+1},$$
(4-178)

which shows that

$$X_{i+1} = (a^i)_* (\operatorname{Im} b) \,\square\, X_i.$$
(4-179)

As a consequence, since $\operatorname{Im} b$ contains e_X, it follows that

$$X_i \subset X_{i+1},$$
(4-180)

$i = 1, 2, \ldots$. This is pictured in Fig. 4.10. Note that this result does not depend upon $\square$ being commutative. ∎

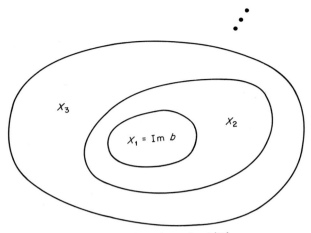

Fig. 4.10. Nested growth of $\{X_i\}$.

Exercises

4.6-1.   If X_i is a subgroup of X, show that

$$a_*(X_i)$$

is a subgroup.

4.6-2.   Give a careful demonstration that the subgroup equality in (4-179) is a direct consequence of (4-178).

4.7   REACHABILITY REVISITED

Most of the emphasis in the SDS elaborations of Chapter 3 was placed upon the concept of observability. This was to aid in the buildup of appropriate background for Section 3.10 on the internal model principle. This section adds a number of ideas to those which were mentioned in Section 3.4.

Recall that a state x in X is said to be reachable from a state x_0 in X if there is an input sequence segment $u_*[0, k]$ in $SS(U^N)$ such that the resulting state sequence segment

$$x_*[0, k + 1] = \pi(x_0, u_*[0, k])$$ (4-181a)

in $SS(X^N)$ satisfies

$$x_{k+1} = x.$$ (4-181b)

The original definition, in terms of the global transition function

$$\pi: X \times SS(U^{\mathbb{N}}) \to SS(X^{\mathbb{N}})$$

(4-182)

was restated in terms of the finite transition function

$$\phi: X \times SS(U^{\mathbb{N}}) \to X$$

(4-183a)

with action

$$\phi(x_0, u_*[0, k]) = x_{k+1}.$$

(4-183b)

In terms of ϕ, the SDS is reachable from a state x if and only if the action

$$\phi_x(u_*[k_1, k_2]) = \phi(x, u_*[k_1, k_2])$$

(4-184)

of $\phi_x : SS(U^{\mathbb{N}}) \to X$ is surjective.

For the group morphic system, special attention is paid to the notion of reachability from the unit state e_X in X. In fact, the usage is so common that a GRMPS is said to be *reachable* if it is reachable from the unit state.

When a GRMPS starts in the unit state e_X, the term

$$a^{i+1}x_0 = a^{i+1}e_X = e_X$$

(4-185)

effectively vanishes from (4-163). Then the mechanism of finite transition has been described in detail as part of Section 4.6. The governing function is (4-164), and Example 4.6-2 has shown that the states x_i which can be reached from the unit state at index i are described by subsets X_i of X which satisfy

$$X_0 \subset X_1 \subset X_2 \cdots,$$

(4-186)

where the set

$$X_0 = \{e_X\}$$

(4-187)

has been adjoined to indicate what happens if the sequence u in $SS(U^{\mathbb{N}})$ is applied. From (4-186), and from Section 3.1, it is apparent that

$$(\{X_i \mid i \in \mathbb{N}\}, \subset)$$

(4-188)

is a chain. Because the subsets increase with their index, (4-188) is sometimes called an *ascending chain*. Notice that this ascending chain is a partially ordered subset of the lattice

$$(\mathbb{P}(X), \subset, \cap, \cup),$$

(4-189)

which is complete. Thus the chain has a least upper bound X^*, where

$$X^* = \sup\{X_i \mid i \in \mathbb{N}\}.$$

(4-190)

Then the GRMPS is reachable only if

$$X^* = X. \tag{4-191}$$

Turn now to the case in which the state group $(X, \Box, e_X)$ is commutative. It has been shown in the section preceding that each of the X_i is then a subgroup of X. It is interesting to ask whether

$$X^* = X_i \tag{4-192}$$

for some $i \in \mathbb{N}$. An answer can be given here to this question, but its justification will have to be postponed.

The following definition is needed. Let $(G, \Box, e_G)$ be a commutative group. From an element g in G, and from its inverse $\hat{g}$ which must also be in G, it is possible to generate the lists

$$
\begin{array}{cc}
g & \hat{g} \\
g \Box g & \hat{g} \Box \hat{g} \\
g \Box g \Box g & \hat{g} \Box \hat{g} \Box \hat{g} \\
\vdots & \vdots
\end{array}
\tag{4-193}
$$

A shorthand for the left column of (4-193), already mentioned in Section 1.4, is

$$(1)g, \quad (2)g, \quad (3)g, \quad \ldots, \tag{4-194a}$$

while a similar shorthand for the right column would be

$$(-1)g, \quad (-2)g, \quad (-3)g, \quad \ldots. \tag{4-194b}$$

Denoting $(0)g$ by e_G, any of the elements in (4-193) or e_G can be written

$$(z)g, \quad z \in \mathbb{Z}. \tag{4-195}$$

This is actually a common practice, and the notation

$$(\mathbb{Z})g = \{(z)g \, | \, z \in \mathbb{Z}\} \tag{4-196}$$

covers all elements of type (4-195). With these preliminaries, the group is said to be of *finite type* if there exists a natural number n in $\mathbb{N}$ and elements

$$g_i, \quad i = 1, 2, \ldots, n, \tag{4-197}$$

in G such that

$$G = (\mathbb{Z})g_1 \Box (\mathbb{Z})g_2 \Box \cdots \Box (\mathbb{Z})g_n. \tag{4-198}$$

When $(X, \Box, e_X)$ is a commutative group of finite type, (4-192) will hold. In that case, the condition (4-191) is also sufficient for reachability. Then X^* is called the *reachable subgroup* and is sometimes denoted by $\langle a \, | \, \text{Im} \, b \rangle$.

It can be instructive to picture some of the aspects associated with X^*. To do this, make the finite type assumption on $(X, \Box, e_X)$, and recall that the commutative assumption on $\Box$ makes every subgroup normal. Then establish the projection

$$p_{X^*}: X \to X/X^* \tag{4-199}$$

onto the quotient group X/X^*. Use (4-192), and calculate

$$a_*(X^*) = a_* X_i \subset a_* X_i \Box \operatorname{Im} b = X_{i+1} = X^*, \tag{4-200}$$

where the results of Example 4.6-1 have been employed. From Example 2.6-2, then, it follows that the diagram of Fig. 4.11 can be constructed, where

$$\bar{b} = p_{X^*} \circ b. \tag{4-201}$$

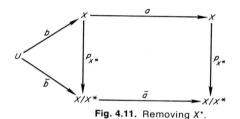

Fig. 4.11. Removing X^*.

Consider the local transition equation

$$x_{k+1} = ax_k \Box bu_k, \tag{4-202}$$

and apply the morphism p_{X^*} to both members. This gives

$$p_{X^*}(x_{k+1}) = p_{X^*}(ax_k \Box bu_k) \tag{4-203a}$$
$$= p_{X^*}(ax_k) \Box p_{X^*}(bu_k) \tag{4-203b}$$
$$= (p_{X^*} \circ a)x_k \Box (p_{X^*} \circ b)u_k \tag{4-203c}$$
$$= (\bar{a} \circ p_{X^*})x_k \Box \bar{b}u_k \tag{4-203d}$$
$$= \bar{a}(p_{X^*}x_k) \Box \bar{b}u_k, \tag{4-203e}$$

which can be simplified to

$$\bar{x}_{k+1} = \bar{a}\bar{x}_k \Box \bar{b}u_k, \tag{4-204}$$

if the previously employed convention

$$p_{X^*}x = \bar{x} \tag{4-205}$$

is used once again. The reader should note that, in keeping with the convention established at the end of Section 4.5, $(X/X^*, \Box, e_{X/X^*})$ represents the quotient group. Thus, although the binary operation symbol in (4-203a) is used in X, the same symbol in (4-203b) is used in X/X^*. Now, for u in U,

$$\bar{b}u = (p_{X^*} \circ b)u = p_{X^*}(bu) = e_{X/X^*}. \tag{4-206}$$

This is a consequence of the fact that

$$\text{Im } b = X_1 \subset X^* = \text{Ker } p_{X^*}. \tag{4-207}$$

Thus

$$\text{Im } \bar{b} = e_{X/X^*}, \tag{4-208}$$

and (4-204) may as well be rewritten

$$\bar{x}_{k+1} = \bar{a}\bar{x}_k, \tag{4-209}$$

which shows clearly that, once the reachable states of a GRMPS are removed in the quotient sense, then the inputs have no effect whatsoever on the resulting cosets. In other words, there is no way that inputs can reach those cosets. This is why it is desirable to have the quotient group be trivial, containing only the unit element. This occurs when (4-191) is satisfied, and may be denoted

$$X/X^* = e. \tag{4-210}$$

There is, of course, another concept related to reachability. A state x in X is *controllable to the unit state* e_X in X if there exists an input sequence segment $u_*[0,k]$ in $\text{SS}(U^{\mathbb{N}})$ such that

$$\phi_x(u_*[0,k]) = e_X. \tag{4-211}$$

A GRMPS is *controllable* if each of its states is controllable to the unit state.

An easy way to see the difference between reachability and controllability is to consider a GRMPS with

$$\text{Im } a = e_X. \tag{4-212}$$

Then

$$X^* = X_1 = \text{Im } b. \tag{4-213}$$

Such a GRMPS is controllable with the choice

$$u_*[0,k] = u_*[0,0] = u_0 = e_U, \tag{4-214}$$

while it may or may not be reachable.

In fact, it follows from the calculation in (4-200), when the state group is commutative and of finite type, that if any element x in X^* can be reached from the unit state, then so can $a^i x$, where i is chosen so that X_i is equal to X^*; and so can $\widehat{a^i x}$, because X_i is a subgroup. Thus there exists $u_*[0, i-1]$

such that

$$\widehat{a^i x} = \bigsqcup_{j=0}^{i-1} a^{i-1-j} b u_j, \qquad (4\text{-}215a)$$

from which it follows that

$$e_X = a^i x \bigsqcup_{j=0}^{i-1} a^{i-1-j} b u_j. \qquad (4\text{-}215b)$$

Under these assumptions, then, if x in X is reachable from the unit state, then it is also controllable to the unit state.

EXAMPLE 4.7

For the GRMPS of Example 4.2-1, it should be noted that the state group is not commutative. On the other hand, it has only eight elements. When a group has only a finite number of elements, it is clear that (4-192) must hold for a finite i in $\mathbb{N}$ also. Now

$$X_1 = \operatorname{Im} b = \{g_2, e\}. \qquad (4\text{-}216)$$

X_1 is always a subgroup, because $\operatorname{Im} b$ is a subgroup. To calculate X_2, use the result of Example 4.6-1:

$$a_* X_1 = \{g_1 \,\square\, g_2, e\}; \qquad (4\text{-}217)$$

$$X_2 = a_* X_1 \,\square\, \operatorname{Im} b = \{g_1 \,\square\, g_2, e\} \,\square\, \{g_2, e\} = \{g_1 \,\square\, g_2, g_1, g_2, e\}. \qquad (4\text{-}218)$$

X_2 is not a subgroup. Next calculate

$$a_* X_2 = a_* X_1, \qquad \text{from which} \qquad X^* = X_2. \qquad (4\text{-}219a,b)$$

Exercises

4.7-1. In this section, attention has been placed primarily upon the notion of reachability from the unit state e_X in X. Explain how the results in this regard can be related to the question of reachability from other, nonunit, states in X.

4.7-2. With reference to Example 4.7, prove that (4-192) must hold for a finite i in $\mathbb{N}$ whenever the group X has only a finite number of elements.

4.8 OBSERVABILITY REVISITED

Section 4.7 preceding put to use the fact that the state group had a unit e_X. Then concepts of reachability and controllability were established relative to the unit state. This section makes the same adjustments for observability.

The basic function for the discussion is the global output function

$$h: X \times U^{\mathbb{N}} \to Y^{\mathbb{N}}, \tag{4-220a}$$

which has action represented by

$$h(x_0, u_*) = y_*. \tag{4-220b}$$

Fixing the input sequence u_* makes h into a function

$$X \to Y^{\mathbb{N}}, \tag{4-221}$$

which can be denoted by h_{u_*}. The system is observable relative to the input sequence u_* if h_{u_*} is injective. This was the statement for the general SDS case.

For GRMPSs, however, there is the possibility of specializing the input sequence to

$$u_* = (e_U, e_U, e_U, \dots), \tag{4-222}$$

which is denoted by e_*. A GRMPS is said to be *observable* if it is observable relative to the input sequence u_* in (4-222).

The discussion on autonomous set-dynamical systems of Section 3.5 can now be handled much more naturally. Consider the local dynamical equations

$$x_{k+1} = ax_k \,\square\, bu_k, \qquad y_k = cx_k * du_k. \tag{4-223a,b}$$

When the input sequence is e_*, the passage through b produces

$$(e_X, e_X, e_X, \dots) \tag{4-224a}$$

while that through d yields

$$(e_Y, e_Y, e_Y, \dots). \tag{4-224b}$$

Accordingly, (4-223) reduces to

$$x_{k+1} = ax_k, \qquad y_k = cx_k, \tag{4-225a,b}$$

which is the *autonomous* GRMPS

$$(X, \square; Y, *; a, c). \tag{4-226}$$

For the autonomous system (4-226), establish the extended global output function

$$h: X \to \mathrm{SS}(Y^{\mathbb{N}}). \tag{4-227}$$

Section 3.8 has established that the fundamental construction involved in (4-227) is that of the morphisms

$$ca^i: X \to Y, \qquad i \in \mathbb{N}. \tag{4-228}$$

The question of observability was then tied up with that of

$$E(ca^i), \qquad i \in \mathbb{N}. \tag{4-229}$$

For Example 4.5-2, it follows that the kernel of the projection morphism

$$p_{E(ca^i)} : X \rightarrow X/\text{Ker}(ca^i) \tag{4-230}$$

is the kernel of ca^i. Thus, two states x_1 and x_2 in X satisfy

$$x_1 \equiv x_2(ca^i) \qquad \text{if and only if} \qquad x_1 = x \,\square\, x_2 \tag{4-231a,b}$$

for some x in $\text{Ker}\, ca^i$. Then

$$x_1\, E(c, a)\, x_2 \quad \text{holds if and only if} \quad x_1 = \tilde{x} \,\square\, x_2 \tag{4-232a,b}$$

for some

$$\tilde{x} \in \bigcap_{j=0}^{\infty} \text{Ker}\, ca^j. \tag{4-232c}$$

The construction in the right member of (4-232c) may seem unusual at first sight. However, it is easily understood as

$$\inf\{\text{Ker}\, ca^j \,|\, j \in \mathbb{N}\} \tag{4-233}$$

in the subgroup lattice

$$(\mathbb{S}(X),\, \subset,\, \cap,\, \vee_{\mathbb{S}(X)}) \tag{4-234}$$

introduced in Section 4.6. Define

$$N(c, a) = \bigcap_{j=0}^{\infty} \text{Ker}\, ca^j. \tag{4-235}$$

It is straightforward to show that $N(c, a)$ is a normal subgroup of X.
Observe that

$$N(c, a) \subset \text{Ker}\, c. \tag{4-236}$$

Moreover,

$$a_* N(c, a) \subset N(c, a). \tag{4-237}$$

To check this, suppose that $\tilde{x}$ is in $N(c, a)$. Then

$$ca^i \tilde{x} = e_Y, \qquad i \in \mathbb{N}. \tag{4-238}$$

Now

$$(ca^i)(a\tilde{x}) = ca^{i+1}\tilde{x} = e_Y, \qquad i \in \mathbb{N}. \tag{4-239}$$

This establishes (4-237). From Example 2.6-2, the commutative diagram of

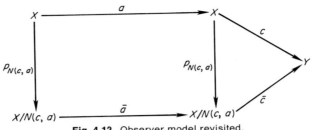

Fig. 4.12. Observer model revisited.

Fig. 4.12 can be drawn, where the existence of $\bar{c}$ is a consequence of (4-236). An earlier version of this diagram was that of Fig. 3.14, without group assumptions.

Starting from (4-225), the meaning of $\bar{a}$ and $\bar{c}$ can be made clearer. Apply the projection morphism $p_{N(c,a)}$ to both members of (4-225a). This gives

$$\bar{x}_{k+1} = p_{N(c,a)}x_{k+1} = (p_{N(c,a)} \circ a)x_k$$
$$= (\bar{a} \circ p_{N(c,a)})x_k = \bar{a}\bar{x}_k. \tag{4-240a}$$

In (4-225b), the calculation

$$y_k = cx_k = (\bar{c} \circ p_{N(c,a)})x_k = \bar{c}\bar{x}_k \tag{4-240b}$$

succeeds. Thus there is an autonomous observer model

$$(X/N(c,a), \square; Y, *; \bar{a}, \bar{c}) \tag{4-241}$$

analogous to (3-163). It is an interesting exercise to show that

$$N(\bar{c}, \bar{a}) = e_{X/N(c,a)}. \tag{4-242}$$

By reason of these conclusions, the normal subgroup $N(c,a)$ is called the *unobservable* subgroup of the GRMPS. When $N(c,a)$ is the unit state, then the GRMPS is observable.

EXAMPLE 4.8

It is not hard to see what effect is produced on observer model cosets by a nonunit input sequence. Simply make the calculations

$$x_{k+1} = ax_k \square bu_k,$$
$$\bar{x}_{k+1} = p_{N(c,a)}(ax_k \square bu_k)$$
$$= (p_{N(c,a)} \circ a)x_k \square (p_{N(c,a)} \circ b)u_k \tag{4-243}$$
$$= \bar{a}\bar{x}_k \square \bar{b}u_k,$$

where

$$\bar{b} = p_{N(c,\,a)} \circ b. \tag{4-244}$$

Thus the input is simply projected into $X/N(c,a)$. ∎

If the global output function is extended in the manner

$$h_{e*}: X \to \mathrm{SS}(Y^{\mathbb{N}}), \tag{4-245}$$

then it is possible to study the effect on knowledge of x_0 brought about by knowledge of $y_*[0,i]$. Then interest centers on the normal subgroups

$$N_i = \bigcap_{j=0}^{i} \mathrm{Ker}\, ca^j, \tag{4-246a}$$

which satisfy

$$\cdots \subset N_3 \subset N_2 \subset N_1 \subset N_0, \tag{4-246b}$$

where N_0 is $\mathrm{Ker}\, c$. Figure 4.13 indicates the situation.

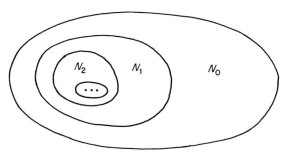

Fig. 4.13. Partial ordering of N_i.

Exercises

4.8-1.   Establish that the unobservable subgroup $N(c,a)$ of (4-235) is a normal subgroup of the state group X.

4.8-2.   Relative to the discussion of Eqs. (4-240)–(4-242), carry out the exercise of showing that

$$N(\bar{c}, \bar{a}) = e_{X/N(c,\,a)}.$$

4.9   THE STATE GROUP

A great deal of interesting behavior has been associated in earlier sections with the difference between products and coproducts. Before passing on to more detailed applications of the group morphic system idea, it is well to spend some time explaining how these issues resolve under reasonable group morphic assumptions.

Begin by recalling the points which have already been made on this subject. The product of two sets was introduced in Section 2.1. However, the important observation that a product set had projections to the factors came up in Section 2.5, and was pictured in Fig. 2.10. These projections

$$p_S : S \times T \to S \quad \text{and} \quad p_T : S \times T \to T \tag{4-247a,b}$$

had actions

$$p_S(s, t) = s, \qquad p_T(s, t) = t. \tag{4-247c,d}$$

At that time, the issue of putting s or t into the product was finessed.

Section 3.10 raised this very issue, in its discussion of ESA, the exosubset assumption. A basic problem was shown to exist, namely, that S and T are not subsets of $S \times T$. Consider S, for example. If S were a subset of $S \times T$, there would be an insertion function on S into $S \times T$. Assignment of a pair (s, t) in $S \times T$ to an element s in S, however, is made nontrivial by the question of determining the t in (s, t).

A feature of the product of two sets can be described as follows. Let

$$f : W \to S, \qquad g : W \to T \tag{4-248a,b}$$

be any two functions. It is always possible to define a function

$$h : W \to S \times T \quad \text{by the action} \quad h(w) = (f(w), g(w)). \tag{4-249a,b}$$

From this definition of h, it follows that

$$p_S \circ h = f, \qquad p_T \circ h = g, \tag{4-250}$$

which may be pictured as in Fig. 4.14 in the form of a commutative diagram. The nature of the product projections establishes that h is the only function which could make the diagram commute.

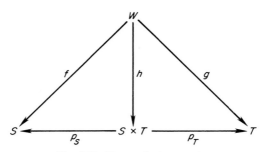

Fig. 4.14. The product property.

EXAMPLE 4.9-1

Section 2.5, which dealt with a common approach to order reduction, provides a good example of the type of construction involved with (4-249). Recall that the state x in X was broken into two parts (x_1, x_2) with the aid of a bijection b. Then x_2 was eliminated from the local dynamical equations, under the assumption that it was changing slowly. The resulting problem then was arranged to contain x_2 as a second output. In this example, then, the role of (4-248a) is played by

$$\tilde{g} \circ \tilde{h} : X_1 \times U \to Y, \tag{4-251}$$

the role of (4-248b) is played by

$$h : X_1 \times U \to X_2, \tag{4-252}$$

and the place of (4-249) is taken by

$$\tilde{g} : X_1 \times U \to Y \times X_2. \tag{4-253}$$

∎

EXAMPLE 4.9-2

The same construction might have been used in Section 3.3, in association with (3-44). Here the role of W is assumed by

$$T \times T \times T \times \cdots, \tag{4-254}$$

and the roles of f and g by

$$p_T[n] : T \times T \times T \times \cdots \to T. \tag{4-255}$$

In this illustration, the construction of Fig. 4.14 must be carried out m times to yield the final function

$$p_T[n_1, n_2] : T \times T \times T \times \cdots \to T^m. \tag{4-256}$$

∎

The idea of the coproduct of two sets was brought forward in Section 2.8, and was pictured in Fig. 2.23. The comparison is striking, in that the coproduct has insertions

$$i_S : S \to S \cup T, \qquad i_T : T \to S \cup T \tag{4-257a,b}$$

with the actions

$$i_S(s) = s, \qquad i_T(t) = t, \tag{4-257c,d}$$

while it has no projection. Moreover, if functions

$$f:S \to W, \qquad g:T \to W \qquad\qquad \text{(4-258a,b)}$$

are given, it is possible to define a function

$$h:S \cup T \to W \qquad\qquad \text{(4-259a)}$$

with the actions

$$h|S = f, \qquad h|T = g. \qquad\qquad \text{(4-259b)}$$

It follows that

$$h \circ i_S = f, \qquad h \circ i_T = g, \qquad\qquad \text{(4-260)}$$

which is indicated in the commutative diagram of Fig. 4.15. This is the only way that such a function can be defined so as to make the diagram commute.

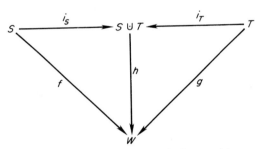

Fig. 4.15. The coproduct property.

What happens when S and T are allowed to become groups $(S, \square, e_S)$ and $(T, *, e_T)$, respectively? In this case it is possible to have the advantages of both insertions and projections. Start with the product group structure on $S \times T$, as developed in Section 4.2. Denote this by

$$(S \times T, **, (e_S, e_T)). \qquad\qquad \text{(4-261)}$$

For an insertion

$$i_S:S \to S \times T, \qquad\qquad \text{(4-262a)}$$

provide the action

$$i_S(s) = (s, e_T). \qquad\qquad \text{(4-262b)}$$

This insertion is a morphism, by the calculation

$$i_S(s_1 \square s_2) = (s_1 \square s_2, e_T) = (s_1 \square s_2, e_T * e_T)$$
$$= (s_1, e_T) ** (s_2, e_T) = i_S(s_1) ** i_S(s_2). \qquad \text{(4-263)}$$

In a similar way, define the insertion morphism

$$i_T:T \rightarrow S \times T \qquad\qquad (4\text{-}264a)$$

by

$$i_T(t) = (e_S, t). \qquad\qquad (4\text{-}264b)$$

Though the same notation i_S has been used for both (4-257) and (4-262) with (4-264), the reader will observe that they are not the same function at all, inasmuch as they have different codomains. Projections are also available in the form (4-247). These turn out to be morphisms as well. For illustration, calculate

$$p_S((s_1, t_1) \ast\ast (s_2, t_2)) = p_S(s_1 \square s_2, t_1 \ast t_2) = s_1 \square s_2$$
$$= p_S(s_1, t_1) \square p_S(s_2, t_2). \qquad (4\text{-}265)$$

A similar calculation works for p_T. In view of this, it might be expected that the diagram of Fig. 4.14 can be specialized right away to the group case. To do this, f and g would be assumed morphisms; then it is necessary to show that h becomes a morphism.

As a test for this hypothesis, consider Fig. 4.14. Let the group description of W be $(W, \triangle, e_W)$. From (4-249b),

$$h(w_1 \triangle w_2) = (f(w_1 \triangle w_2), g(w_1 \triangle w_2))$$
$$= (f(w_1) \square f(w_2), g(w_1) \ast g(w_2))$$
$$= (f(w_1), g(w_1)) \ast\ast (f(w_2), g(w_2))$$
$$= h(w_1) \ast\ast h(w_2). \qquad (4\text{-}266)$$

Thus Fig. 4.14 does specialize to the group case.

Now the diagram of Fig. 4.15 is not directly applicable to the product group case, because $S \cup T$ and $S \times T$ are different sets. Thus, Fig. 4.16 is provided, where (4-262) and (4-264) provide the definitions for the insertions, and the action of h is given by

$$h(s, t) = (f \circ p_S)(s, t) \triangle (g \circ p_T)(s, t). \qquad (4\text{-}267)$$

It is a modest exercise to demonstrate that the diagram in Fig. 4.16 now commutes. A bit less obvious to show, but equally true, is that this is the only way to define such a function to make the diagram commute. Attention turns here to whether or not (4-267) defines a morphism. Now f, p_S, g, and p_T are morphisms—f and g by assumption and p_S and p_T by construction. It follows from Section 4.3 that $f \circ p_S$ and $g \circ p_T$ are morphisms. However, also from that section, the combination

$$f \circ p_S \triangle g \circ p_T \qquad\qquad (4\text{-}268)$$

is not necessarily a morphism unless $(W, \triangle, e_W)$ is a commutative group.

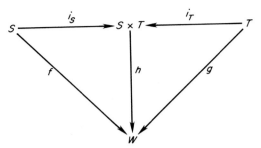

Fig. 4.16. A substitute for Fig. 4.15.

Equipped with both its insertion and projection morphisms, the product group $S \times T$ is called a *biproduct* and is denoted by $S \oplus T$. The diagram of Fig. 4.14 is available at the group level; that of Fig. 4.16 is available at the group level, with the exception that the assumption that W is a commutative group is made so as to establish that h is a morphism.

EXAMPLE 4.9-3

The fact that W is assumed to be a commutative group so that the diagram Fig. 4.16 has h a morphism is at the heart of the constructions of Section 4.2. There it was assumed that the local transition function $f: X \times U \to X$ was a morphism, with the resulting appearance of morphisms $a: X \to X$ and $b: U \to X$. That situation can be visualized in the manner of Fig. 4.16. The role of h is played by f, and that of f and g by a and b, respectively; W is X. Given morphisms a and b, it is seen that the local transition function always exists, but is a morphism if the state group is commutative. Moreover, the definitions for f_1, f_2, g_1, and g_2 in Section 4.2 are seen in the manner that

$$f_1 = f \circ i_X, \qquad f_2 = f \circ i_U,$$
$$g_1 = g \circ i_X, \qquad g_2 = g \circ i_U \qquad (4\text{-}269)$$

when f and g are morphisms. ∎

EXAMPLE 4.9-4

One nice feature about the biproduct $X \oplus U$ is that it permits pleasing figurative representations of local dynamical activity. In fact, the transition equation

$$x_{k+1} = f(x_k, u_k) \qquad (4\text{-}270)$$

can be pictured as in Fig. 4.17, even if f is not a morphism. ∎

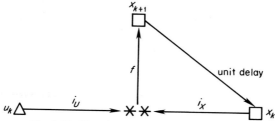

Fig. 4.17. Biproduct picture of local transition.

The feature of state group commutativity is thus brought out as a very fundamental issue for group morphic systems.

Exercises

4.9-1. Reconsider Fig. 4.14, and the definition (4-249) for h, which makes the diagram commute. Show that h is the only function which could make the diagram commute.

4.9-2. Fill in the missing details in Example 4.9-2.

4.9-3. Reexamine the illustrative calculation of (4-265). Perform the corresponding computation for p_T.

4.9-4. Refer to Fig. 4.16, and to the definitions (4-262), (4-264), and (4-267). Demonstrate that the diagram commutes.

4.9-5. Extend Exercise 4.9-4 to show that (4-267) is the only way to define the function h so that the diagram commutes.

4.10 DISCUSSION

A great deal of the intuition growing out of many readers' background can be brought to bear usefully in the group morphic system. One reason for this is that a great number of readers have received their early exposure to system theory in a context of vectors; and the algebraic description of a set of vectors has come to be accepted as a commutative group.

When the input, state, and output groups of an SDS are endowed with group structure, it becomes possible to talk about the local dynamical functions as morphisms. If the local dynamical functions are morphisms, on the product group of states and inputs, then the well-known 4-tuple representation (a, b, c, d) of morphisms develops naturally; and this leads to a basic notion of group morphic system (GRMPS). Curiously enough, a given 4-tuple of morphisms does not have to lead to local dynamical functions having the morphism property. This turns out to be the case if the state group is commutative. Thus the idea of a group morphic system has a generality slightly greater than its originating discussion.

Though some systems topics can be discussed solely in the context of one system, most of the interest in, and applications of, system theory are a consequence of connecting several systems together. A very basic question then concerns whether the interconnection of two GRMPSs is another GRMPS. The answer has been shown to be a qualified yes for series and parallel connections. The series connection requires a commutative assumption on the state and output groups of the driven system. The parallel connection requires the commutative assumption on the output group. Feedback connections are considered in the next chapter.

The key triangle of Chapter 2 has been specialized to the group case. As part of this discussion, it became clear that kernels of morphisms have the special property of normality, which is crucial to the triangle, and that the type of equivalence relations which could be used in the group morphism version of the triangle was completely specified.

With the key triangle adjusted to the group morphic case, the concepts of reachability and observability were specialized also. This led to the notions of reachable subgroup and unobservable subgroup.

Finally, the issue of products and coproducts was reexamined, with the creation of a biproduct, which is similar to the coproduct in the presence of its insertions, but different from the coproduct because of its use of pairs. This discussion then established just why it was that a 4-tuple (a, b, c, d) of morphisms might not lead to a local transition morphism. The issue turned out to be commutativity of the state group.

In closing this chapter, it can be noted that a number of properties of functions and morphisms have been developed, when their domains and their codomains admit group structure. It is time to gather together these notions; this leads to the idea of a ring.

5 INVERSES
OF GROUP MORPHIC SYSTEMS

The reader may recall from Section 4.1 that a function

$$f : S \to T \tag{5-1}$$

presents some new possibilities when its codomain T becomes a group $(T, *, e_T)$. In Example 4.1-3, it was explained that the binary operation $*$ on T induced a corresponding binary operation on T^S. Indeed, for any pair of functions f_1 and f_2 in T^S, the action of the induced operation is given by

$$(f_1 * f_2)(s) = f_1(s) * f_2(s). \tag{5-2}$$

Notice that $*$ in the right member of (5-2) has domain $T \times T$, whereas the same symbol in the left member has domain $T^S \times T^S$. If S is $\mathbb{R}$, and if $(T, *, e_T)$ is $(\mathbb{R}, +, 0)$ as in Example 4.1-1, then an example of (5-2) could be

$$(\sin + \cos)(r) = \sin(r) + \cos(r). \tag{5-3}$$

Here it is easy to see why the notation (5-2) is customary. The operation $*$ on T^S is associative because $*$ on T is associative. It has a unit e_{T^S} defined by (4-19). Moreover, each element in T^S has an inverse defined by (4-20). Accordingly,

$$(T^S, *, e_{T^S}) \tag{5-4}$$

becomes a group, which is a natural consequence of the group structure on T.

Now composition represents an additional way of combining two functions to make another function. This issue was brought forward in the opening paragraphs of Chapter 3. To make this combination a binary operation, it is appropriate to select S equal to T. Composition is easily demonstrated to be associative, and it has a unit, which is the identity function 1_T on T. However, it fails to establish group structure, because not every function is bijective. The structure

$$(T^T, \circ, 1_T),\tag{5-5}$$

having an associative binary operation, together with unit, is a monoid, as defined in Section 4.2.

Together, (5-4) and (5-5) have a structure

$$(T^T, *, e_{TT}, \circ, 1_T).\tag{5-6}$$

Sometimes, the structure (5-6) is an example of a ring. One case in which this is true occurs when interest is specialized to the subset of functions which are morphisms.

Constructions utilizing such structures play an important role in system theory. This chapter illustrates the point by considering such a notion in the GRMP context: invertibility of group morphic systems and the construction of GRMP inverses.

5.1 RINGS

To construct a ring, begin with a set R, and equip it with two binary operations

$$\square : R \times R \to R \qquad \text{and} \qquad \triangle : R \times R \to R.\tag{5-7a,b}$$

Both of these operations are required to have units. The unit for one of them, say $\square$, is designated by the usual symbol e_R. The unit for the other, $\triangle$, is written 1_R. The structure

$$(R, \square, e_R)\tag{5-8a}$$

must be a commutative group; the structure

$$(R, \triangle, 1_R)\tag{5-8b}$$

is taken as a monoid. But the essential final step in the construction is to arrange that $\triangle$ distributes over $\square$, and that this be true from both sides, in the sense

$$r_1 \triangle (r_2 \square r_3) = (r_1 \triangle r_2) \square (r_1 \triangle r_3),\tag{5-9a}$$
$$(r_2 \square r_3) \triangle r_1 = (r_2 \triangle r_1) \square (r_3 \triangle r_1)\tag{5-9b}$$

for all r_1, r_2, and r_3 in R.

A structure

$$(R, \Box, e_R, \triangle, 1_R), \tag{5-10}$$

in which (5-8a) is a commutative group, (5-8b) is a monoid, and the distributive rules (5-9) hold, is a *ring*. The ring is commutative if $\triangle$ is a commutative binary operation. For historical reasons, the binary operation $\Box$ is often called addition, while $\triangle$ is called multiplication. These identifications can certainly be helpful in understanding the notion of ring. They can also be quite a hindrance in seeing new ways to apply the concept.

EXAMPLE 5.1-1

Denote the binary operation of real multiplication by

$$\cdot : \mathbb{R} \times \mathbb{R} \to \mathbb{R}. \tag{5-11}$$

Then

$$(\mathbb{R}, +, 0, \cdot, 1) \tag{5-12}$$

is a commutative ring, where 0 and 1 are the real numbers zero and one, respectively. Likewise, the structures

$$(\mathbb{Z}, +, 0, \cdot, 1), \qquad (\mathbb{Q}, +, 0, \cdot, 1), \qquad (\mathbb{C}, +, 0, \cdot, 1) \tag{5-13a,b,c}$$

are commutative rings. In (5-13), + is defined in the usual way, as is $\cdot$ in each case.

∎

EXAMPLE 5.1-2

The prefatory paragraphs of this chapter have provided an example of a structure that does not quite measure up to ring standards. This is that of

$$(T^T, *, e_{TT}, \circ, 1_T). \tag{5-14}$$

It turns out that (5-14) fails to meet the requirement that $\circ$ distribute over $*$ from the left, although it can meet all other requirements if $(T, *, e_T)$ is a commutative group. Note that $*$ on T^T is commutative if $*$ on T is commutative. It is not difficult to establish that $\circ$ distributes over $*$ from the right. Indeed,

$$\begin{aligned}
((f_1 * f_2) \circ f_3)(t) &= (f_1 * f_2)(f_3(t)) \\
&= f_1(f_3(t)) * f_2(f_3(t)) \\
&= (f_1 \circ f_3)(t) * (f_2 \circ f_3)(t) \\
&= ((f_1 \circ f_3) * (f_2 \circ f_3))(t). \tag{5-15}
\end{aligned}$$

On the other hand, the same calculation from the left,

$$(f_3 \circ (f_1 * f_2))(t) = f_3(f_1(t) * f_2(t)) \tag{5-16}$$

encounters difficulty unless f_3 is a morphism of the group $(T, *, e_T)$. ∎

A great deal of information has been gathered about morphisms of groups. In Section 4.1, it was shown that a morphism of groups carries the unit of its domain to the unit of its codomain. Also established was the fact that the inverse for a morphism of groups f in the group $(T^S, *, e_{TS})$ had action

$$\hat{f}(s) = f(\hat{s}), \tag{5-17}$$

where $\hat{s}$ was the inverse of s in the group $(S, \square, e_S)$. Section 4.3 demonstrated that the composition of morphisms of groups is a morphism of groups, and that the binary operation $*$ in the group $(T^S, *, e_{TS})$, when applied to a pair of morphisms f_1 and f_2 of groups, yields another morphism of groups $f_1 * f_2$, provided that $*$ is commutative.

These facts lead to some interesting conclusions. Let $(T, *, e_T)$ be a commutative group. Then $(T^S, *, e_{TS})$ is a commutative group, for S just a set. Now let $(S, \square, e_S)$ be a group, not necessarily commutative as yet. Consider the subset $\mathbb{M}(T^S)$ of T^S consisting of all those functions which are morphisms of groups. $\mathbb{M}(T^S)$ is closed under $*$ according to Section 4.3, because $(T, *, e_T)$ is a commutative group. $\mathbb{M}(T^S)$ is not empty, because e_{TS} is in $\mathbb{M}(T^S)$. This may be verified by the calculation

$$e_{TS}(s_1 \square s_2) = e_T = e_T * e_T = e_{TS}(s_1) * e_{TS}(s_2). \tag{5-18}$$

Moreover, if $f \in T^S$ is a morphism, so is $\hat{f} \in T^S$, by the calculation

$$\hat{f}(s_1 \square s_2) = \widehat{f(s_1 \square s_2)} \tag{5-19a}$$
$$= \widehat{f(s_1) * f(s_2)} \tag{5-19b}$$
$$= \hat{f}(s_2) * \hat{f}(s_1) \tag{5-19c}$$
$$= \hat{f}(s_1) * \hat{f}(s_2). \tag{5-19d}$$

Notice that (5-19c) is easily an inverse for $f(s_1) * f(s_2)$, by verification, and in a group the inverse is unique—because the group binary operation is associative. Accordingly, then, by the terms of Section 4.1,

$$(\mathbb{M}(T^S), *, e_{TS}) \tag{5-20}$$

is a subgroup of $(T^S, *, e_{TS})$. Next select $(S, \square, e_S)$ equal to $(T, *, e_T)$. Then 1_T is an element of $\mathbb{M}(T^T)$, and the composition of two morphisms is another morphism. Thus

$$(\mathbb{M}(T^T), \circ, 1_T) \tag{5-21}$$

is a monoid.

By Example 5.1-2,

$$(\mathbb{M}(T^T), *, e_{TT}, \circ, 1_T) \tag{5-22}$$

is a ring, because $\circ$ distributes over $*$ from the right even if the functions are not morphisms of groups, and (5-16) can be completed to

$$
\begin{aligned}
(f_3 \circ (f_1 * f_2))(t) &= f_3(f_1(t) * f_2(t)) \\
&= f_3(f_1(t)) * f_3(f_2(t)) \\
&= (f_3 \circ f_1)(t) * (f_3 \circ f_2)(t) \\
&= ((f_3 \circ f_1) * (f_3 \circ f_2))(t) \tag{5-23}
\end{aligned}
$$

when f_3 is a morphism. Using the endomorphism definition of Section 4.1, $\mathbb{M}(T^T)$ might be denoted $\mathbb{EM}(T)$ instead. Then (5-22) appears as

$$(\mathbb{EM}(T), *, e_{TT}, \circ, 1_T). \tag{5-24}$$

Up until this point, the only type of morphism discussed has been the morphism of groups. But when a function

$$f : R_1 \to R_2 \tag{5-25}$$

has rings for both its domain and codomain, it is possible to inquire now both as to its compatibility with the group structures and as to its compatibility with the monoid structures. A function (5-25) from one ring to another is said to be a *morphism of rings* if it is both a morphism of groups, in the sense of Section 4.1, and a morphism of monoids as defined in Section 4.2. If the rings are described by

$$(R_i, \square_i, e_{R_i}, \triangle_i, 1_{R_i}), \qquad i = 1, 2, \tag{5-26}$$

then f of (5-25) is a morphism of rings if

$$f(1_{R_1}) = 1_{R_2}, \tag{5-27a}$$

$$f(r_1 \triangle_1 \tilde{r}_1) = f(r_1) \triangle_2 f(\tilde{r}_1), \tag{5-27b}$$

$$f(r_1 \square_1 \tilde{r}_1) = f(r_1) \square_2 f(\tilde{r}_1), \tag{5-27c}$$

the latter two equations holding for all r_1 and $\tilde{r}_1$ in R_1.

The algebraic structures available on sets of functions and morphisms make it possible to describe many interesting system theoretic behaviors. This is much in the same spirit as the lattice structures on relations—with the exception that these function structures depend upon domain or codomain structure.

The ease with which morphisms can be manipulated creates the possibility that their domains and codomains could be forgotten. It is to be hoped that the discussions of Chapter 1 have made clear why such a loss of memory is highly undesirable.

Exercises

5.1-1. The structure

$$(T^T, \circ, 1_T)$$

of (5-5) has been introduced with the assertion that the binary operation $\circ$ of composition is associative. Show that this is the case.

5.1-2. In studying (5-16), it was asserted that distribution from the left encounters difficulty unless f_3 is a morphism. Clearly, f_3 a morphism is sufficient for the desired distributive law. Can you show that f_3 a morphism is a necessary condition as well?

5.1-3. Study the image of a morphism of rings. Is such an image also a ring? Explain in detail.

5.1-4. Is the composition of two morphisms of rings another morphism of rings?

5.2 GRMP GLOBAL OUTPUT CALCULATIONS

Global functions for an SDS have been introduced in Section 3.3. In particular, the global output function was defined in the manner

$$h: X \times U^{\mathbb{N}} \to Y^{\mathbb{N}} \qquad (5\text{-}28)$$

and was subsequently extended according to

$$h: X \times SS(U^{\mathbb{N}}) \to SS(Y^{\mathbb{N}}). \qquad (5\text{-}29)$$

A later chapter will examine some algebraic structures which can be attached to these sets of sequences. The aim of this section is to examine the rules of calculation which attach to the global output function, in the GRMP case.

Section 4.6 on finite transitions has laid the groundwork for these calculations. From (4-163), it follows that

$$x_{i+1} = a^{i+1}x_0 \prod_{j=0}^{i} a^{i-j}bu_j. \qquad (5\text{-}30)$$

It should be recalled that the composition symbol $\circ$ has been suppressed in (5-30), for simplicity. From this expression, it is easy to calculate the output, namely,

$$y_{i+1} = cx_{i+1} * du_{i+1}$$

$$= c\left(a^{i+1}x_0 \underset{j=0}{\overset{i}{\square}} a^{i-j}bu_j \right) * du_{i+1}$$

$$= c(a^{i+1}x_0) * c\left(\underset{j=0}{\overset{i}{\square}} a^{i-j}bu_j \right) * du_{i+1}. \qquad (5\text{-}31)$$

Before completing this development, recall from (4-160) that

$$\underset{j=0}{\overset{i}{\square}} a^{i-j}bu_j = a^i bu_0 \,\square\, a^{i-1}bu_1 \,\square\, \cdots \,\square\, abu_{i-1} \,\square\, bu_i; \qquad (5\text{-}32\text{a})$$

therefore

$$c\left(\underset{j=0}{\overset{i}{\square}} a^{i-j}bu_j \right) = c(a^i bu_0 \,\square\, a^{i-1}bu_1 \,\square\, \cdots \,\square\, abu_{i-1} \,\square\, bu_i)$$

$$= c(a^i bu_0) * c(a^{i-1}bu_1) * \cdots * c(abu_{i-1}) * c(bu_i)$$

$$= ca^i bu_0 * ca^{i-1}bu_1 * \cdots * cabu_{i-1} * cbu_i$$

$$= \underset{j=0}{\overset{i}{*}} \, ca^{i-j}bu_j, \qquad (5\text{-}32\text{b})$$

where the last step has been taken in the same spirit as (4-161). Combining (5-31) and (5-32) then leads quickly to the completed calculation

$$y_{i+1} = ca^{i+1}x_0 * \left(\underset{j=0}{\overset{i}{*}} \, ca^{i-j}bu_j \right) * du_{i+1} \qquad (5\text{-}33\text{a})$$

$$= ca^{i+1}x_0 \underset{j=0}{\overset{i}{*}} \, ca^{i-j}bu_j * du_{i+1}, \qquad (5\text{-}33\text{b})$$

where the passage from (5-33a) to (5-33b) is made along the same lines as that made from (4-162) to (4-163). Equation (5-32) is useful for i in $\mathbb{N}$; for calculation of y_0, it must be supplemented by the equation

$$y_0 = cx_0 * du_0. \qquad (5\text{-}34)$$

EXAMPLE 5.2-1

Reconsider the GRMPS of Example 4.2-1. The morphism

$$cb: U \rightarrow Y \qquad (5\text{-}35\text{a})$$

is defined by

$$(cb)(0) = 0, \qquad (cb)(1) = 1. \qquad (5\text{-}35\text{b,c})$$

Inasmuch as $(U, \triangle, e_U)$ and $(Y, *, e_Y)$ are the same group $(\mathbb{Z}_2, +, 0)$, it is clear that

$$cb = 1_{\mathbb{Z}_2}, \tag{5-36}$$

the identity morphism. Continuing, it follows that

$$cab : U \to Y \tag{5-37a}$$

has the action

$$(cab)(0) = 0, \qquad (cab)(1) = 1, \tag{5-37b,c}$$

so that

$$cab = 1_{\mathbb{Z}_2} \tag{5-38a}$$

as well. In fact, the reader may verify that

$$ca^ib = 1_{\mathbb{Z}_2}, \qquad i \in \mathbb{N}. \tag{5-38b}$$

∎

EXAMPLE 5.2-2

When (5-36) holds, a very interesting possibility occurs. Suppose that x_0 is the unit state e_X. Then, in Example 4.2-1,

$$d = e_{YU}, \qquad \text{so that} \qquad y_0 = e_Y, \tag{5-39a,b}$$

irrespective of the value of u_0. The next operation of the system yields

$$y_1 = cx_1 * du_1 = cx_1 * e_{YU}(u_1) = cx_1 * e_Y = cx_1$$
$$= c(ax_0 \,\square\, bu_0) = cae_X * cbu_0 = e_Y * 1_{\mathbb{Z}_2}u_0 = u_0. \tag{5-40}$$

From (5-40) it is seen that the control input u_0 is recoverable from the output y_1, which in this example is exactly equal to it. The next system operation, by (5-33), gives

$$y_2 = ca^2x_0 \mathop{*}_{j=0}^{1} ca^{1-j}bu_j * du_2 \tag{5-41a}$$

$$= ca^2e_X * (cabu_0 * cbu_1) * e_{YU}(u_2) \tag{5-41b}$$

$$= cabu_0 * cbu_1. \tag{5-41c}$$

But u_0 is known from (5-40) to be y_1, and cb is $1_{\mathbb{Z}_2}$; thus

$$y_2 = caby_1 * u_1, \tag{5-42a}$$

which provides

$$cab\hat{y}_1 * y_2 = u_1. \tag{5-42b}$$

By now a property is becoming clearer. When the initial state is the unit, inputs to the GRMPS of the example can be determined one operation later from the outputs. This type of behavior was brought up earlier in Section 3.5, in connection with (3-80). It suggests a notion of an inverse GRMPS, which is to be pursued in this chapter. ∎

Notice that statements of the form (5-36) and (5-38) are made in the type of context cited in Section 5.1. These amount to the observation that ca^ib, $i \in \mathbb{N}$, is equal to the unit for composition in the ring

$$(\mathbb{EM}(\mathbb{Z}_2), +, e_{YU}, \circ, 1_{\mathbb{Z}_2}). \tag{5-43}$$

More generally, suppose that there exists a function

$$f : Y^2 \to U \tag{5-44a}$$

which is able to apply the action

$$f(y_0, y_1) = u_0. \tag{5-44b}$$

Then (5-44) could be used to find out the value of x_1 by

$$x_1 = ax_0 \,\square\, bf(y_0, y_1) = bf(y_0, y_1) \tag{5-45}$$

when x_0 is the unit state. Moreover, since

$$x_2 = ax_1 \,\square\, bu_1, \tag{5-46}$$

and since (5-45) provides a ready expression for x_1, it follows that

$$x_2 = abf(y_0, y_1) \,\square\, bu_1, \tag{5-47a}$$

and that

$$\overline{abf(y_0, y_1)} \,\square\, x_2 = bu_1. \tag{5-47b}$$

In this way the state x_2 can be purged of its dependence upon u_0. The manner in which this purging occurs, however, suggests the use of feedback, along the lines of (5-45). Chapter 4 did not take up the question of feedback interconnections of GRMPSs. Such feedback connections are the topic of the section following.

Exercise

5.2.   In Example 5.2-1, verify that

$$ca^ib = 1_{\mathbb{Z}_2}$$

for all i in $\mathbb{N}$. Notice that this may require an induction.

5.3 STATE FEEDBACK INTERCONNECTION

Examples 3.5-1 and 3.5-2 have made an early study of the concept of applying feedback to a system, in the SDS context. Then, in Section 3.10, some features of feedback were shown, in introducing the internal model principle.

This section places its emphasis along the lines of state feedback, as in Example 3.5-1. The question asked here is somewhat different than the question of that example. There it was the goal to show that feedback led to the formation of a new SDS. Here, it is of interest whether state feedback around a GRMPS leads to a new GRMPS. The discussion is in the spirit of Sections 4.3 and 4.4.

Begin by recalling the state feedback construction in SDS terms. The local transition equation takes the form

$$x_{k+1} = f(x_k, u_k). \tag{5-48}$$

The idea is to construct the input u_k from the state x_k and from a set of exogenous inputs V. Such a construction can be represented as a function

$$\mathrm{fb}: X \times V \to U \tag{5-49a}$$

with the action

$$\mathrm{fb}(x_k, v_k) = u_k. \tag{5-49b}$$

Substitution of (5-49b) into (5-48) provides the local transition equation

$$x_{k+1} = f(x_k, \mathrm{fb}(x_k, v_k)) \tag{5-50}$$

of the overall SDS with feedback included. The equation (5-50) can be used to define the action of a new function

$$\alpha: X \times V \to X, \tag{5-51a}$$

in the manner

$$\alpha(x, v) = f(x, \mathrm{fb}(x, v)). \tag{5-51b}$$

The local transition is expressed by

$$x_{k+1} = \alpha(x_k, v_k). \tag{5-52}$$

A similar approach can be taken in regard to the local output equation

$$y_k = g(x_k, u_k) = g(x_k, \mathrm{fb}(x_k, v_k)) = \beta(x_k, v_k), \tag{5-53a,b,c}$$

where

$$\beta: X \times V \to Y \tag{5-54a}$$

has the action

$$\beta(x, v) = g(x, \mathrm{fb}(x, v)). \tag{5-54b}$$

The main point to be noted here is this. Given an arbitrary SDS

$$(J, \leq; U, X, Y; f, g), \tag{5-55}$$

and an arbitrary state feedback rule (5-49), the resulting SDS

$$(J, \leq; V, X, Y; \alpha, \beta) \tag{5-56}$$

is defined in every case.

Do any restrictions have to be placed when the SDS admits the GRMP structure (a, b, c, d)? In this situation, (5-48) is replaced by

$$x_{k+1} = ax_k \,\square\, bu_k. \tag{5-57}$$

Applying (5-49b) gives

$$x_{k+1} = ax_k \,\square\, b\,\mathrm{fb}(x_k, v_k), \tag{5-58}$$

and it is clear that development of group morphic structure will depend upon the way in which the feedback function interacts with the various group operations. One approach is to place group structure on the exogenous input set V, to develop product group structure on $X \times V$, and to require fb to be a morphism of groups. This is the approach which was taken in Section 4.2. The reader will recall, however, that the structure developed from that assumption was actually more general in its interpretation. To be specific, the assumption that fb is a morphism of groups leads to

$$\mathrm{fb}(x, v) = a_\mathrm{f} x \,\triangle\, b_\mathrm{f} v, \tag{5-59}$$

for

$$a_\mathrm{f} : X \to U \qquad \text{and} \qquad b_\mathrm{f} : V \to U \tag{5-60a,b}$$

morphisms of groups. On the other hand, not every pair of morphisms of groups (5-60) has to lead to an fb function which is also a morphism of groups. This remarkable state of affairs has been discussed in several sections, but especially in Section 4.9, and in Example 4.9-3. Thus another approach is simply to assume the action (5-59), with a_f and b_f morphisms of groups, and not to make any assumption about the morphic nature of fb. The group structure on V may be taken to be of the form $(V, \cdot, e_V)$.

Proceeding, then, develop (5-58) with the aid of (5-59) so as to obtain

$$x_{k+1} = ax_k \,\square\, b(a_\mathrm{f} x_k \,\triangle\, b_\mathrm{f} v_k). \tag{5-61a}$$

Because b is a morphism of groups, (5-61a) expands in the manner

$$x_{k+1} = ax_k \,\square\, (b \circ a_\mathrm{f})x_k \,\square\, (b \circ b_\mathrm{f})v_k. \tag{5-61b}$$

It is clear that definition of a morphism of groups

$$\beta : V \to X \qquad \text{is possible by} \qquad \beta = b \circ b_f. \qquad (5\text{-}62a,b)$$

Moreover,

$$ax_k \square (b \circ a_f)x_k = (a \square (b \circ a_f))x_k \qquad (5\text{-}63)$$

in every case, and the function

$$a \square (b \circ a_f) : X \to X \qquad (5\text{-}64)$$

becomes a morphism of groups if the state group $(X, \square, e_X)$ is commutative. Denote the morphism (5-64) by α.

Turning then to the local output function, with subscripts suppressed,

$$
\begin{aligned}
y &= cx * d(a_f x \triangle b_f v) \\
&= cx * (d \circ a_f)x * (d \circ b_f)v \\
&= (c * (d \circ a_f))x * (d \circ b_f)v.
\end{aligned}
\qquad (5\text{-}65)
$$

Here it is possible to define a morphism of groups

$$\delta : V \to Y \qquad \text{by} \qquad \delta = d \circ b_f, \qquad (5\text{-}66a,b)$$

and a morphism of groups

$$\gamma : X \to Y \qquad \text{by} \qquad \gamma = c * (d \circ a_f), \qquad (5\text{-}67a,b)$$

if the output group $(Y, *, e_Y)$ is commutative.

Then the GRMPS

$$(U, \triangle; X, \square; Y, *; a, b, c, d), \qquad (5\text{-}68)$$

when equipped with the feedback function (5-59), becomes another GRMPS

$$(V, \cdot; X, \square; Y, *; \alpha, \beta, \gamma, \delta), \qquad (5\text{-}69)$$

provided that the state and output groups of (5-68) are commutative. The GRMPS (5-68) is often called the open loop system, while that of (5-69) is frequently called the closed loop system, for reasons which have a certain intuitive appeal.

The technicalities involved with construction of α and γ, as well as β and δ, may be reviewed in Section 4.3. Notice that the restrictions on the open loop GRMPS in this section, namely, commutativity of state and output groups, are the same as those placed upon the driven system in the series interconnection treatment of Section 4.3.

EXAMPLE 5.3

Suppose that the input

$$u_k = a_f x_k \triangle b_f v_k \qquad (5\text{-}70)$$

could itself be regarded as the output of another GRMPS. This would then be an aid in understanding why the commutative group restrictions of this section are the same as those for the series interconnection section of the preceding chapter. Now consider the manner in which the states are evolving, namely,

$$x_{k+1} = \alpha x_k \,\square\, \beta v_k. \tag{5-71}$$

Combination of (5-70) and (5-71) suggests the interpretation

$$(V, \cdot\,; X, \square\,; U, \triangle\,; \alpha, \beta, a_f, b_f) \tag{5-72}$$

as a GRMPS to drive (5-68). This way of thinking can be visualized in Figs. 5.1 and 5.2. The first of these figures comprises the usual way of thinking. The second is drawn along the lines of this example.   ∎

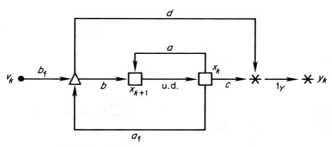

Fig. 5.1. GRMP feedback; u.d. = unit delay.

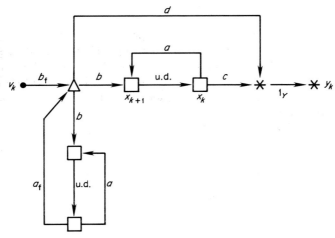

Fig. 5.2. Another view of feedback.

Exercises

5.3-1.   Supply a complete explanation to support the statement that fb as a morphism of groups leads to (5-59) with the definitions (5-60).

5.3-2.   Explain in careful detail how the assumptions made in this section result in α of (5-64) and γ of (5-67) becoming morphisms of groups.

5.3-3.   To the feedback configuration of this section, add another GRMPS in series. Develop conditions for this series connection to be a GRMPS. Explain and define the morphisms associated with the total system.

5.3-4.   Place a feedback connection around two GRMPSs which are in series. Develop conditions for the total configuration to be a GRMPS. Explain and define the morphisms associated with the total system.

5.4   FEEDBACK AND THE REACHABLE SUBGROUP

In Section 4.7, a GRMPS was defined to be reachable if it is reachable from the unit state e_X in the state group $(X, \square, e_X)$. The set X_i of states x_i which can be reached in the manner

$$x_i = \phi(e_X, u_*[0, i-1])$$ (5-73)

for an appropriate choice of sequence segment $u_*[0, i-1]$ in $SS(U^{\mathbb{N}})$ is a member of an ascending chain

$$X_0 \subset X_1 \subset X_2 \subset \cdots.$$ (5-74)

Because of the properties of the complete lattice

$$(\mathbb{P}(X), \subset, \cap, \cup),$$ (5-75)

it is known that the chain (5-74) has a supremum X^*. If the state group is commutative, all the X_i are normal subgroups. If also the state group is of finite type, then there exists a natural number i in $\mathbb{N}$ such that X_i is equal to X^*. In this case, X^* is said to be the reachable subgroup of the GRMPS.

At this time, it is appropriate to discuss the effect of feedback upon the reachable subgroup.

The basic GRMPS is described by

$$x_{k+1} = ax_k \square bu_k, \qquad y_k = cx_k * du_k.$$ (5-76a,b)

Feedback is expressed by the equation

$$u_k = a_f x_k \triangle b_f v_k,$$ (5-76c)

where

$$a_f : X \rightarrow U \qquad \text{and} \qquad b_f : V \rightarrow U$$ (5-77a,b)

are morphisms of groups, the latter with domain group $(V, \cdot, e_V)$ of exogenous inputs.

The preceding section has established that, if the state and output groups are commutative, the set of three equations (5-76) amounts to the description of a closed loop GRMPS

$$x_{k+1} = \alpha x_k \,\square\, \beta v_k, \qquad y_k = \gamma x_k \,*\, \delta v_k, \qquad (5\text{-}78\text{a,b})$$

where the morphisms of groups are given by

$$\alpha = a \,\square\, (b \circ a_f), \qquad \beta = b \circ b_f, \qquad (5\text{-}79\text{a,b})$$
$$\gamma = c \,*\, (d \circ a_f), \qquad \delta = d \circ b_f. \qquad (5\text{-}79\text{c,d})$$

To discuss the effect of a_f and b_f upon the reachable subgroup, it is assumed also that the state group is of finite type. For the open loop GRMPS (a, b, c, d), the reachable set notation $X_0, X_1, X_2, \ldots, X^*$ is used. For the closed loop GRMPS $(\alpha, \beta, \gamma, \delta)$, the reachable sets are denoted $X_0^f, X_1^f, \ldots, (X^f)^*$.

The ground-level observation is that

$$X_0 = X_0^f = \{e_X\}. \qquad (5\text{-}80)$$

In the case of the open loop system, this is obtained by u, whereas in the case of the closed loop system, it is obtained by v. Next, it is easy to see that

$$X_1 = \operatorname{Im} b, \qquad X_1^f = \operatorname{Im} \beta. \qquad (5\text{-}81\text{a,b})$$

In general, of course, (5-81a) and (5-81b) can range from being identical to having only the unit state e_X in common. The way this is usually addressed is to require the morphism b_f of (5-77b) to be epic. Then

$$\operatorname{Im} \beta = \operatorname{Im} b, \qquad \text{and} \qquad X_1 = X_1^f. \qquad (5\text{-}82\text{a,b})$$

Make this assumption on b_f, and proceed to the next step.

From Example 4.6-1, it is known that

$$X_{i+1}^f = \alpha_* X_i^f \,\square\, \operatorname{Im} \beta. \qquad (5\text{-}83)$$

From (5-79a), and from (5-82), it follows therefore that

$$X_2^f = (a \,\square\, (b \circ a_f))_* X_1 \,\square\, \operatorname{Im} b. \qquad (5\text{-}84)$$

This brings the discussion to a minor technical digression, which is attacked by way of example.

EXAMPLE 5.4-1

Suppose that $(X, \square, e_X)$ is a group, and that f and g are functions on the group to itself. What can be said about

$$(f \,\square\, g)_*(S), \qquad (5\text{-}85)$$

where S is a subset of X? Well, examine the construction

$$(f \,\square\, g)(s) = f(s) \,\square\, g(s), \tag{5-86}$$

for arbitrary s in S. This makes it clear that

$$(f \,\square\, g)_*(S) \subset f_*(S) \,\square\, g_*(S). \tag{5-87}$$

But the reverse inclusion is not necessarily true. This may be seen by selecting g to be $\hat{f}$ in the group

$$(X^X, \square, e_{XX}). \tag{5-88}$$

So the inclusion in (5-87) cannot be replaced in general by an equality. This completes the technical digression associated with Example 5.4-1, namely to establish the general inclusion (5-87).   ∎

Return now to (5-84); then

$$(a \,\square\, (b \circ a_f))_* X_1 \subset a_* X_1 \,\square\, (b \circ a_f)_* X_1, \tag{5-89a}$$

which yields

$$X_2^f \subset a_* X_1 \,\square\, (b \circ a_f)_* X_1 \,\square\, \operatorname{Im} b \tag{5-89b}$$

$$= a_* X_1 \,\square\, \operatorname{Im} b \tag{5-89c}$$

$$= X_2. \tag{5-89d}$$

Passage from (5-89b) to (5-89c) is possible because

$$(b \circ a_f)_* X_1 \subset \operatorname{Im} b; \tag{5-90}$$

and so

$$X_2^f \subset X_2. \tag{5-91}$$

Now, is it possible to reverse the inclusion in (5-91)? Begin by writing

$$X_2 = a_* X_1 \,\square\, \operatorname{Im} b, \tag{5-92}$$

and notice that

$$a = a \,\square\, (b \circ a_f) \,\square\, \widehat{(b \circ a_f)}. \tag{5-93}$$

We come now to a second technical digression, which is also attacked by way of example.

EXAMPLE 5.4-2

Let $f : X \to U$ and $g : U \to X$ be morphisms of groups. Then

$$\widehat{(g \circ f)}(x) = (g \circ f)(\hat{x}) = g(\hat{f}(x)) = (g \circ \hat{f})(x), \tag{5-94a,b,c}$$

from which it follows that

$$\widehat{g \circ f} = g \circ \hat{f}. \tag{5-95}$$

Thus the second technical digression is complete, with the establishment of the identity in (5-95). Return now to the development. ∎

The idea of (5-93) can now be developed another step to

$$a = a \,\square\, (b \circ a_f) \,\square\, (b \circ \hat{a}_f) \tag{5-96a}$$
$$= \alpha \,\square\, (b \circ \hat{a}_f). \tag{5-96b}$$

With this knowledge, (5-92) advances to

$$X_2 = (\alpha \,\square\, b \circ \hat{a}_f)_* X_1^f \,\square\, \operatorname{Im} \beta \tag{5-97a}$$
$$\subset \alpha_* X_1^f \,\square\, (b \circ \hat{a}_f)_* X_1^f \,\square\, \operatorname{Im} \beta \tag{5-97b}$$
$$= \alpha_* X_1^f \,\square\, \operatorname{Im} \beta \tag{5-97c}$$
$$= X_2^f, \tag{5-97d}$$

which gives

$$X_2 \subset X_2^f. \tag{5-98}$$

From (5-91) and (5-98),

$$X_2 = X_2^f. \tag{5-99}$$

These ideas lead to the general conclusion

$$X_j = X_j^f, \qquad j = 0, 1, 2, \dots, \tag{5-100}$$

by induction. To see this, suppose that (5-100) is true for $j = 0, 1, 2, \dots, i$. Then

$$\check{X}_{i+1}^f = \alpha_* X_i^f \,\square\, \operatorname{Im} \beta \tag{5-101a}$$
$$= (a \,\square\, (b \circ a_f))_* X_i \,\square\, \operatorname{Im} b \tag{5-101b}$$
$$\subset a_* X_i \,\square\, (b \circ a_f)_* X_i \,\square\, \operatorname{Im} b \tag{5-101c}$$
$$= a_* X_i \,\square\, \operatorname{Im} b \tag{5-101d}$$
$$= X_{i+1}, \tag{5-101e}$$

and similarly for the reverse inclusion.
  In view of (5-100), it is also true that

$$X^* = (X^f)^*, \tag{5-102}$$

by the uniqueness of a supremum. This establishes the observation that feedback does not change the reachable subgroup of a GRMPS.

Exercises

5.4-1. The feedback discussion at the end of this section led to the construction

$$X^f_{i+1} \subset X_{i+1}$$

of (5-101). Present the argument for the reverse inclusion, and so establish the equality asserted.

5.4-2. In Example 5.4-2, show that (5-95) could be replaced by

$$\widehat{g \circ f} = \hat{g} \circ \hat{f}.$$

5.4-3. Let f and g be defined as in Example 5.4-2. Develop three alternative expressions for

$$\widehat{g \circ (f \triangle f)}.$$

5.5 INVERTIBILITY WITH DELAY L

Earlier sections have made evident the fact that connections, in an appropriate sense, of more than one system depend upon a notion of binary operation on the set associated with the point at which the connection is made. In introductory treatments, these binary operations are almost always associative. As a result, the most frequent issues to arise after the establishment of a binary operation concern the existence of a unit and existence of inverse elements under the operation.

Section 5.1 has brought out the idea of ring structure on the set of endomorphisms of a commutative group. This ring structure addresses itself to the interrelation of two binary operations, one induced by the cummutative group serving as domain and codomain and one arising from composition of morphisms. The former establishes a commutative group structure of its own, with unit e_{GG} if $(G, \square, e_G)$ is the domain and codomain group. The latter establishes a monoid, with unit 1_G the identity morphism. In general, this monoid is not a group, because not every element has an inverse under composition. This engenders the question, just what elements do have inverses under composition and what elements do not.

It turns out that quite similar questions can be asked for GRMPSs, and this leads up to the point of this section.

Consider a group morphic system

$$(U, \triangle; X, \square; Y, *; a, b, c, d) \tag{5-103}$$

with the local dynamical equations

$$x_{k+1} = ax_k \,\square\, bu_k, \qquad y_k = cx_k * du_k. \tag{5-104a,b}$$

Let L be an element of $\mathbb{N}$. Such a GRMPS is said to be *invertible with delay L* if, given x_0, the sequence segment

$$u_*[0, k] \tag{5-105a}$$

can be determined from the sequence segment

$$y_*[0, k + L] \tag{5-105b}$$

for all k in $\mathbb{N}$.

An initial step in studying invertibility with delay L is to observe that a GRMPS will satisfy the definition if and only if the sequence segment (5-105a) can be determined from the sequence segment (5-105b) for all k in $\mathbb{N}$ when the initial state x_0 is the unit state e_X.

The necessity of this statement is apparent, because e_X is one possible x_0 which could be given. For sufficiency, recall the output calculations of Section 5.2, which determined that

$$y_0 = cx_0 * du_0, \qquad y_{i+1} = ca^{i+1}x_0 \overset{i}{\underset{j=0}{\text{\Large $\ast$}}} ca^{i-j}bu_j * du_{i+1}, \tag{5-106a,b}$$

i in $\mathbb{N}$, and set the initial state equal to e_X. Then (5-106) becomes

$$y_0 = du_0, \qquad y_{i+1} = \overset{i}{\underset{j=0}{\text{\Large $\ast$}}} ca^{i-j}bu_j * du_{i+1}, \tag{5-107a,b}$$

i in $\mathbb{N}$. Assume that (5-105a) can be determined from (5-105b) for all k in $\mathbb{N}$ when (5-107) is satisfied. When x_0 is not e_X, define

$$\tilde{y}_0 = c\hat{x}_0 * y_0, \qquad \tilde{y}_{i+1} = ca^{i+1}\hat{x}_0 * y_{i+1}, \tag{5-108a,b}$$

i in $\mathbb{N}$, and note that

$$\tilde{y}_0 = du_0, \qquad \tilde{y}_{i+1} = \overset{i}{\underset{j=0}{\text{\Large $\ast$}}} ca^{i-j}bu_j * du_{i+1}, \tag{5-109a,b}$$

i in $\mathbb{N}$. Knowledge of $y_*[0, k + L]$, together with knowledge of x_0, implies knowledge of $\tilde{y}_*[0, k + L]$. But by assumption, this implies knowledge of $u_*[0, k]$.

Thus attention may as well be focused upon (5-109). It will be helpful to combine the equations (5-109a) and (5-109b) into one entity. To do this define

$$m_0 = d, \qquad m_i = ca^{i-1}b, \quad i = 1, 2, 3, \ldots. \tag{5-110a,b}$$

The sequence

$$(m_0, m_1, m_2, \ldots) \in (Y^U)^{\mathbb{N}} \tag{5-111}$$

constitutes the *Markov morphisms* of the GRMPS (5-103). In terms of the Markov morphisms, (5-109) becomes

$$\tilde{y}_i = \overset{i}{\underset{j=0}{*}}\, m_{i-j} u_j, \qquad i \in \mathbb{N}. \tag{5-112}$$

A next step in studying invertibility with delay L is to establish that a GRMPS is invertible with delay L if and only if the input sequence segment

$$u_*[0,0] = u_0 \tag{5-113a}$$

can be determined from the sequence segment

$$\tilde{y}_*[0,L]. \tag{5-113b}$$

It is of course straightforward that this is necessary, simply by setting k equal to zero in (5-105). For sufficiency, an inductive argument can be pursued. The first step in the induction is to show that $u_*[0,1]$ can be found from $\tilde{y}_*[0, 1 + L]$. Because (5-113a) is assumed known from (5-113b), it is of course also known from $\tilde{y}_*[0, 1 + L]$. Thus it remains to see that u_1 is also known from $\tilde{y}_*[0, 1 + L]$. Begin by removing the effect of u_0 from (5-112) in the manner

$$y_i' = m_i \hat{u}_0 * \tilde{y}_i, \qquad i \in \mathbb{N}, \tag{5-114a}$$

$$= m_i \hat{u}_0 \overset{i}{\underset{j=0}{*}}\, m_{i-j} u_j, \qquad i \in \mathbb{N}, \tag{5-114b}$$

$$= e_Y, \qquad i = 0, \tag{5-114c}$$

$$= \overset{i}{\underset{j=1}{*}}\, m_{i-j} u_j, \qquad i = 1, 2, \ldots . \tag{5-114d}$$

Certainly y_0' contains no information about u_1, so consider (5-114d). Make shifts of variable

$$k = j - 1, \qquad n = i - 1, \tag{5-115}$$

which give

$$y_n'' = y_{n+1}' \tag{5-116a}$$

$$= \overset{n}{\underset{k=0}{*}}\, m_{n-k} u_{k+1}, \qquad n \in \mathbb{N}, \tag{5-116b}$$

$$= \overset{n}{\underset{k=0}{*}}\, m_{n-k} u_k'', \qquad n \in \mathbb{N}, \tag{5-116c}$$

where

$$u_k'' = u_{k+1}. \tag{5-117}$$

A comparison of (5-112) with (5-116c), together with the induction hypothesis, leads to the conclusion that

$$u_*''[0,0] = u_0'' = u_1 \tag{5-118a}$$

can be determined from

$$y_*''[0,L] = y'_*[1, 1+L], \tag{5-118b}$$

which is available from $y'_*[0, 1 + L]$. This completes the base step of the induction. The general step follows in the same spirit, and is therefore left as an exercise.

Thus a GRMPS is invertible with delay L if and only if u_0 can be determined from $\tilde{y}_*[0, L]$.

Recall the discussion of Section 4.3, on the subject of two GRMPSs connected in series. Figure 4.3 pictures the situation. Suppose that GRMPS 2 is a system which is invertible with delay L; then interest now turns to finding another system, GRMPS 1, which actually constructs (5-105a) from (5-105b). It is helpful to use the basic notation of (4-77), (4-78), and (4-79). Notice, however, that the present discussion also justifies the following conclusion:

$$(U_2, \triangle_2, e_{U_2}) = (Y_1, *_1, e_{Y_1}). \tag{5-119}$$

Thus subscripts on input and output groups may as well be suppressed. GRMPS 2 is therefore described by

$$(U, \triangle; X_2, \square_2; Y, *; a_2, b_2, c_2, d_2) \tag{5-120}$$

and GRMPS 1 by

$$(Y, *; X_1, \square_1; U, \triangle; a_1, b_1, c_1, d_1). \tag{5-121}$$

Moreover, the assumptions that $(U, \triangle, e_U)$ and $(X_1, \square_1, e_{X_1})$ are commutative groups will be sufficient to assure that the series connection is itself a group morphic system.

Suppose, then, that any input sequence

$$u_0, \quad u_1, \quad u_2, \quad \ldots, \tag{5-122a}$$

enters GRMPS 2 which is initially in the unit state; the task of GRMPS 1 is to produce an output sequence

$$\tilde{u}_0, \quad \tilde{u}_1, \quad \tilde{u}_2, \quad \ldots, \quad \tilde{u}_{L-1}, \quad u_0, \quad u_1, \quad u_2, \quad \ldots, \tag{5-122b}$$

in which the sequence (5-122a) reappears, with delay L. For brevity, attention is oriented to the case in which GRMPS 1 is initially in the unit state and in which

$$\tilde{u}_j = e_U, \quad j = 0, 1, 2, \ldots, L - 1. \tag{5-123}$$

If GRMPS 1 achieves this effect, it is called an *inverse with delay L* for GRMPS 2. Necessary conditions for one GRMPS to be an inverse with delay L for another GRMPS can be deduced without much trouble. Sufficiency of any set of conditions depends, of course, upon construction of the inverse system. Generally speaking, there is no unique way to make such a construction.

The following section considers one way to construct inverse GRMPSs.

Exercise

5.5. In studying invertibility with delay L, an important step was to establish that a GRMPS is invertible with delay L if and only if the input sequence segment (5-113a) can be determined from the sequence segment (5-113b). The argument in this section completes the base step of the induction for sufficiency. Refer to (5-118). Complete the general step of the induction.

5.6   INVERSE GRMP SYSTEMS

The preceding section has introduced the notion of invertibility with delay L, for L a natural number in $\mathbb{N}$. Associated with such a system, say GRMPS 2 in Fig. 4.3, is the idea of another group morphic system, GRMPS 1, which actually performs the inversion. Such a system is called an inverse with delay L. This section establishes necessary and sufficient conditions for the existence of inverses with delay L, subject only to various commutativity assumptions on the signal groups. These assumptions are made in the spirit of Sections 4.3 and 5.3, having to do with the GRMP nature of interconnected GRMPSs.

It is assumed that there is available a GRMPS 2, which is invertible with delay L. GRMPS 2 takes the form (5-120). The proposed inverse GRMPS is described in (5-121). At the outset, it is assumed that $(X_1, \square_1, e_{X_1})$ and $(U, \triangle, e_U)$ are commutative groups, according to Section 4.3. This will ensure that, regardless of how GRMPS 1 is constructed, the series connection of the inverse with the original system has group morphic nature.

Surprisingly, an appropriate set of necessary and sufficient conditions can be discovered by activating GRMPS 2 with a very simple sequence. Consider now the impulse input sequence

$$u_0, \quad e_U, \quad e_U, \quad e_U, \quad \ldots. \tag{5-124}$$

At the output of GRMPS 2, this produces the sequence

$$m_0^2 u_0, \quad m_1^2 u_0, \quad m_2^2 u_0, \quad \ldots, \tag{5-125}$$

where m_i^2 is the ith Markov morphism of GRMPS 2. Denote the Markov morphisms of GRMPS 1 by

$$m_0^1, \quad m_1^1, \quad m_2^1, \quad \ldots. \tag{5-126}$$

When (5-125) becomes the input sequence for GRMPS 1, the output sequence of GRMPS 1 becomes

$$
\begin{aligned}
&(m_0^1 \circ m_0^2)u_0, \\
&(m_1^1 \circ m_0^2)u_0 \, \triangle \, (m_0^1 \circ m_1^2)u_0, \\
&(m_2^1 \circ m_0^2)u_0 \, \triangle \, (m_1^1 \circ m_1^2)u_0 \, \triangle \, (m_0^1 \circ m_2^2)u_0, \\
&\qquad\qquad \vdots \\
&(m_{L-1}^1 \circ m_0^2)u_0 \, \triangle \, (m_{L-2}^1 \circ m_1^2)u_0 \, \triangle \, \cdots \, \triangle \, (m_0^1 \circ m_{L-1}^2)u_0, \\
&(m_L^1 \circ m_0^2)u_0 \, \triangle \, (m_{L-1}^1 \circ m_1^2)u_0 \, \triangle \, \cdots \, \triangle \, (m_0^1 \circ m_L^2)u_0,
\end{aligned}
\tag{5-127}
$$

up to the $(L+1)$st element. If (5-122b) and (5-123) are to hold for all u_0 in U up to the $(L+1)$st element described in (5-127), then, for GRMPS 1 to be an inverse with delay L for GRMPS 2, it is necessary that the Markov morphisms m_i^1, $i = 0, 1, \ldots, L$, of GRMPS 1 satisfy the equations

$$
\begin{aligned}
m_0^1 \circ m_0^2 &= e_{UU}, \\
(m_1^1 \circ m_0^2) \, \triangle \, (m_0^1 \circ m_1^2) &= e_{UU}, \\
&\vdots \\
(m_{L-1}^1 \circ m_0^2) \, \triangle \, \cdots \, \triangle \, (m_0^1 \circ m_{L-1}^2) &= e_{UU}, \\
(m_L^1 \circ m_0^2) \, \triangle \, \cdots \, \triangle \, (m_0^1 \circ m_L^2) &= 1_U.
\end{aligned}
\tag{5-128}
$$

In a more compact notation, (5-128) can be written

$$
\begin{aligned}
&\mathop{\triangle}_{j=0}^{i} (m_{i-j}^1 \circ m_j^2) = e_{UU}, \qquad i = 0, 1, 2, \ldots, L - 1, \\
&\mathop{\triangle}_{j=0}^{L} (m_{L-j}^1 \circ m_j^2) = 1_U.
\end{aligned}
\tag{5-129}
$$

The reader should note that each

$$m_k^1 \circ m_n^2 \tag{5-130}$$

is an element of the ring of endomorphisms acting on the commutative group $(U, \triangle, e_U)$. This was the ring discussed in Section 5.1. The necessary condition (5-129) is expressed naturally in this ring.

Suppose next that there exist morphisms

$$m_i^1 : Y \rightarrow U, \qquad i = 0, 1, 2, \ldots, L, \tag{5-131}$$

which satisfy the condition (5-129). Is it then possible to construct a GRMPS 1 which can serve as an inverse with delay L for GRMPS 2? The answer is yes, with minor qualifying conditions on group commutativity.

Begin by constructing a group morphic system which will achieve the first $L + 1$ elements of the sequence (5-122b), with (5-123) in force. Such a GRMP system is shown in Fig. 5.3. Denote it by GRMPS 1′; intuitively, it is called the "feedforward" portion of the inverse. The input group of GRMPS 1′ is $(Y, * , e_Y)$; the state group is the product

$$Y^L, \tag{5-132a}$$

equipped with the binary operation

$$(y_1, y_2, \ldots, y_L) ** (\tilde{y}_1, \tilde{y}_2, \ldots, \tilde{y}_L) = (y_1 * \tilde{y}_1, y_2 * \tilde{y}_2, \ldots, y_L * \tilde{y}_L), \tag{5-132b}$$

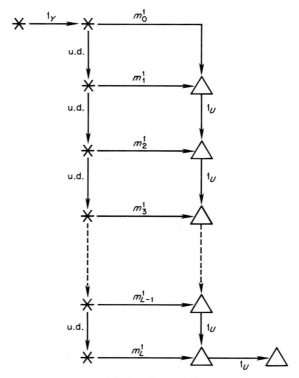

Fig. 5.3. GRMPS 1′.

which has unit $(e_Y, e_Y, \ldots, e_Y)$. The inverse of $(y_1, y_2, \ldots, y_L)$ is $(\hat{y}_1, \hat{y}_2, \ldots, \hat{y}_L)$. And the output group is $(U, \triangle, e_U)$. The 4-tuple (a', b', c', d') characterizing GRMPS 1′ has the actions

$$a'(y_1, y_2, \ldots, y_L) = (y_2, y_3, \ldots, y_L, e_Y),$$
$$b'(y) = (e_Y, e_Y, \ldots, e_Y, y), \qquad (5\text{-}133)$$
$$c'(y_1, y_2, \ldots, y_L) = m_L^1 y_1 \triangle m_{L-1}^1 y_2 \triangle \cdots \triangle m_1^1 y_L,$$
$$d'(y) = m_0^1 y.$$

It is a straightforward exercise to check that the first $L + 1$ elements of the sequence (5-122b), with (5-123) in force, are produced by GRMPS 1′, when it starts in the unit state. For i greater than L, however, GRMPS 1′ produces outputs

$$\bigtriangleup_{j=0}^{L} (m_{L-j}^1 \circ m_{j+i-L}^2) u_0. \qquad (5\text{-}134)$$

This contribution can be nullified by the GRMPS 1″ structure shown in Fig. 5.4. The idea is to connect GRMPS 1″ in series with GRMPS 1′, as shown in Fig. 5.5. GRMPS 1″ is described by

$$(U, \triangle; X_2, \square_2; U, \triangle; a'', b'', c'', d''), \qquad (5\text{-}135a)$$

where the actions are given by

$$a'' = a_2 \square_2 b_2 \hat{\tilde{c}}, \qquad b'' = b_2, \qquad c'' = \hat{\tilde{c}}, \qquad d'' = 1_U, \qquad (5\text{-}135b,c,d,e)$$

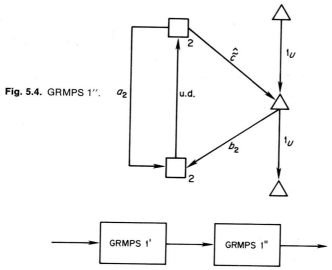

Fig. 5.4. GRMPS 1″.

Fig. 5.5. GRMPS 1.

with $\tilde{c}: X_2 \rightarrow U$ defined by

$$\tilde{c} = \bigwedge_{j=0}^{L} (m_{L-j}^1 \circ c_2 a_2^j). \tag{5-136}$$

Again, from Section 4.3, assume that $(X_2, \square_2, e_{X_2})$ is a commutative group. This assumption, together with that already made on $(U, \triangle, e_U)$, ensures that the series connection of Fig. 5.5 can be understood as a group morphic system. When i is equal to $L + 1$, Fig. 5.4 and (5-134) show that the output of GRMPS 1'' is

$$\bigwedge_{j=0}^{L} (m_{L-j}^1 \circ m_{j+1}^2) u_0 \bigwedge_{j=0}^{L} (m_{L-j}^1 \circ c_2 a_2^j) b_2 \hat{u}_0, \tag{5-137}$$

which is equal to e_U because

$$c_2 a_2^j b_2 = m_{j+1}^2. \tag{5-138}$$

Moving on to an induction, suppose that the output of GRMPS 1'' has been e_U for

$$i = L + 1, L + 2, \ldots, L + n; \tag{5-139}$$

then the output at $L + n + 1$ is

$$\bigwedge_{j=0}^{L} (m_{L-j}^1 \circ m_{j+n+1}^2) u_0 \bigwedge_{j=0}^{L} (m_{L-j}^1 \circ c_2 a_2^j) a_2^n b_2 \hat{u}_0, \tag{5-140}$$

which is once again e_U because

$$c_2 a_2^{j+n} b_2 = m_{j+n+1}^2. \tag{5-141}$$

These arguments establish that GRMPS 1 is an inverse with delay L for GRMPS 2 input sequences of the form (5-124).

The case for the general input sequence (5-122a) is an easy extension. The main difference is that the output sequence of GRMPS 2 becomes

$$\mathop{\ast}_{j=0}^{i} m_{i-j}^2 u_j, \qquad i \in \mathbb{N}. \tag{5-142}$$

The zeroth output of GRMPS 1' is then

$$m_0^1(m_0^2 u_0) = e_U; \tag{5-143}$$

and the first output is

$$m_1^1(m_0^2 u_0) \triangle m_0^1(m_1^2 u_0 \ast m_0^2 u_1)$$
$$= [(m_1^1 \circ m_0^2) \triangle (m_0^1 \circ m_1^2)] u_0 \triangle (m_0^1 \circ m_0^2) u_1 = e_U. \tag{5-144}$$

Continuing, the ith output is

$$\bigtriangleup_{j=0}^{i} m_{i-j}^{1}\left(\underset{k=0}{\overset{j}{*}}\, m_{j-k}^{2}u_{k}\right) = \bigtriangleup_{j=0}^{i}\bigtriangleup_{k=0}^{j} (m_{i-j}^{1} \circ m_{j-k}^{2})u_{k} \qquad (5\text{-}145)$$

$$= \bigtriangleup_{k=0}^{i}\bigtriangleup_{j=k}^{i} (m_{i-j}^{1} \circ m_{j-k}^{2})u_{k} \qquad (5\text{-}146)$$

$$= \bigtriangleup_{k=0}^{i}\bigtriangleup_{q=0}^{i-k} (m_{i-k-q}^{1} \circ m_{q}^{2})u_{k}, \qquad (5\text{-}147)$$

where the change of variable

$$q = j - k \qquad (5\text{-}148)$$

has been made. From the condition (5-129), as applied to the inner expression in (5-147), it follows that the ith output is e_U until $i - k$ is equal to L, which cannot occur before L. At the stage when i is equal to L, $i - k$ equal to L implies k equal to 0, which means that u_0 becomes the output of GRMPS 1′. For i greater than L, the GRMPS 1′ output is

$$\bigtriangleup_{j=0}^{L} m_{L-j}^{1}\left(\underset{k=0}{\overset{j+i-L}{*}}\, m_{j+i-L-k}^{2}u_{k}\right). \qquad (5\text{-}149)$$

Consider i equal to $L + 1$. Then GRMPS 1′ outputs

$$\bigtriangleup_{j=0}^{L} m_{L-j}^{1}\left(\underset{k=0}{\overset{j+1}{*}}\, m_{j+1-k}^{2}u_{k}\right)$$

$$= \bigtriangleup_{j=0}^{L} m_{L-j}^{1}\left\{m_{j+1}^{2}u_{0}\underset{k=1}{\overset{j+1}{*}}\, m_{j+1-k}^{2}u_{k}\right\} \qquad (5\text{-}150)$$

$$= \bigtriangleup_{j=0}^{L} (m_{L-j}^{1} \circ m_{j+1}^{2})u_{0}\bigtriangleup_{j=0}^{L} m_{L-j}^{1}\left\{\underset{q=0}{\overset{j}{*}}\, m_{j-q}^{2}u_{q+1}\right\}, \qquad (5\text{-}151)$$

where q has replaced $k - 1$. From a brief reconsideration of (5-137), it is clear that GRMPS 1″ will nullify the first term in (5-151). Moreover, the analysis of (5-145), (5-146), and (5-147), when applied to the second term of (5-151) yields

$$\bigtriangleup_{q=0}^{L}\bigtriangleup_{m=0}^{L-q} (m_{L-q-m}^{1} \circ m_{m}^{2})u_{q+1}, \qquad (5\text{-}152)$$

from which it follows that u_1 is the output of GRMPS 1″ when i is equal to $(L + 1)$. The induction here is more complicated. Suppose that the output of GRMPS 1″ has been

$$u_1, \quad u_2, \quad \ldots, \quad u_n \tag{5-153}$$

for the values of i given in (5-139). It is important to know what is produced by $\hat{\tilde{c}}$ when i is equal to $L + n + 1$. Calculate the state x'' of GRMPS 1″ at the interesting values of i:

$$x''(i) = e_{x_2}, \quad i = 0, 1, 2, \ldots, L, \tag{5-154a}$$

$$x''(L + 1) = b_2 u_0,$$

$$x''(L + 2) = a_2 b_2 u_0 \,\square_2\, b_2 u_1,$$

$$\vdots$$

$$x''(L + n) = \bigsqcup_{2\,j=0}^{n-1} a_2^{n-1-j} b_2 u_j, \tag{5-154b}$$

$$x''(L + n + 1) = \bigsqcup_{2\,j=0}^{n} a_2^{n-j} b_2 u_j.$$

Accordingly,

$$\tilde{c}(x''(L + n + 1)) = \bigwedge_{j=0}^{n} \tilde{c} a_2^{n-j} b_2 u_j. \tag{5-155}$$

With the aid of (5-136), (5-155) becomes

$$\bigwedge_{j=0}^{n} \left(\bigwedge_{k=0}^{L} (m_{L-k}^1 \circ c_2 a_2^k) \right) a_2^{n-j} b_2 u_j = \bigwedge_{j=0}^{n} \bigwedge_{k=0}^{L} (m_{L-k}^1 \circ m_{k+n-j+1}^2) u_j. \tag{5-156}$$

From (5-149), the output of GRMPS 1′ at $L + n + 1$ is

$$\bigwedge_{j=0}^{L} m_{L-j}^1 \left(\underset{k=0}{\overset{j+n+1}{\text{\Large ✳}}} m_{j+n+1-k}^2 u_k \right)$$

$$= \bigwedge_{j=0}^{L} m_{L-j}^1 \left\{ \underset{k=0}{\overset{n}{\text{\Large ✳}}} m_{j+n+1-k}^2 u_k \right\} \bigwedge_{j=0}^{L} m_{L-j}^1 \left\{ \underset{k=n+1}{\overset{j+n+1}{\text{\Large ✳}}} m_{j+n+1-k}^2 u_k \right\}$$

$$= \bigwedge_{j=0}^{L} \bigwedge_{k=0}^{n} (m_{L-j}^1 \circ m_{j+n+1-k}^2) u_k \bigwedge_{j=0}^{L} m_{L-j}^1 \left\{ \underset{q=0}{\overset{j}{\text{\Large ✳}}} m_{j-q}^2 u_{q+n+1} \right\}, \tag{5-157a}$$

where q has replaced $k - n - 1$. Compare the first term with (5-156), to see that GRMPS 1″ nullifies its effect. Moreover, the second term converts, according to the same arguments, to

$$\bigwedge_{q=0}^{L} \bigwedge_{m=0}^{L-q} (m_{L-q-m}^1 \circ m_m^2) u_{q+n+1}, \tag{5-157b}$$

from which only u_{n+1} appears. The induction is therefore accomplished.

To complete the discussion, notice that the state group for GRMPS 1 will be the product group

$$X_1 = X_2 \times Y^L, \tag{5-158}$$

as discussed in Section 4.3. By assumption, this is to be commutative. Thus it is now appropriate to assume that $(Y, *, e_Y)$ is a commutative group as well.

Accordingly, all three groups associated with GRMPS 2 have been assumed commutative. A group morphic system in which all signal groups are commutative may be called a *commutative group morphic system.*

For commutative GRMPSs, there exists an inverse with delay L if and only if Eq. (5-129) can be solved for $L + 1$ Markov morphisms (5-131). The inverse GRMPS constructed in this section is also commutative.

Exercise

5.6. Show that the first $L + 1$ elements of the sequence (5-122b), with (5-123) in force, are produced by GRMPS 1′, when it starts in the unit state.

5.7 EXAMPLE

The purpose of this section is to offer illustration of the concepts of invertibility with delay L, as well as the construction of inverse GRMPSs.

EXAMPLE 5.7

Select for the input and output groups

$$(U, \triangle, e_U) = (Y, *, e_Y) = (\mathbb{Z}_2, +, 0). \tag{5-159}$$

For the state group X_2, select the product group

$$X^4, \tag{5-160}$$

where

$$(X, \square, e_X) = (\mathbb{Z}_2, +, 0), \tag{5-161}$$

which is also a commutative group. Define the actions of the 4-tuple (a_2, b_2, c_2, d_2) in the manner

$$a_2(x_1, x_2, x_3, x_4) = (x_2, x_3, x_4, e_X), \tag{5-162a}$$
$$b_2(z) = (e_X, e_X, e_X, z), \tag{5-162b}$$
$$c_2(x_1, x_2, x_3, x_4) = x_1, \tag{5-162c}$$
$$d_2(z) = 0. \tag{5-162d}$$

An initial construction of the Markov morphisms is appropriate. These are, if 0_{UV} is denoted simply by 0 and if 1_U is denoted simply by 1,

$$m_0^2 = d_2 = 0, \tag{5-163a}$$

$$m_1^2 = c_2 b_2 = 0, \tag{5-163b}$$

$$m_2^2 = c_2 a_2 b_2 = 0, \tag{5-163c}$$

$$m_3^2 = c_2 a_2^2 b_2 = 0, \tag{5-163d}$$

$$m_4^2 = c_2 a_2^3 b_2 = 1, \tag{5-163e}$$

$$\vdots$$

$$m_i^2 = c_2 a_2^{i-1} b_2 = 0, \qquad i = 5,6,7,\ldots. \tag{5-163f}$$

The reader has probably already foreseen a very simple solution to this problem. For the sake of completeness, however, examine the fundamental condition (5-128). With L equal to 0, it would be required to solve

$$m_0^1 \circ m_0^2 = 1, \tag{5-164}$$

which is impossible in view of (5-163a). If L is equal to 1, the conditions are

$$m_0^1 \circ m_0^2 = 0, \qquad (m_1^1 \circ m_0^2) + (m_0^1 \circ m_1^2) = 1. \tag{5-165a,b}$$

Consider (5-163a); this means that (5-165a) is satisfied trivially and that (5-165b) amounts to

$$m_0^1 \circ m_1^2 = 1. \tag{5-165c}$$

But this is impossible by (5-163b). In like manner, L equal to 2 or 3 is also impossible. If, however, L is 4, then the condition (5-129) is

$$m_0^1 \circ m_0^2 = 0, \tag{5-166a}$$

$$(m_1^1 \circ m_0^2) + (m_0^1 \circ m_1^2) = 0, \tag{5-166b}$$

$$(m_2^1 \circ m_0^2) + (m_1^1 \circ m_1^2) + (m_0^1 \circ m_2^2) = 0, \tag{5-166c}$$

$$(m_3^1 \circ m_0^2) + (m_2^1 \circ m_1^2) + (m_1^1 \circ m_2^2) + (m_0^1 \circ m_3^2) = 0, \tag{5-166d}$$

$$(m_4^1 \circ m_0^2) + (m_3^1 \circ m_1^2) + (m_2^1 \circ m_2^2) + (m_1^1 \circ m_3^2) + (m_0^1 \circ m_4^2) = 1. \tag{5-166e}$$

Equations (5-163a)–(5-163d) ensure that (5-166a)–(5-166d) are satisfied trivially. Moreover, (5-166e) reduces to

$$m_0^1 \circ m_4^2 = 1. \tag{5-166f}$$

From (5-163e), this implies the choice

$$m_0^1 = 1. \tag{5-167}$$

Thus the GRMPS is invertible with delay 4, because the remaining morphisms

$$m_i^1, \qquad i = 1,2,3,4, \tag{5-168}$$

are arbitrary. For simplicity, choose them equal to the zero morphism. GRMPS 1′ may then be pictured as in Fig. 5.6. Turn now to GRMPS 1″. One has

$$a'' = a_2 \,\square_2\, b_2 \tilde{\hat{c}}, \qquad b'' = b_2, \qquad c'' = \hat{\tilde{c}}, \qquad d'' = 1. \qquad (5\text{-}169)$$

Examine $\tilde{c}$. The general expression for $\tilde{c}$ is found in (5-136). For the case at hand, this becomes

$$\tilde{c} = (m_4^1 \circ c_2) + (m_3^1 \circ c_2 a_2) + (m_2^1 \circ c_2 a_2^2) + (m_1^1 \circ c_2 a_2^3) + (m_0^1 \circ c_2 a_2^4)$$
$$= m_0^1 \circ c_2 a_2^4 = c_2 a_2^4. \qquad (5\text{-}170)$$

But

$$a_2^4 = e_{X_2 x_2} \qquad (5\text{-}171)$$

so that

$$\tilde{c} = e_{U x_2}. \qquad (5\text{-}172)$$

Accordingly, the feedback loop in GRMPS 1″, as may be seen in Fig. 5.4, produces only units and may be disconnected. The result is Fig. 5.7. Moreover, from Fig. 5.5, the total inverse with delay 4 can be pictured in Fig. 5.8.

∎

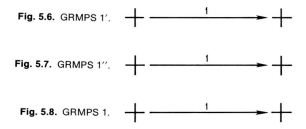

Fig. 5.6. GRMPS 1′.

Fig. 5.7. GRMPS 1″.

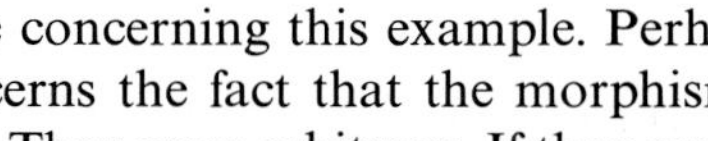

Fig. 5.8. GRMPS 1.

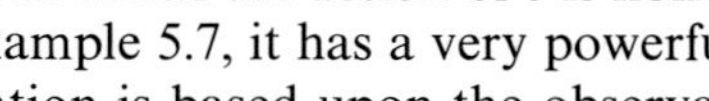

    Several points can be made concerning this example. Perhaps the most intriguing of these points concerns the fact that the morphisms in (5-168) did not have to be chosen zero. They were arbitrary. If they are chosen nonzero, inverse systems can result in which the action of $\tilde{c}$ is nontrivial.

    Despite the simplicity of Example 5.7, it has a very powerful generalization. The key to this generalization is based upon the observation that the first nonzero morphism in the sequence

$$m_i^2, \qquad i \in \mathbb{N}, \qquad (5\text{-}173)$$

is monic. Whenever this is the case, an argument similar to that above can be completed. Input, state, and output groups are assumed to be commutative, so that all the various series interconnections have group morphic interpretation.

Suppose, therefore, that

$$m_i^2 = e_{YU}, \qquad i = 0, 1, 2, \ldots, L - 1, \qquad (5\text{-}174a)$$

and that

$$\mathrm{Ker}\, m_L^2 = e_U. \qquad (5\text{-}174b)$$

Then condition (5-129) is satisfied trivially for

$$i = 0, 1, 2, \ldots, L - 1, \qquad (5\text{-}175a)$$

and reduces to

$$m_0^1 \circ m_L^2 = 1_U \qquad (5\text{-}175b)$$

for i equal to L. Because m_L^2 is monic, as may be recalled from (4-41), it has at least one left inverse

$$(m_L^2)^{-L} : Y \to U. \qquad (5\text{-}176)$$

For the moment, regard (5-176) as the function discussed in Section 2.1. This function satisfies

$$(m_L^2)^{-L} \circ m_L^2 = 1_U. \qquad (5\text{-}177)$$

Is (5-176) a morphism? The answer can be yes, if the function is properly defined. First, examine the action of (5-176) on $\mathrm{Im}\, m_L^2$. Here (5-177) implies, because m_L^2 is a morphism, that

$$(m_L^2)^{-L}\{m_L^2(u_1) * m_L^2(u_2)\} = u_1 \triangle u_2 \qquad (5\text{-}178)$$

for all u_1 and u_2 in U. But (5-177) also implies that

$$u_1 \triangle u_2 = ((m_L^2)^{-L} \circ m_L^2)u_1 \triangle ((m_L^2)^{-L} \circ m_L^2)u_2. \qquad (5\text{-}179)$$

Together, (5-178) and (5-179) provide

$$(m_L^2)^{-L}\{m_L^2(u_1) * m_L^2(u_2)\} = (m_L^2)^{-L}\{m_L^2(u_1)\} \triangle (m_L^2)^{-L}\{m_L^2(u_2)\}, \quad (5\text{-}180)$$

which is the desired statement of morphism on $\mathrm{Im}\, m_L^2$. Second, observe that (5-177) places no constraint whatever on the manner in which the function $(m_L^2)^{-L}$ is defined for elements y in Y but not in $\mathrm{Im}\, m_L^2$. To show the existence of a morphism (5-176) on the domain Y involves demonstrating that there exists a way to extend the morphic action of $(m_L^2)^{-L}$ on $\mathrm{Im}\, m_L^2$ to a morphic action on Y. There are broad classes of groups in which this is possible; but they will not be discussed at this point.

Therefore assume such an extension, so that (5-176) becomes a morphism. Then choose

$$m_0^1 = (m_L^2)^{-L} \quad \text{and} \quad m_i^1 = e_{UY}, \quad i = 1, 2, 3, \ldots, L. \quad (5\text{-}181\text{a,b})$$

Unlike Example 5.7, $\tilde{c}$ is this time in general nontrivial, with the definition

$$\tilde{c} = (m_L^2)^{-L} \circ c_2 a_2^L. \quad (5\text{-}182)$$

So the feedback section GRMPS 1″ is operative in this case.

It often happens that m_0^2 is monic. If the existence of a morphism

$$(m_0^2)^{-L} : Y \to U \quad (5\text{-}183)$$

is assumed, then there is an alternative way to look at an inverse with delay zero. To do this, begin with the local dynamical equations

$$x_2(k + 1) = a_2 x_2(k) \,\square_2\, b_2 u(k), \quad (5\text{-}184\text{a})$$

$$y(k) = c_2 x_2(k) * d_2 u(k) \quad (5\text{-}184\text{b})$$

of GRMPS 2. Operate on both members of (5-184b) with the morphism (5-183) to obtain

$$(m_0^2)^{-L} y(k) = (m_0^2)^{-L} \{ c_2 x_2(k) * d_2 u(k) \}$$

$$= (m_0^2)^{-L} \circ c_2 x_2(k) \,\triangle\, (m_0^2)^{-L} \circ d_2 u(k). \quad (5\text{-}185)$$

This expression can be simplified, inasmuch as

$$m_0^2 = d_2, \quad (5\text{-}186)$$

and by rearrangement of terms, to give

$$u(k) = d_2^{-L} \circ \hat{c}_2 x_2(k) \,\triangle\, d_2^{-L} y(k). \quad (5\text{-}187)$$

Now substitute (5-187) into the local transition equation (5-184a) and find

$$x_2(k + 1) = a_2 x_2(k) \,\square_2\, b_2 \{ d_2^{-L} \circ \hat{c}_2 x_2(k) \,\triangle\, d_2^{-L} y(k) \}$$

$$= (a_2 \,\square_2\, b_2 \circ d_2^{-L} \circ \hat{c}_2) x_2(k) \,\square_2\, b_2 \circ d_2^{-L} y(k). \quad (5\text{-}188)$$

From (5-187) and (5-188), the GRMPS

$$(a_2 \,\square_2\, b_2 \circ d_2^{-L} \circ \hat{c}_2, b_2 \circ d_2^{-L}, d_2^{-L} \circ \hat{c}_2, d_2^{-L}) \quad (5\text{-}189)$$

describes an inverse with delay zero.

Similar calculations can be carried out with delays greater than zero, but only at the expense of adding a symbol for the unit delay. Though this symbol is going to become available in a later chapter, it would needlessly complicate things here.

One case in which (5-176) is certainly a morphism occurs when m_L^2 is an isomorphism. In this case,

$$\mathrm{Im}\, m_L^2 = Y, \tag{5-190}$$

and the above discussion goes through without any question about the manner in which (5-176) is defined for y not in $\mathrm{Im}\, m_L^2$. For a hint of an example in which (5-176) cannot be a morphism, see the exercise.

Exercise

5.7. Give an example of a monomorphism of groups with the property that it has a left inverse function which cannot be extended to a morphism. *Hint*: Consider the morphism

$$f:\mathbb{Z} \to \mathbb{Z} \qquad \text{with action} \qquad f(z) = 2z.$$

A left inverse

$$f^{-L}:\mathbb{Z} \to \mathbb{Z}$$

will have to have the property

$$f^{-L}(1) + f^{-L}(1) = f^{-L}(2)$$

if it is a morphism. Show that this is impossible.

5.8 DECOUPLING WITH STATE MEASUREMENT

The three sections preceding have discussed the general question of what it means for a GRMPS to be invertible with delay L, a necessary and sufficient condition for the existence of another GRMPS which can act as an inverse with delay L, including a construction for one such inverse system, and a major class of invertible systems, namely, those whose first nonunit Markov morphism has a left-inverse morphism.

This section briefly examines one of the most interesting control problems relating to the question of invertibility. This problem is part of a broad class of studies which have come to be described by the adjective *decoupling*. Generally speaking, a decoupling problem begins with a given GRMPS (a, b, c, d), having local dynamical equations

$$x_{k+1} = ax_k \,\square\, bu_k, \qquad y_k = cx_k * du_k. \tag{5-191a,b}$$

The typical situation can be sketched as follows. A commutative group $(V, \circ, e_V)$ is given, and the input and output groups are induced from V in the usual product group form

$$U = V^m, \qquad Y = V^p. \qquad (5\text{-}192\text{a,b})$$

Often the state group has a similar structure, though there is no need to so assume here. Now suppose that the input sequence in $U^\mathbb{N}$ satisfies

$$u_i = (e_V, \ldots, e_V, v_k(i), e_V, \ldots, e_V), \qquad (5\text{-}193)$$

$i \in \mathbb{N}$.

In other words, there is but one nonunit entry $v_k(i)$ in (5-193), in the kth position. It will ordinarily be the case that the global output function produces a sequence in $Y^\mathbb{N}$ whose typical element is

$$y_j = (v_1(j), v_2(j), \ldots, v_p(j)), \qquad (5\text{-}194)$$

$j \in \mathbb{N}$. Thus, even though but one input channel has been excited, all output channels respond. This phenomenon is described as *coupling* between the outputs and the inputs. Decoupling, then, consists in adjoining another GRMPS to the original, in such a way that coupling vanishes, that is, in a manner such that the outputs of the interconnected system take the form

$$(e_V, \ldots, e_V, v_k(j), e_V, \ldots, e_V). \qquad (5\text{-}195)$$

For decoupling, it is not enough to establish a bijection between input sequences and output sequences of the interconnected system. Essentially, it is desired that one input channel affect one and only one output channel. In other words, what is sought is more in the line of an identity function. In the context arranged up to this point, decoupling can be arranged by choosing the second GRMPS to be an inverse with delay L for the first GRMPS. Then the relation between input sequences and output sequences of the interconnected system is in the same spirit as an identity function, with adjustment for the delay.

The reader should note that the previous sections have considered inverses which made their constructions upon the output sequence of the GRMPS in question. This section has to do with what can be done when the construction can make use of the state sequence as well.

Basically, the whole idea proceeds from an observation which can be made by Fig. 5.4. If the tandem connection of Fig. 5.5 is operating as an inverse with delay L, then it is true that the output sequence of GRMPS $1''$ is L copies of e_U followed by the input sequence to GRMPS 2. An examination of Fig. 5.4 then shows that the state sequence of GRMPS $1''$ is $L + 1$ copies of e_{X_2} followed by the state sequence of GRMPS 2. This raises the possibility of connecting $\hat{c}$ in an appropriate way to GRMPS 2, and avoiding GRMPS $1''$ altogether.

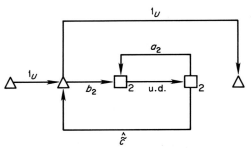

Fig. 5.9. Another view of Fig. 5.4.

To see this, it is helpful to redraw Fig. 5.4 from an alternative viewpoint. This has been done in Fig. 5.9. Next, compare Fig. 5.9 with a corresponding figure of GRMPS 2 itself, as sketched in Fig. 5.10. The morphism b_2 in Fig. 5.10 processes the input sequence

$$u_0, \quad u_1, \quad u_2, \quad \ldots ; \tag{5-196}$$

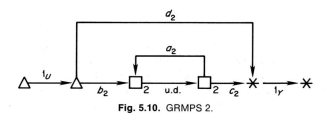

Fig. 5.10. GRMPS 2.

the morphism b_2 in Fig. 5.9 processes the sequence

$$e_U, \quad e_U, \quad \ldots, \quad e_U, \quad u_0, \quad u_1, \quad u_2, \quad \ldots, \tag{5-197}$$

which is the same as (5-196) but preceded by L copies of the unit input e_U in U. The group morphic systems of Figs. 5.9 and 5.10 both begin their operations in the unit state. If the state sequence in Fig. 5.10 is

$$e_{X_2}, \quad x_2(1), \quad x_2(2), \quad \ldots, \tag{5-198}$$

then the state sequence of Fig. 5.9 is

$$e_{X_2}, \quad e_{X_2}, \quad \ldots, \quad e_{X_2}, \quad x_2(1), \quad x_2(2), \quad \ldots, \tag{5-199}$$

with L copies of e_{X_2} preceding the sequence of (5-198).

Accordingly, $\hat{c}$ of Fig. 5.9 may as well be connected to a corresponding point in Fig. 5.10, provided that it is preceded by L unit delays. Such a

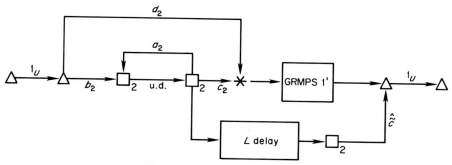

Fig. 5.11. GRMPS and its inverse.

scheme is indicated in Fig. 5.11. Like GRMPS 2 and GRMPS 1', the L delays are initially loaded with units from X_2.

The GRMPS of Fig. 5.11, taken as an interconnected whole, accepts the input sequence (5-196) and produces the output sequence (5-197). In the sense described above, then, these series constructions have achieved a decoupling of the original GRMPS. Of course, the GRMPS to be decoupled is assumed to be commutative, which is sufficient to assure that the inverses of the foregoing sections are applicable.

Consider again now the series connection of Fig. 4.3. The discussion of invertibility and inverses has assumed that GRMPS 1 is the inverse and that GRMPS 2 is the system having an inverse. A moment's reflection upon the inverse system constructed in Section 5.6, however, shows that the connection can also be made in reverse order, as in Fig. 5.12. What effect does this new positioning of the inverse system have upon the resulting series interconnection? This topic is explored in the remainder of the section.

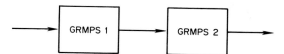

Fig. 5.12. Alternative connection of inverse system.

Recall the global output function

$$h: X \times U^{\mathbb{N}} \to Y^{\mathbb{N}} \tag{5-200}$$

associated with an arbitrary GRMPS

$$(U, \triangle; X, \square; Y, *; a, b, c, d). \tag{5-201}$$

Denote by

$$h_{ex}: U^{\mathbb{N}} \to Y^{\mathbb{N}} \tag{5-202a}$$

the restriction of (5-200) to the unit state e_X, with corresponding action

$$h_{ex}(u_*) = h(e_X, u_*). \tag{5-202b}$$

In order to simplify the notation for this discussion, write

$$\tilde{h} = h_{ex}. \tag{5-203}$$

For GRMPSs 1 and 2, then, the corresponding notations are

$$\tilde{h}_1: Y^{\mathbb{N}} \to U^{\mathbb{N}}, \tag{5-204}$$

and

$$\tilde{h}_2: U^{\mathbb{N}} \to Y^{\mathbb{N}}. \tag{5-205}$$

Define a function

$$f_L: U^{\mathbb{N}} \to U^{\mathbb{N}} \tag{5-206a}$$

with the action

$$f_L(u_0, u_1, u_2, \dots) = (e_U, \dots, e_U, u_0, u_1, u_2, \dots), \tag{5-206b}$$

the right member being prefixed by L copies of the unit e_U. According to the developments of Section 5.6,

$$\tilde{h}_1 \circ \tilde{h}_2 = f_L. \tag{5-207}$$

As a function, f_L is injective and therefore possesses left inverses

$$f_L^{-L}: U^{\mathbb{N}} \to U^{\mathbb{N}}. \tag{5-208}$$

Select one of these and apply it to both members of (5-207). Then

$$(f_L^{-L} \circ \tilde{h}_1) \circ \tilde{h}_2 = 1_{U^{\mathbb{N}}}, \tag{5-209}$$

which states that

$$f_L^{-L} \circ \tilde{h}_1: Y^{\mathbb{N}} \to U^{\mathbb{N}} \tag{5-210}$$

is a left inverse for $\tilde{h}_2$, which thus must be injective.

The question concerning what happens when GRMPSs 1 and 2 are interchanged is then fundamentally involved with the possibility of changing the order of the composition in the left member of (5-209).

One way to attack this issue is to make $\tilde{h}_2$ into a bijection by redefining its codomain to be its image. Thus,

$$\bar{\tilde{h}}_2: U^{\mathbb{N}} \to \text{Im } \tilde{h}_2, \tag{5-211}$$

with the same action. Then (5-209) becomes

$$(f_L^{-L} \circ \tilde{h}_1 | \operatorname{Im} \tilde{h}_2) \circ \overline{h}_2 = 1_{U^{\mathbb{N}}}. \tag{5-212}$$

Now,

$$f_L^{-L} \circ \tilde{h}_1 | \operatorname{Im} \tilde{h}_2 \tag{5-213}$$

has become the left inverse of a bijection, namely, $\overline{h}_2$. It is therefore also a right inverse satisfying

$$\overline{h}_2 \circ (f_L^{-L} \circ \tilde{h}_1 | \operatorname{Im} \tilde{h}_2) = 1_{\operatorname{Im} \tilde{h}_2}. \tag{5-214}$$

This equation contains the essence of the desired result, but is not yet in its most recognizable form. Define a function

$$\tilde{f}_L : \operatorname{Im} \tilde{h}_2 \to \operatorname{Im} \tilde{h}_2 \tag{5-215a}$$

with action

$$\tilde{f}_L(y_0, y_1, y_2, \dots) = (e_Y, \dots, e_Y, y_0, y_1, y_2, \dots), \tag{5-215b}$$

and apply it to both members of (5-214). This provides

$$\overline{h}_2 \circ (f_L^{-L} \circ \tilde{h}_1 | \operatorname{Im} \tilde{h}_2 \circ \tilde{f}_L) = \tilde{f}_L. \tag{5-216}$$

Notice that

$$f_L^{-L} \circ \tilde{h}_1 | \operatorname{Im} \tilde{h}_2 \circ \tilde{f}_L = \tilde{h}_1 | \operatorname{Im} \tilde{h}_2. \tag{5-217}$$

Thus,

$$\overline{h}_2 \circ \tilde{h}_1 | \operatorname{Im} \tilde{h}_2 = \tilde{f}_L, \tag{5-218}$$

which states that the interchanged connection acts as a delay L inversion as well, but only for input sequences which are possible output sequences of GRMPS 2. This is an entirely reasonable restriction when GRMPSs 1 and 2 have changed places in Fig. 4.3.

The reverse connection of Fig. 5.12 has also, therefore, a decoupling effect.

One of the best-known examples of this situation occurs when d_2 is an isomorphism, and corresponds to the discussion of Section 5.7, with L equal to zero. In this case

$$\operatorname{Im} \tilde{h}_2 = Y^{\mathbb{N}}, \tag{5-219}$$

and (5-218) becomes a statement of inversion with delay zero, namely,

$$\overline{h}_2 \circ \tilde{h}_1 = 1_{Y^{\mathbb{N}}}. \tag{5-220}$$

The details of this inverse are shown in Fig. 5.13.

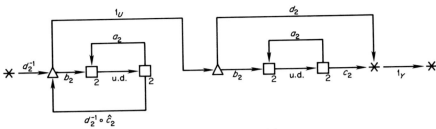

Fig. 5.13. Alternative connection, delay zero case.

If, as was the case in the earlier part of the section, the state of GRMPS 2 is available for measurement, then the diagram can be redrawn in the style of Fig. 5.14. This is a somewhat classic picture of what is now known as state feedback decoupling.

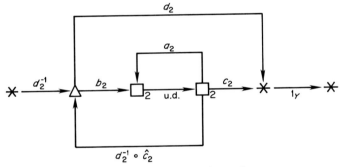

Fig. 5.14. State feedback decoupling.

It is quite straightforward to verify the arguments above for the diagram of Fig. 5.14. The key is the local output equation, which is

$$
\begin{aligned}
y_2 &= c_2 x_2 * d_2 u_2 \\
&= c_2 x_2 * d_2 (d_2^{-1} \circ \hat{c}_2 x_2 \triangle d_2^{-1} \tilde{y}_2) \qquad\qquad (5\text{-}221)\\
&= (c_2 * \hat{c}_2) x_2 * \tilde{y}_2 = \tilde{y}_2,
\end{aligned}
$$

where arguments have been suppressed and where $\tilde{y}_2$ represents the input to the system of Fig. 5.14, which may be visualized as a requested output of GRMPS 2.

Exercises

5.8-1. Why is (5-217) true?

5.8-2. Explain the statement (5-219).

5.9 DISCUSSION

When a set has been equipped with an associative binary operation, it becomes natural in a variety of systems contexts to inquire whether or not the operation has a unit. If it does have a unit, then an essential question follows involving the possible existence of inverses for various set elements under the operation. Chapter 4 has made extensive use of these ideas for the introduction to group morphic systems.

In this chapter, a similar question has been put forward for group morphic systems. Such a system has been defined to be invertible with delay L, provided that its input sequence can be recovered after L operations of the system from a knowledge of the initial condition and of the output sequence. Necessary and sufficient conditions have been stated for the existence of another GRMPS which acts as the inverse. The sufficiency condition was constructive.

The inverse system makes use of state feedback around a model of the state dynamics of the system which is invertible. Because of this fact, a new type of GRMPS interconnection has been studied. Together with the studies of Chapter 4, this study has provided the beginnings of a reasonably comprehensive picture of GRMPS interconnections. While in the context of discussing the state feedback interconnection, the chapter has established that the reachable subgroup is not changed by the process of state feedback.

Almost all of the calculations associated with inverses take place in terms of morphisms. This has provided an opportunity to introduce the notion of a ring, which is then motivated by all the inverse calculations.

A brief introduction to the idea of decoupling with state measurement has been given, and its close relationship to the theory of inverse systems has been illustrated. Calculations to indicate the effect of interchanging the system and its inverse have been made, and these led to a classical diagram involved with state feedback decoupling.

In the last analysis, all the signal groups have been assumed commutative in order to assure various group morphic interpretations. This motivated the definition of a commutative group morphic system.

The structures now in place, namely, the commutative group and the ring of endomorphisms on it, generate an interplay which leads naturally to many of the remaining algebraic structures in common use in system theory. The next chapter begins to develop these structures.

6 INTERCONNECTED SYSTEMS

The idea of connections in system theory arises in a variety of ways. On an intuitive basis, Section 1.3 has pointed out the relationship between connections and the binary operation on a set at the point of connection. Such an idea is at the heart of the local transition equation

$$x_{k+1} = ax_k \,\square\, bu_k \qquad (6\text{-}1)$$

for a GRMPS. The binary operation $\square$ appearing in the right member of (6-1) serves in such a capacity, and the reader should have no difficulty recalling pictorial instances of connection corresponding to (6-1) in the figures of Chapter 5.

Another notion of connection is more involved with the input and output sets of the system. Examples of such connections have been given in Sections 4.3 and 4.4, which dealt with the frequently occurring series and parallel cases. Attention in those sections was focused upon the local dynamical equations of the interconnection, in order to ascertain conditions sufficient to assure GRMP representation of the resulting interconnection. Nonetheless, the connection was established by means of the binary operations on input sets and output sets for the system.

The role of the input and output groups in studies of connections began to make itself more evident in Section 5.8, where the functions

$$h_{ex} : U^{\mathbb{N}} \to Y^{\mathbb{N}} \qquad (6\text{-}2)$$

were used to study the decoupling possibilities associated with two different series connections of a GRMPS and its inverse.

At this point, it is not yet convenient to establish in what sense (6-2) is a morphism. This matter will be resolved in Chapter 7. In the early sections of this chapter, a system will be understood simply as a morphism. The idea will be to refine further the key triangle of Section 4.5 and to apply the result to the problem of simplifying large-scale systems with a view toward preservation of their connection patterns.

6.1   KEY TRIANGLE REVISITED

The status of triangle discussion, as left off by Section 4.5, may be seen in Fig. 6.1. Given are groups $(S, \square, e_S)$ and $(T, *, e_T)$, together with a morphism of groups

$$f : S \to T. \tag{6-3}$$

W is a normal subgroup of S, and

$$p_W : S \to S/W \tag{6-4}$$

is the morphism of groups from S onto its quotient group S/W. The diagram is involved with the question whether or not there exists a unique morphism of groups

$$\bar{f} : S/W \to T \tag{6-5}$$

in such a way that the diagram commutes, that is,

$$f = \bar{f} \circ p_W. \tag{6-6}$$

The answer is yes, provided that

$$W \subset \mathrm{Ker}\, f. \tag{6-7}$$

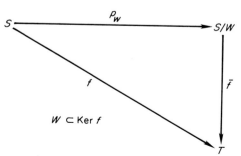

Fig. 6.1. Present status of key triangle.

The reader may find it useful to recall briefly the nature of the quotient group. Basic to the issue was the idea of locating equivalence relations which would make it possible for the key triangle of Section 2.6, which is discussed solely in terms of sets and functions, to be refined to the case in which all the sets admit group structure and all the functions are morphisms of groups. By assuming that the quotient set was a group, and by assuming that the quotient set projection was a morphism of groups, it was determined that the appropriate type of equivalence relation was given by

$$s_1 \, E \, s_2 \qquad \text{if} \qquad s_1 = w \, \square \, s_2 \qquad\qquad \text{(6-8a,b)}$$

for some element w in a normal subgroup W.

Thus the morphism (6-4) is really assigning to each element s in S its equivalence class in S/W. Moreover, S/W is just a subset of $\mathbb{P}(S)$, equipped with a special type of binary operation.

This section looks at the case in which the sets take on ring structure, and in which the functions become morphisms of rings.

The notation for the ring is chosen to be

$$(S, \square, e_S, \triangle, 1_S), \qquad\qquad \text{(6-9)}$$

which is an amalgam of (5-10) with Fig. 6.1. Recall that

$$(S, \square, e_S) \qquad\qquad \text{(6-10)}$$

is a commutative group, if (6-9) is a ring. This suggests that the discussion of key triangles for rings can and must proceed from the key triangle for groups. Moreover, it proceeds from the triangle for commutative groups, which means that every subgroup is normal. Accordingly, it is assumed that W is a subgroup of the commutative group (6-10), and that the kernel of f, understood as a morphism of groups, includes W.

To see what adjustments have to be added for the ring situation, it helps to take a closer look at $\mathrm{Ker}\, f$, where f is a morphism of rings. Adopt the notation

$$(T, *, e_T, \circ, 1_T) \qquad\qquad \text{(6-11)}$$

for the ring which is codomain for f. The statement that f is a morphism of rings means that it is a morphism of groups in the sense of Section 4.1 and a morphism of monoids in the sense of Section 4.2. In concrete terms, this means

$$f(s_1 \, \square \, s_2) = f(s_1) * f(s_2), \qquad\qquad \text{(6-12a)}$$
$$f(s_1 \, \triangle \, s_2) = f(s_1) \circ f(s_2), \qquad\qquad \text{(6-12b)}$$
$$f(1_S) = 1_T, \qquad\qquad \text{(6-12c)}$$

for all s_1 and s_2 in S.

From the fact that f is a morphism of groups, according to (6-12a), it is known that $\operatorname{Ker} f$ is a subgroup. But the fact that f is also a morphism of monoids, and thus satisfies (6-12b), means also that

$$s_1 \in \operatorname{Ker} f \quad \text{and} \quad s_2 \in S \qquad (6\text{-}13a,b)$$

imply

$$s_1 \triangle s_2 \in \operatorname{Ker} f \quad \text{and} \quad s_2 \triangle s_1 \in \operatorname{Ker} f. \qquad (6\text{-}14a,b)$$

To establish this basic result requires showing that

$$e_R \triangle r = r \triangle e_R = e_R \qquad (6\text{-}15)$$

for every element r in a ring

$$(R, \square, e_R, \triangle, 1_R). \qquad (6\text{-}16)$$

It is easy to do this, by means of the distributive properties of (6-16). Begin with the construction

$$r = r \square e_R, \qquad (6\text{-}17a)$$

which follows because of the group structure on $(R, \square, e_R)$. From (6-17a), develop by multiplication

$$r \triangle r = r \triangle (r \square e_R) = (r \triangle r) \square (r \triangle e_R). \qquad (6\text{-}17b,c)$$

Then add the inverse of $r \triangle r$ to both members, so that

$$e_R = r \triangle e_R. \qquad (6\text{-}17d)$$

A similar procedure achieves the other part of (6-15). Armed with the result (6-15), compute (6-14a) by a calculation

$$f(s_1 \triangle s_2) = f(s_1) \circ f(s_2) = e_R \circ f(s_2) \qquad (6\text{-}18a)$$

$$= e_R, \qquad (6\text{-}18b)$$

and likewise for (6-14b). Thus, $\operatorname{Ker} f$ is a subset of S that is a subgroup and is closed under left and right ring multiplication.

As the reader may well have suspected, such subsets receive special designation in the theory of rings. An *ideal I* in the ring (6-16) is a subgroup of $(R, \square, e_R)$ with the property that

$$i \in I \quad \text{and} \quad r \in R \qquad (6\text{-}19a,b)$$

imply

$$r \triangle i \in I \quad \text{and} \quad i \triangle r \in I. \qquad (6\text{-}20a,b)$$

Return now to a discussion of the key triangle of Fig. 6.1. When $f : S \to T$ is a morphism of rings, it is now known that $\operatorname{Ker} f$ is an ideal contained in S.

The reader probably has already conjectured that the natural way to adapt the triangle to the ring case is to require W to be an ideal as well. Indeed, this turns out to be the case.

Recall from Section 4.5 that S/W already has group structure

$$(S/W, \Box, e_{S/W}), \tag{6-21}$$

with

$$e_{S/W} = p_W(e_S) \tag{6-22}$$

and

$$\bar{s}_1 \Box \bar{s}_2 = p_W(s_1 \Box s_2) \tag{6-23a}$$

for $\bar{s}_1$ and $\bar{s}_2$ elements in S/W and s_i in S any elements satisfying

$$p_W(s_i) = \bar{s}_i, \qquad i = 1, 2. \tag{6-23b}$$

If (6-21) can be developed into a ring

$$(S/W, \Box, e_{S/W}, \triangle, 1_{S/W}) \tag{6-24}$$

with (6-4) a morphism of rings, it is certainly necessary that

$$1_{S/W} = p_W(1_S) \tag{6-25}$$

and

$$\bar{s}_1 \triangle \bar{s}_2 = p_W(s_1 \triangle s_2) \tag{6-26}$$

for any s_i satisfying (6-23b).

The verifications are a bit taxing. Nonetheless, they are hardly obvious and thus will be included here.

Begin by checking that (6-26) represents a valid definition of binary operation on S/W. The situation begins in the same spirit as (4-128), as pictured in Fig. 4.8. In place of (4-129a), there is

$$\triangle_E = p_E \circ \triangle \circ (p_E \times p_E)^{-R}; \tag{6-27}$$

but the step (4-132) is replaced by

$$(p_E \circ \triangle)(w_1 \Box s_1, w_2 \Box s_2) = p_E((w_1 \Box s_1) \triangle (w_2 \Box s_2)), \tag{6-28}$$

the argument of the right member of which develops to

$$((w_1 \Box s_1) \triangle w_2) \Box ((w_1 \Box s_1) \triangle s_2)$$
$$= (w_1 \triangle w_2) \Box (s_1 \triangle w_2) \Box (w_1 \triangle s_2) \Box (s_1 \triangle s_2)$$
$$= \tilde{w} \Box (s_1 \triangle s_2), \tag{6-29}$$

where

$$\tilde{w} = (w_1 \triangle w_2) \,\square\, (s_1 \triangle w_2) \,\square\, (w_1 \triangle s_2). \tag{6-30}$$

Now w_1 and w_2 are in the ideal W and s_1 and s_2 are in S. Because an ideal is closed under ring multiplication, the three terms in the right member of (6-30) are also in W; and because an ideal is a subgroup, $\tilde{w}$ is in W. It follows that (6-26) represents a well-defined binary operation $\triangle$ on S/W. $\triangle$ on S/W is associative because $\triangle$ on S is associative. Now let $\bar{s}$ be any element in S/W; then

$$\bar{s} \triangle 1_{S/W} = \bar{s} \triangle p_W(1_S) = p_W(s \triangle 1_S) = p_W(s) = \bar{s}, \tag{6-31}$$

and similarly for $1_{S/W} \triangle \bar{s}$. Thus (6-25) is the unique unit for $\triangle$ on S/W.

Next the distributive laws must be checked on S/W. Multiplication from the left gives

$$\bar{s}_1 \triangle (\bar{s}_2 \,\square\, \bar{s}_3) = p_W(s_1 \triangle (s_2 \,\square\, s_3)) \tag{6-32a}$$
$$= p_W((s_1 \triangle s_2) \,\square\, (s_1 \triangle s_3)) \tag{6-32b}$$
$$= (p_W(s_1 \triangle s_2)) \,\square\, (p_W(s_1 \triangle s_3)) \tag{6-32c}$$
$$= (\bar{s}_1 \triangle \bar{s}_2) \,\square\, (\bar{s}_1 \triangle \bar{s}_3), \tag{6-32d}$$

which establishes distribution of $\triangle$ over $\square$ from the left. A similar calculation shows distribution of $\triangle$ over $\square$ from the right. In following (6-32), it is well to bear in mind that $\square$ on S/W is already known to establish group structure, with p_W a morphism of groups. Then (6-32a) grows out of (6-26) and (4-128). The passage from (6-32a) to (6-32b) is a consequence of ring structure on S.

It is now known that (6-24) is indeed a ring, and that p_W is a morphism of rings. The last step is to verify that

$$\bar{f}: S/W \to T \tag{6-33}$$

is a morphism of rings. Of course, it is already known to be a morphism of groups, from Section 4.5. It remains to show that $\bar{f}$ is a morphism of monoids.

Consider the calculation

$$\bar{f}(1_{S/W}) = \bar{f}(p_W(1_S)) \tag{6-34a}$$
$$= (\bar{f} \circ p_W)(1_S) \tag{6-34b}$$
$$= f(1_S) \tag{6-34c}$$
$$= 1_T. \tag{6-34d}$$

This shows that $\bar{f}$ carries the unit $1_{S/W}$ of $\triangle$ on S/W to the unit 1_T of $\circ$ on T. Next make the computation

$$\bar{f}(\bar{s}_1 \triangle \bar{s}_2) = \bar{f}(p_W(s_1 \triangle s_2)) \qquad (6\text{-}35\text{a})$$
$$= (\bar{f} \circ p_W)(s_1 \triangle s_2) \qquad (6\text{-}35\text{b})$$
$$= f(s_1 \triangle s_2) \qquad (6\text{-}35\text{c})$$
$$= f(s_1) \circ f(s_2) \qquad (6\text{-}35\text{d})$$
$$= (\bar{f} \circ p_W)(s_1) \circ (\bar{f} \circ p_W)(s_2) \qquad (6\text{-}35\text{e})$$
$$= \bar{f}(p_W(s_1)) \circ \bar{f}(p_W(s_2)) \qquad (6\text{-}35\text{f})$$
$$= \bar{f}(\bar{s}_1) \circ \bar{f}(\bar{s}_2). \qquad (6\text{-}35\text{g})$$

This makes it clear that $\bar{f}$ is compatible with $\triangle$ on S/W and $\circ$ on T. Accordingly, $\bar{f}$ is also a morphism of monoids. And so $\bar{f}$ becomes a morphism of rings.

The diagram of Fig. 6.1 has been adapted therefore to the case of rings. The only change in the condition (6-7) is that W must be an ideal contained in Ker f, and not just a subgroup.

Just as $(S/W, \square, e_{S/W})$ is known as the quotient group of S by the normal subgroup W, so (6-24) is called the *quotient ring* of S by the ideal W.

It is important to realize that an ideal is not a ring in its own right. This is a consequence of its failure to meet the monoid requirements with respect to ring multiplication. In turn, this follows because interesting ideals do not include the ring unit 1_R for multiplication. If they did, the ideal would be equal to the ring.

Other types of subsets of rings, however, can have the structure of a ring. A *subring* of the ring (6-16) is a subgroup of $(R, \square, e_R)$ that is closed under $\triangle$ and includes 1_R.

The discussion now passes on to the application of ideals, subrings, and quotient rings to the problem of simplifying large-scale interconnected systems. As an example class, these studies will clarify such issues as the difference between subrings and ideals.

Exercises

6.1-1. Establish that

$$e_R \triangle r = e_R.$$

6.1-2. Under the assumptions (6-13), show that

$$s_2 \triangle s_1 \in \text{Ker } f.$$

6.1-3. In view of the discussion following (6-30), why does it follow that (6-26) represents a well-defined binary operation $\triangle$ on S/W? Explain also how $\triangle$ on S being associative leads to $\triangle$ on S/W being associative.

6.1-4. Corresponding to the calculation (6-31), make the analogous steps for

$$1_{S/W} \triangle \bar{s}.$$

6.1-5. The calculations of (6-32) establish that $\triangle$ on S/W distributes from the left over $\square$ on S/W. Make the related calculation for checking that $\triangle$ distributes over $\square$ from the right.

6.2 NOTIONS OF INTERCONNECTION

In this section, the discussion uses groups and rings, but proceeds much in the same spirit as Section 1.3. For the sake of convenience, however, input and output groups of all systems are chosen to be the same commutative group $(G, \square \, e_G)$. A system, for present purposes, is represented by an endomorphism

$$h: G \to G. \tag{6-36}$$

Of course, h is in essence an input–output system description. According to the covenant of the preceding section, the exact way in which such global functions acquire the morphism property will be considered in the following chapter.

The parallel connection of two systems can be visualized as in Fig. 6.2. The two systems are represented by the endomorphisms

$$h_i: G \to G, \qquad i = 1, 2, \tag{6-37}$$

and their parallel connection is made possible by the binary operation on G, regarded as codomain. Now select an element g in G, regarded as domain. This element acts as an input to both systems, which produce outputs

$$h_i(g), \qquad i = 1, 2, \tag{6-38}$$

that are then combined in the codomain by the action

$$h_1(g) \, \square \, h_2(g). \tag{6-39}$$

Fig. 6.2. Parallel interconnection.

Fig. 6.3. Equivalent for Fig. 6.2.

But the reader has undoubtedly foreseen the desirability of rephrasing (6-39) in the manner

$$(h_1 \,\Box\, h_2)(g),\qquad\qquad(6\text{-}40)$$

from which is suggested the parallel equivalent of Fig. 6.3.

This idea was proposed already in Chapter 1. But its meaning is now much enriched by the algebraic structures which have developed in the interim. Recall at the outset that the step from (6-39) to (6-40) does not depend upon the fact that the systems are described by morphisms. In fact, it could be taken for any functions h_1 and h_2 on G to itself. That was a fundamental message of Section 1.3. This message was developed in Example 4.1-3 and paragraphs following, where it was seen that group structure on the codomain T of T^S induces group structure on T^S in a natural way. The context of this section would make the statement by declaring

$$(\mathbb{EM}(G), \Box, e_{GG})\qquad\qquad(6\text{-}41)$$

a commutative group, where $\mathbb{EM}(G)$ denotes the set of endomorphisms of G. Though a similar assertion can indeed be made for G^G, it will not be studied in the subsequent discussion.

The series connection of two systems can be visualized as in Fig. 6.4. This type of connection is meaningful even when the set G has no algebraic structure whatsoever. It is governed by the composition operation, and in the present case, when G admits the structure of a commutative group, amounts to

$$(\mathbb{EM}(G), \circ, 1_G)\qquad\qquad(6\text{-}42)$$

being a monoid. As a result, there is a series equivalent for Fig. 6.4, which is sketched in Fig. 6.5. Again, the same statement could be made for G^G; but this will not be studied in the sequel.

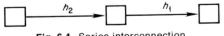

Fig. 6.4. Series interconnection.

Fig. 6.5. Equivalent for Fig. 6.4.

It is worth mentioning that the validity of $\Box$ in (6-41) and $\circ$ in (6-42) being binary operations on $\mathbb{EM}(G)$ depends upon showing that

$$h_1 \,\Box\, h_2 \quad \text{and} \quad h_1 \circ h_2 \qquad \text{(6-43a,b)}$$

are morphisms when h_1 and h_2 are morphisms. These arguments were provided in Section 5.1.

The essential reason for assuming (6-37) to be morphisms has to do with the idea of developing $\mathbb{EM}(G)$ into a ring. Inasmuch as (6-41) is a commutative group even if $\mathbb{EM}(G)$ is replaced by G^G, and inasmuch as (6-42) remains a monoid when $\mathbb{EM}(G)$ is replaced by G^G, whether or not there is a binary operation on G, it is apparent that the morphism assumption has to do with the distributive features of $\circ$ relative to $\Box$. Now Example 5.1-2 has shown that $\circ$ will distribute over $\Box$ from the right, without any use of the morphism property. Moreover, the calculation (5-23) has established that $\circ$ also distributes over $\Box$ from the left, *if* the morphism assumption is employed.

In a sense, then, the morphism assumption is made only so that $\circ$ will distribute over $\Box$ from the left. But this does have the positive result of making

$$(\mathbb{EM}(G), \Box, e_{G^G}, \circ, 1_G) \qquad \text{(6-44)}$$

a ring, and the ring is a very convenient place to begin to develop an intuition for interconnected systems.

The reader has probably surmised by now that a great deal of the essence of the ring structure is involved with the distributive requirements that relate the two binary operations to each other. It is useful to visualize the issue of distributivity in diagrammatic terms. First, consider the fact that $\circ$ distributes over $\Box$ from the right. This can be pictured as in Figs. 6.6 and 6.7. The first of these two figures pictures the left member of

$$(h_1 \,\Box\, h_2) \circ h_3 = (h_1 \circ h_3) \,\Box\, (h_2 \circ h_3), \qquad \text{(6-45)}$$

while the second depicts the right member. Second, consider the fact that $\circ$ distributes over $\Box$ from the left. This is pictured in Figs. 6.8 and 6.9. The

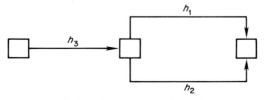

Fig. 6.6. Left member of (6-45).

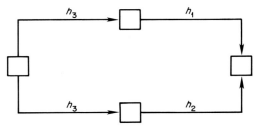

Fig. 6.7. Right member of (6-45).

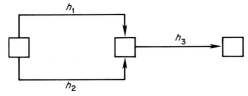

Fig. 6.8. Left member of (6-46).

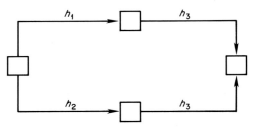

Fig. 6.9. Right member of (6-46).

first of these figures pictures the left member of

$$h_3 \circ (h_1 \,\square\, h_2) = (h_3 \circ h_1) \,\square\, (h_3 \circ h_2), \tag{6-46}$$

with the right member represented by the second.

It is, perhaps, a bit surprising that the diagrams are so closely related to the algebraic expressions. In fact, with proper use of parentheses, very many of these diagrams can be represented by an algebraic expression in the ring (6-44).

EXAMPLE 6.2-1

Consider a diagram such as that shown in Fig. 6.10. Part (a) of the diagram pictures the morphism

$$h_1 \circ (h_2 \,\square\, h_3) \circ h_4, \tag{6-47a}$$

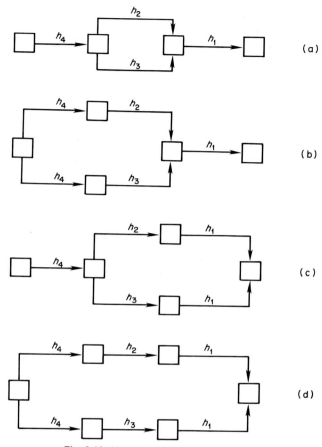

Fig. 6.10. Versions of the ring viewpoint.

in the ring (6-44) built upon the set $\mathbb{EM}(G)$ of endomorphisms of G. Part (b) regards the morphism (6-47a) from another viewpoint, by distributing $\circ$ over $\square$ from the right, namely,

$$h_1 \circ ((h_2 \circ h_4) \,\square\, (h_3 \circ h_4)). \tag{6-47b}$$

Part (c) is similar in spirit to (6-47b), but takes its essence from a distribution of $\circ$ over $\square$ from the left, which converts (6-47a) to

$$((h_1 \circ h_2) \,\square\, (h_1 \circ h_3)) \circ h_4. \tag{6-47c}$$

Finally, there is part (d) of the figure, which may be obtained either from (6-47b), by distribution from the left, or from (6-47c), by distribution from the right. In either case, the result is

$$(h_1 \circ h_2 \circ h_4) \,\Box\, (h_1 \circ h_3 \circ h_4). \tag{6-47d}$$

It is apparent that the diagrams of Fig. 6.10 and the expressions of (6-47) constitute logical alternatives to one another. ∎

EXAMPLE 6.2-2

A different type of diagram is that of Fig. 6.11, which is patterned after the classical notion of feedback. The signal constructed from the binary operation at the right of the loop is denoted g_1, while that constructed from the binary operation at the left is g_2. The signal injected at the left is g_3. The thing to remember about such a configuration is that it need not give rise to a meaningful relationship between g_1 and g_3. This can be seen as follows. The ring equations for Fig. 6.11 are

$$g_1 = h_1 g_2, \qquad g_2 = \hat{h}_2 g_1 \,\Box\, g_3. \tag{6-48a,b}$$

Substitute (6-48b) into (6-48a), obtaining

$$g_1 = h_1(\hat{h}_2 g_1 \,\Box\, g_3) = h_1(\hat{h}_2 g_1) \,\Box\, h_1 g_3$$
$$= h_1(h_2 \hat{g}_1) \,\Box\, h_1 g_3 = (h_1 \circ h_2)\hat{g}_1 \,\Box\, h_1 g_3, \tag{6-49}$$

and rearrange in the manner

$$h_1 g_3 = (h_1 \circ h_2) g_1 \,\Box\, g_1 = g_1 \,\Box\, (h_1 \circ h_2) g_1 = (1_G \,\Box\, (h_1 \circ h_2)) g_1. \tag{6-50}$$

Whether (6-50) has a solution for g_1 in terms of g_3, and whether such a solution will be unique, depends upon whether the morphism

$$1_G \,\Box\, (h_1 \circ h_2) \tag{6-51}$$

has certain image and kernel properties. To avoid these particular issues and assure a general interpretation, it is agreed here that the configuration of Fig. 6.11 is *well defined* if (6-51) is an isomorphism. In that case, there is an inverse morphism

$$(1_G \,\Box\, (h_1 \circ h_2))^{-1} : G \to G \tag{6-52}$$

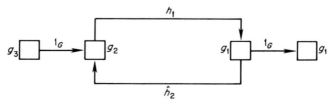

Fig. 6.11. A loop diagram.

which effectively solves (6-50) and establishes

$$g_1 = (1_G \,\square\, (h_1 \circ h_2))^{-1} \circ h_1 g_3. \tag{6-53}$$

∎

It should be carefully noted that (6-52), when it exists, is just another member of $\mathbb{EM}(G)$. Therefore, when Fig. 6.11 is well defined, it is understandable within the ring context.

This section has been designed to bring out the conceptual notion that interconnections in a diagrammatic context bear a close resemblance to carefully articulated elements of rings. The systems discussed insofar as their interconnections are concerned have been described in an input-to-output sense by endomorphisms of commutative groups. It has been shown that series and parallel connections fit naturally into this viewpoint. Moreover, feedback loops are also included when they are well defined.

An interesting special case of the ring interconnection idea occurs when all the endomorphisms are the same. Before the reader writes this remark off as implausible, he or she is invited to examine the next section.

Exercise

6.2. Why does the fact that G admits the structure of a commutative group lead to (6-42) being a monoid?

6.3 IDENTICAL SYSTEMS: SERIES, PARALLEL, INTERPLAYS

The remarks of this section proceed from the viewpoint that a fixed endomorphism

$$h : G \to G \tag{6-54}$$

is given on the commutative group $(G, \square, e_G)$ to itself. A glimpse of the discussion which follows was given in Section 1.4. In that earlier section, however, the symbol **a** was used in place of the symbol h. Moreover, of course, h is an endomorphism, whereas **a** was not so assumed.

It is helpful to begin by establishing some structures which arise when only hs can be used as building blocks.

EXAMPLE 6.3-1

One way, of course, to generate new systems from a supply of given systems described by h is to connect various numbers of them in series.

Thus, a series connection of two *h*s leads to the composite endomorphism

$$h \circ h, \tag{6-55a}$$

which can be written more briefly as

$$h^2, \tag{6-55b}$$

if the composition symbol $\circ$ is suppressed. Along inductive lines, this approach extends quite readily. Suppose that *i* systems have been connected in series in the manner

$$h \circ h \circ \cdots \circ h, \tag{6-56}$$

where (6-56) contains *i* factors of *h*, and the composite endomorphism has been briefly designated by

$$h^i. \tag{6-57}$$

Then a series connection of $i + 1$ systems can be written as

$$h^{i+1} = h \circ h^i = h^i \circ h, \tag{6-58}$$

where the associative nature of the binary operation $\circ$ has been used. This establishes the meaning of the systems

$$h, \quad h^2, \quad h^3, \quad \ldots, \tag{6-59}$$

and explains the notion (6-57) for $i = 1, 2, 3, \ldots$. Can the list (6-59) be extended? The answer is yes. Define

$$h^0 = 1_G; \tag{6-60}$$

then (6-57) is defined for $i \in \mathbb{N}$. For arbitrary *h*, the list does not extend further. However, if *h* is an automorphism, h^{-1} is defined; and an induction

$$h^{-(i+1)} = h^{-1} \circ h^{-i} = h^{-i} \circ h^{-1} \tag{6-61}$$

can be completed. In such a case, the meaning of (6-57) extends further to $i \in \mathbb{Z}$. ∎

Of course, it should be remembered that the construction of h^{-1} from *h* is an interesting issue in itself, as the reader may recall from Chapter 5. However, it is quite useful to assume here that suitable constructions are indeed available for this purpose.

Example 6.3-1 makes use of the binary operation of composition on the ring of endomorphisms. There is an alternative way to construct new systems from *h*s. This alternative method corresponds to parallel connection, and makes use of the binary operation induced on the ring of endomorphisms by the binary operation $\square$ on $G \times G$ to G.

EXAMPLE 6.3-2

Another way to produce new systems from a supply of given systems described by h is to connect them in parallel. In this way, a parallel combination of two hs leads to a composite endomorphism

$$h \square h, \tag{6-62a}$$

which can be designated as

$$(2)h. \tag{6-62b}$$

Here the reader may wish to review briefly the related discussion contained in Section 4.7. Along inductive lines, this idea proceeds naturally. If i systems have been connected in parallel so as to have the composite endomorphism

$$h \square h \square \cdots \square h, \tag{6-63a}$$

where h appears exactly i times as a factor, and if this composite has been designated by

$$(i)h, \tag{6-63b}$$

then a parallel connection of $i + 1$ systems can be constructed by

$$(i + 1)h = h \square (i)h = (i)h \square h, \tag{6-64}$$

where the associative property of $\square$ is in evidence. As in the preceding example, this discussion establishes the meaning of the notation (6-63b) for $i = 1, 2, 3, \ldots$. Again, the notation (6-63b) can be extended to $i \in \mathbb{N}$, by the agreement

$$(0)h = e_{GG}. \tag{6-65}$$

In fact, without further assumption, it can be extended immediately to $i \in \mathbb{Z}$. To accomplish this, define

$$(-1)h = \hat{h}, \tag{6-66}$$

and proceed recursively to

$$(-(i + 1))h = \hat{h} \square (-i)h = (-i)h \square \hat{h}, \tag{6-67}$$

for $i = 1, 2, 3, \ldots$. ∎

Examples 6.3-1 and 6.3-2 employ the two binary operations $\square$ and $\circ$ on the ring

$$(\mathbb{EM}(G), \square, e_{GG}, \circ, 1_G). \tag{6-68}$$

This suggests an intriguing possibility. Can the ring structure (6-68) in some way carry over to interconnections of identical systems? The answer is a qualified yes. A glimpse of the idea is available if the ideas of Examples 6.3-1 and 6.3-2 are interleaved, so as to write down a system of the form

$$(-3)h \,\Box\, (5)h^2 \,\Box\, h^3. \tag{6-69}$$

Such an expression has an intuitive connotation of polynomial in h, which the reader may wish to compare with similar ideas broached in Section 1.4.

To develop an intuitive feeling for the interplays possible in a polynomic context, it will be helpful to discuss in some detail a few examples.

EXAMPLE 6.3-3

Consider a series connection of two hs which is placed in parallel with a single h. The series hs have representation

$$h^2 \qquad \text{and} \qquad h. \tag{6-70a,b}$$

Regard (6-70a) in the manner

$$h_1 = (0)h^0 \,\Box\, (0)h^1 \,\Box\, (1)h^2 \tag{6-71a}$$

and (6-70b) in the manner

$$h_2 = (0)h^0 \,\Box\, (1)h^1 \,\Box\, (0)h^2. \tag{6-71b}$$

Notice that the notations

$$(i)h^j, \qquad j \neq 1, \tag{6-72}$$

are slight extensions of the method of Example 6.3-2 to the case in which a collection of systems h^j, for a fixed j which is not equal to 1, is available. The details of this extension are left to the reader as an exercise. Notice also that the binary operation symbol $\Box$ in the right members of (6-71) describes a function

$$\mathbb{EM}(G) \times \mathbb{EM}(G) \to \mathbb{EM}(G). \tag{6-73}$$

It is of interest to propose the use of this binary operation to form $h_1 \,\Box\, h_2$ in the manner

$$[(0)h^0 \,\Box\, (0)h^1 \,\Box\, (1)h^2] \,\Box\, [(0)h^0 \,\Box\, (1)h^1 \,\Box\, (0)h^2]$$
$$= (0)h^0 \,\Box\, (0)h^0 \,\Box\, (0)h^1 \,\Box\, (1)h^1 \,\Box\, (1)h^2 \,\Box\, (0)h^2 \tag{6-74a}$$
$$= (0+0)h^0 \,\Box\, (0+1)h^1 \,\Box\, (1+0)h^2 \tag{6-74b}$$
$$= (0)h^0 \,\Box\, (1)h^1 \,\Box\, (1)h^2 \tag{6-74c}$$
$$= h \,\Box\, h^2. \tag{6-74d}$$

The step (6-74b) evolves, of course, from Example 6.3-2. It suggests a straightforward way to form a group operation on polynomials in h. ∎

EXAMPLE 6.3-4

Suppose that a pair of polynomials

$$h_1 = h \,\square\, h^2 \qquad \text{and} \qquad h_2 = h \,\square\, h^3 \qquad (6\text{-}75\text{a,b})$$

are given. Consider the construction of $h_1 \circ h_2$ along the lines

$$h_1 \circ h_2 = [h \,\square\, h^2] \circ [h \,\square\, h^3] \qquad (6\text{-}76\text{a})$$

$$= [(0)h^0 \,\square\, (1)h^1 \,\square\, (1)h^2] \circ [(0)h^0 \,\square\, (1)h^1 \,\square\, (0)h^2 \,\square\, (1)h^3] \quad (6\text{-}76\text{b})$$

$$= [(0)h^0 \,\square\, (1)h^1 \,\square\, (1)h^2] \circ (0)h^0$$
$$\square\, [(0)h^0 \,\square\, (1)h^1 \,\square\, (1)h^2] \circ (1)h^1$$
$$\square\, [(0)h^0 \,\square\, (1)h^1 \,\square\, (1)h^2] \circ (0)h^2$$
$$\square\, [(0)h^0 \,\square\, (1)h^1 \,\square\, (1)h^2] \circ (1)h^3 \qquad (6\text{-}76\text{c})$$

$$= [(0)h^0] \circ [(0)h^0] \,\square\, [(1)h^1] \circ [(0)h^0] \,\square\, [(1)h^2] \circ [(0)h^0]$$
$$\square\, [(0)h^0] \circ [(1)h^1] \,\square\, [(1)h^1] \circ [(1)h^1] \,\square\, [(1)h^2] \circ [(1)h^1]$$
$$\square\, [(0)h^0] \circ [(0)h^2] \,\square\, [(1)h^1] \circ [(0)h^2] \,\square\, [(1)h^2] \circ [(0)h^2]$$
$$\square\, [(0)h^0] \circ [(1)h^3] \,\square\, [(1)h^1] \circ [(1)h^3] \,\square\, [(1)h^2] \circ [(1)h^3].$$
$$(6\text{-}76\text{d})$$

The presence of terms of the form

$$[(i)h^m] \circ [(j)h^n] \qquad (6\text{-}77)$$

is somewhat of a novelty. However, it is not difficult to see what is happening. For illustrative purposes, calculate

$$[(2)h^3] \circ [(3)h^2]. \qquad (6\text{-}78)$$

By Examples 6.3-1 and 6.3-2 and the extensions suggested in Example 6.3-3, (6-78) means

$$[h^3 \,\square\, h^3] \circ [h^2 \,\square\, h^2 \,\square\, h^2]$$
$$= [h^3 \,\square\, h^3] \circ h^2 \,\square\, [h^3 \,\square\, h^3] \circ h^2 \,\square\, [h^3 \,\square\, h^3] \circ h^2$$
$$= h^3 \circ h^2 \,\square\, h^3 \circ h^2 \,\square\, h^3 \circ h^2 \,\square\, h^3 \circ h^2 \,\square\, h^3 \circ h^2 \,\square\, h^3 \circ h^2$$
$$= (6)h^5, \qquad (6\text{-}79)$$

from the fact that

$$h^3 \circ h^2 = [h \circ h \circ h] \circ [h \circ h], \qquad (6\text{-}80)$$

together with the observation that composition is associative. Thus the meaning of (6-77) is not in question, being

$$(i \cdot j)h^{m+n}, \qquad (6\text{-}81)$$

where $(i \cdot j)$ represents the familiar multiplication of natural numbers, the complete proof of which is a worthwhile exercise. With the aid of the pair

(6-77) and (6-81), (6-76d) develops to

$$h_1 \circ h_2 = (0 \cdot 0)h^{0+0} \,\square\, (1 \cdot 0)h^{1+0} \,\square\, (1 \cdot 0)h^{2+0}$$
$$\square\, (0 \cdot 1)h^{0+1} \,\square\, (1 \cdot 1)h^{1+1} \,\square\, (1 \cdot 1)h^{2+1}$$
$$\square\, (0 \cdot 0)h^{0+2} \,\square\, (1 \cdot 0)h^{1+2} \,\square\, (1 \cdot 0)h^{2+2}$$
$$\square\, (0 \cdot 1)h^{0+3} \,\square\, (1 \cdot 1)h^{1+3} \,\square\, (1 \cdot 1)h^{2+3}$$
$$= h^2 \,\square\, h^3 \,\square\, h^4 \,\square\, h^5. \tag{6-82}$$

The details of this example outline a viewpoint which may permit the establishment of a monoid operation on polynomials in h. ∎

This section has considered the idea of Section 6.2 preceding, but applied to the case in which the supply of building blocks for the interconnection is limited to identical systems h. From series connections of these hs, new systems

$$h^i, \qquad i \in \mathbb{N}, \tag{6-83}$$

have been found in general, with extension to $i \in \mathbb{Z}$, where h is an automorphism. From parallel connections of the hs, systems

$$(i)h, \qquad i \in \mathbb{Z}, \tag{6-84}$$

develop in general. The notions (6-83) and (6-84) can be combined in the manner

$$(i)h^j, \tag{6-85}$$

by analogous arguments.

Accordingly, there arises the opportunity to define polynomials in h as even more general classes of interconnections of identical elements. Examples 6.3-3 and 6.3-4 have illustrated a possible way to handle parallel and series connections of systems represented by polynomials in h. Especially in Example 6.3-4, the reader has undoubtedly taken note of the distributive laws from the endomorphism ring being put into play.

All this suggests that Section 6.2 may have an analog for the case in which the connections involve only identical systems. To see the analog, it will be necessary to establish ring structure on the polynomials in h. This is the purpose of the next section.

Exercises

6.3-1. In place of the single fixed endomorphism (6-54), suppose instead that two endomorphisms

$$h_i : G \to G, \qquad i = 1, 2,$$

are available, and suppose further that

$$h_1 \circ h_2 = h_2 \circ h_1.$$

How does Example 6.3-1 generalize in this case? Comment upon the increase in difficulty and complexity which could arise if

$$h_1 \circ h_2 \neq h_2 \circ h_1.$$

6.3-2. Again assume two endomorphisms as in Exercise 6.3-1, and discuss the adjustments which will have to be made in the details of Example 6.3-2.

6.3-3. Extend the discussions of this section to the case (6-72), involving interconnections of like systems h^j for a fixed integer $j > 1$.

6.3-4. Examine in detail the relationship between (6-77) and (6-81).

6.4 IDENTICAL SYSTEMS: THE RING STRUCTURE

Generally expressed, the goal of this section is to complete the technical details omitted in the preceding section and to offer interpretations thereon. The approach is to give an accurate definition of polynomials in h and then to establish that the set of such objects is a subring of the endomorphism ring on the commutative group $(G, \square, e_G)$.

Begin with the definition. Consider the formal construction of a parallel connection of hs and h-composites h^i which may contain more than a finite number of branches. Such a construction could be described by

$$(k_0)h^0 \,\square\, (k_1)h^1 \,\square\, (k_2)h^2 \,\square\, \cdots, \qquad (6\text{-}86)$$

where each $k_i \in \mathbb{Z}$. The object indicated in (6-86) is called a *formal power series* in h over the coefficient ring $\mathbb{Z}$. To recall the ring structure on $\mathbb{Z}$, the reader may wish to re-examine Example 5.1-1. The adjective formal is attached to this series because there is no guarantee that (6-86) actually turns out to be a morphism. To see this, write (6-86) in the alternative form

$$\underset{j \in \mathbb{N}}{\square}\,(k_j)h^j \qquad (6\text{-}87)$$

and consider its potential action

$$\left\{\underset{j \in \mathbb{N}}{\square}\,(k_j)h^j\right\}(g) \qquad (6\text{-}88)$$

on an element g in G. The natural development

$$\underset{j \in \mathbb{N}}{\square}\,\left[\{(k_j)h^j\}(g)\right] \qquad (6\text{-}89)$$

of (6-88) presents a difficulty in computation, because, although each

$$\{(k_j)h^j\}(g) \in G, \qquad (6\text{-}90)$$

there is no convenient way in the present context to assure that the recursive action of $\square$ on a nonfinite number of elements in G turns out to be another element in G. Thus it cannot be directly asserted that (6-87) is an element of $\mathbb{EM}(G)$.

However, there is a standard way to overcome this difficulty, and that is to require that (6-87) have the property of *finite support*, which means that only a finite number of the k_j are nonzero in $\mathbb{Z}$. In this fashion, it then follows directly that (6-87) is an endomorphism of G, if it is agreed that the result of applying (6-87) to an element of G is determined by accounting only for a finite number of terms, including those in which k_j is nonzero. To denote the finite support property on a formal series (6-87), use the notation

$$\boxed{\sim}_{j \in \mathbb{N}} (k_j)h^j. \tag{6-91}$$

For the formal power series in h over $\mathbb{Z}$ which have finite support, employ the notation $\mathbb{Z}[h]$ introduced in Chapter 1. Notice that the formal power series of finite support are then a natural way of discussing polynomials in h. Thus,

$$\mathbb{Z}[h] = \left\{ \boxed{\sim}_{j \in \mathbb{N}} (k_j)h^j \,\middle|\, k_j \in \mathbb{Z} \right\} \tag{6-92}$$

completes the definition mentioned at the start of this section.

Clearly,

$$\mathbb{Z}[h] \subset \mathbb{EM}(G). \tag{6-93}$$

Recall from Section 6.1 that $\mathbb{Z}[h]$ must satisfy three properties in order to qualify as a subring of $\mathbb{EM}(G)$. First,

$$(\mathbb{Z}[h], \Box, e_{GG}) \quad \text{must be a subgroup of} \quad (\mathbb{EM}(G), \Box, e_{GG}). \tag{6-94a,b}$$

Second, $\mathbb{Z}[h]$ must be closed under composition; and third, $\mathbb{Z}[h]$ must include 1_G.

Begin with the first condition. Notice that (6-94a) requires that

$$e_{GG} \in \mathbb{Z}[h]. \tag{6-95}$$

Is this the case? Well, as a unit for the binary operation $\Box$ on the commutative group (6-94b), e_{GG} is unique. Moreover, the action of e_{GG} on an element in G is exactly the same as the action of

$$\boxed{\sim}_{j \in \mathbb{N}} (0)h^j \tag{6-96}$$

on that element. Thus, by virtue of the discussion on functions in Chapter 2, (6-95) and (6-96) have the same domain, codomain, and action; therefore they may be identified. Next it is necessary to show that (6-94a) is a subgroup of (6-94b). Because (6-94a) is not empty, it suffices to show (6-94a) closed under $\Box$ and under inversion. These details may be reviewed in (4-34).

Consider two elements in (6-94a), say

$$\underset{j \in \mathbb{N}}{\boxed{\sim}} (k_j^i) h^j, \qquad i = 1, 2. \tag{6-97}$$

It is required to form

$$\left\{ \underset{j \in \mathbb{N}}{\boxed{\sim}} (k_j^1) h^j \right\} \square \left\{ \underset{j \in \mathbb{N}}{\boxed{\sim}} (k_j^2) h^j \right\}. \tag{6-98}$$

It will be helpful to explain in detail how this is done. Because the elements (6-97) have finite support, there is natural number M such that the actions of (6-97) are the same as the actions of

$$\overset{M}{\underset{j=0}{\boxed{}}} (k_j^i) h^j, \qquad i = 1, 2. \tag{6-99}$$

Then (6-98) can be computed by

$$\left\{ \overset{M}{\underset{j=0}{\boxed{}}} (k_j^1) h^j \right\} \square \left\{ \overset{M}{\underset{j=0}{\boxed{}}} (k_j^2) h^j \right\}. \tag{6-100}$$

From the commutative and associative properties of $\square$, (6-100) can be rearranged in the manner

$$\overset{M}{\underset{j=0}{\boxed{}}} \{(k_j^1) h^j \square (k_j^2) h^j\}. \tag{6-101}$$

Extending the ideas of Example 6.3-2 from h to h^j, and making use of the additive group structure on $\mathbb{Z}$, it follows that (6-101) can become

$$\overset{M}{\underset{j=0}{\boxed{}}} (k_j^1 + k_j^2) h^j, \tag{6-102}$$

which identifies easily with an element of $\mathbb{Z}[h]$ by the process of providing additional terms

$$\overset{\infty}{\underset{j=M+1}{\boxed{}}} (0) h^j. \tag{6-103}$$

So (6-94a) is closed under the binary operation $\square$.

To complete the argument that (6-94a) is a subgroup of (6-94b), it remains to show that (6-94a) is closed under inversion. By virtue of the treatment of (6-98), it is now an easy exercise to verify that

$$\underset{j \in \mathbb{N}}{\boxed{\sim}} (-k_j) h^j \tag{6-104}$$

is an inverse for an element (6-91). Notice again that the additive group structure on $\mathbb{Z}$ is employed in (6-104). Because $(\mathbb{Z}, +, 0)$ is a group, $k_j \in \mathbb{Z}$ implies that $-k_j \in \mathbb{Z}$. Thus the inverse (6-104) is in $\mathbb{Z}[h]$ as well.

This means that (6-94a) is indeed a subgroup of (6-94b).

Turn now to the second condition in the subring study. This condition says that $\mathbb{Z}[h]$ must be closed under composition. Reconsider (6-97), and rewrite (6-98) in the manner

$$\left\{ \bigsqcup_{j \in \mathbb{N}} (k_j^1) h^j \right\} \circ \left\{ \bigsqcup_{j \in \mathbb{N}} (k_j^2) h^j \right\}. \qquad (6\text{-}105)$$

Once again make use of (6-99), so that a form

$$\left\{ \prod_{j=0}^{M} (k_j^1) h^j \right\} \circ \left\{ \prod_{i=0}^{M} (k_i^2) h^i \right\} \qquad (6\text{-}106)$$

analogous to (6-100) is obtained. Here it is necessary to make use of the distributive laws on the endomorphism ring. Then (6-106) develops to

$$\prod_{i=0}^{M} \left\{ \prod_{j=0}^{M} (k_j^1) h^j \right\} \circ \{(k_i^2) h^i\}. \qquad (6\text{-}107)$$

Yet another use of the distributive law converts (6-107) to

$$\prod_{i=0}^{M} \prod_{j=0}^{M} \{(k_j^1) h^j\} \circ \{(k_i^2) h^i\}, \qquad (6\text{-}108)$$

which from the ideas of Example 6.3-4 and the multiplicative monoid structure on $\mathbb{Z}$ becomes

$$\prod_{i=0}^{M} \prod_{j=0}^{M} (k_j^1 \cdot k_i^2) h^{j+i}. \qquad (6\text{-}109)$$

Now (6-109) can be identified with an element of the form

$$\prod_{m=0}^{2M} (k_m) h^m \qquad (6\text{-}110)$$

in a straightforward way. Indeed, the coefficient k_m in (6-110) is computed from (6-109) by considering the various ways in which i and j can sum to m. This gives

$$k_m = \sum_{j=0}^{m} (k_j^1 \cdot k_{m-j}^2). \qquad (6\text{-}111)$$

Finally, (6-110) identifies with an element of $\mathbb{Z}[h]$ by adjoining the terms

$$\prod_{m=2M+1}^{\infty} (0) h^m. \qquad (6\text{-}112)$$

Therefore $\mathbb{Z}[h]$ is closed under composition.

It remains to establish the third condition, which is that $\mathbb{Z}[h]$ must include 1_G. This, though last, is not the least interesting. Recall from Example

6.3-1 the identification

$$h^0 = 1_G.$$ (6-113)

Then define an element

$$\underset{j \in \mathbb{N}}{\boxed{\sim}} (k_j) h^j \in \mathbb{Z}[h]$$ (6-114a)

by the conditions

$$k_0 = 1, \qquad k_j = 0, \qquad j \neq 1.$$ (6-114b)

Then (6-114) identifies with (6-113); and the third condition is satisfied.

Accordingly, it has been established that $\mathbb{Z}[h]$ is a subring of $\mathbb{EM}(G)$. More technically,

$$(\mathbb{Z}[h], \square, e_{GG}, \circ, 1_G)$$ (6-115)

is a subring of

$$(\mathbb{EM}(G), \square, e_{GG}, \circ, 1_G).$$ (6-116)

This means that the discussions of Section 6.2 on interconnections and rings carry over with equal force to the case of interconnections of identical systems.

Moreover, though the case of identical systems appears at first glance to be rather special, this type of situation carries with it considerable applicability in systems theory and application. This assertion is advanced a bit farther in the subsequent section.

Exercises

6.4-1. Extend the conclusions of this section from one endomorphism h to two endomorphisms h_1 and h_2. Develop a ring structure $\mathbb{Z}[h_1, h_2]$ of polynomials in h_1 and h_2 and having integer coefficients. For simplicity, assume that

$$h_1 \circ h_2 = h_2 \circ h_1,$$

but comment on the case in which this equality may not hold.

6.4-2. In the light of your results from Exercise 6.4-1, reconsider Example 6.2-2 and determine the conditions under which it can be discussed completely within the ring $\mathbb{Z}[h_1, h_2]$.

6.5 EXTENSION

Section 6.4 presented a reasonably thorough discussion of the assertion that $\mathbb{Z}[h]$ is a subring of $\mathbb{EM}(G)$. However, it turns out that the use of $\mathbb{Z}$ as a coefficient ring for the polynomials is a bit restrictive. This section points out the possibility of generalization on this point and exemplifies an application of the generalization. In particular, it is intended to begin with a focus upon the set

$$\mathbb{EM}(G)[h]. \tag{6-117}$$

Again, h is a fixed endomorphism of the commutative group $(G, \square, e_G)$.

Consider a typical candidate for a term in the polynomial set (6-117). It would take the form

$$\alpha_i \circ h^i, \tag{6-118}$$

where $\alpha_i \in \mathbb{EM}(G)$. Thus (6-118) belongs to $\mathbb{EM}(G)$. Proceed to observe further that

$$\mathbb{EM}(G)[h] = \left\{ \underset{j \in \mathbb{N}}{\boxed{\sim}} \, \alpha_j h^j \,\middle|\, \alpha_j \in \mathbb{EM}(G) \right\}, \tag{6-119}$$

where the composition of (6-118) has been suppressed, is properly defined, and that

$$\mathbb{EM}(G)[h] \subset \mathbb{EM}(G). \tag{6-120}$$

The same task, then, is at hand, namely, to find out whether (6-119) is a subring of the endomorphism ring. The investigations are similar in spirit to those of Section 6.4, and so the generalizations are just sketched here.

With regard to closure under $\square$, consider the situation

$$\left\{ \underset{j \in \mathbb{N}}{\boxed{\sim}} \, \alpha_j^1 h^j \right\} \square \left\{ \underset{j \in \mathbb{N}}{\boxed{\sim}} \, \alpha_j^2 h^j \right\}. \tag{6-121}$$

As before, this can be rewritten as

$$\left\{ \underset{j=0}{\overset{M}{\boxed{\square}}} \, \alpha_j^1 h^j \right\} \square \left\{ \underset{j=0}{\overset{M}{\boxed{\square}}} \, \alpha_j^2 h^j \right\} \tag{6-122}$$

for a suitable $M \in \mathbb{N}$. Again, $\square$ is commutative and associative. Thus, (6-122) rewrites to

$$\underset{j=0}{\overset{M}{\boxed{\square}}} \, \{ \alpha_j^1 h^j \square \alpha_j^2 h^j \}. \tag{6-123}$$

But

$$\begin{aligned} (\alpha_j^1 h^j \square \alpha_j^2 h^j)(g) &= (\alpha_j^1 h^j)(g) \square (\alpha_j^2 h^j)(g) \\ &= \alpha_j^1 (h^j g) \square \alpha_j^2 (h^j g) = (\alpha_j^1 \square \alpha_j^2)(h^j g). \end{aligned} \tag{6-124}$$

So (6-123) becomes

$$\underset{j=0}{\overset{M}{\boxed{\square}}} \, (\alpha_j^1 \square \alpha_j^2) h^j, \tag{6-125}$$

which may then be identified with an element of (6-119) by adjoining the formal series tail

$$\coprod_{j=M+1}^{\infty} e_{GG}h^j. \tag{6-126}$$

The unit e_{GG} is of course identified with

$$\widetilde{\coprod_{j\in\mathbb{N}}} e_{GG}h^j; \tag{6-127}$$

the inverse of an element

$$\widetilde{\coprod_{j\in\mathbb{N}}} \alpha_j h^j \tag{6-128a}$$

is found in the manner

$$\widetilde{\coprod_{j\in\mathbb{N}}} \hat{\alpha}_j h^j, \tag{6-128b}$$

where $\hat{\alpha}_j$ is the inverse of α_j under $\square$ in the endomorphism ring. Notice that the group structure on $(\mathbb{EM}(G), \square, e_{GG})$ assures that (6-128b) is an element of (6-119).

For closure under composition, the key construction is

$$\left\{\coprod_{j=0}^{M} \alpha_j^1 h^j\right\} \circ \left\{\coprod_{i=0}^{M} \alpha_i^2 h^i\right\} = \coprod_{i=0}^{M} \left\{\coprod_{j=0}^{M} \alpha_j^1 h^j\right\} \circ \{\alpha_i^2 h^i\}$$

$$= \coprod_{i=0}^{M} \coprod_{j=0}^{M} \{\alpha_j^1 h^j\} \circ \{\alpha_i^2 h^i\}. \tag{6-129}$$

Here it is seen that the desirable situation

$$\{\alpha_j^1 h^j\} \circ \{\alpha_i^2 h^i\} = (\alpha_j^1 \circ \alpha_i^2)h^{j+i} \tag{6-130}$$

does *not* occur in general. Thus $\mathbb{EM}(G)[h]$ is *not* a subring of the endomorphism ring on G.

Recall, however, that h is fixed. Now examine the set

$$S_h = \{\alpha \mid \alpha \in \mathbb{EM}(G) \text{ and } \alpha \circ h = h \circ \alpha\} \tag{6-131}$$

of endomorphisms which *commute* with h. It follows directly that

$$e_{GG} \in S_h. \tag{6-132}$$

Moreover, if $\alpha_i \in S_h$ for $i = 1, 2$, then

$$(\alpha_1 \square \alpha_2) \circ h = (\alpha_1 \circ h) \square (\alpha_2 \circ h) = (h \circ \alpha_1) \square (h \circ \alpha_2)$$

$$= h \circ (\alpha_1 \square \alpha_2), \tag{6-133}$$

so that S_h is closed under $\square$. Also, if $\alpha \in S_h$, then

$$(\hat{\alpha} \circ h)(g) = \hat{\alpha}(hg) \tag{6-134a}$$
$$= \widehat{\alpha(hg)} \tag{6-134b}$$
$$= \overline{(\alpha \circ h)(g)} \tag{6-134c}$$
$$= (\alpha \circ h)(\hat{g}) \tag{6-134d}$$
$$= (h \circ \alpha)(\hat{g}) \tag{6-134e}$$
$$= h(\alpha \hat{g}) \tag{6-134f}$$
$$= h(\widehat{\alpha g}) \tag{6-134g}$$
$$= h(\hat{\alpha} g) \tag{6-134h}$$
$$= (h \circ \hat{\alpha})(g) \tag{6-134i}$$

for all g in G, so that S_h is closed under inversion relative to $\square$. This establishes that S_h is a subgroup of $\mathbb{EM}(G)$. When $\alpha_i \in S_h$, $i = 1, 2$, it follows that

$$(\alpha_1 \circ \alpha_2) \circ h = \alpha_1 \circ (\alpha_2 \circ h) = \alpha_1 \circ (h \circ \alpha_2)$$
$$= (\alpha_1 \circ h) \circ \alpha_2 = (h \circ \alpha_1) \circ \alpha_2$$
$$= h \circ (\alpha_1 \circ \alpha_2), \tag{6-135}$$

so that S_h is closed under composition; $1_G \in S_h$ by inspection.

This means that S_h is a subring of the endomorphism ring $\mathbb{EM}(G)$. Accordingly, the set

$$S_h[h] \tag{6-136}$$

suggests itself in place of (6-117). The calculation (6-129) would then proceed to

$$\prod_{i=0}^{M} \prod_{j=0}^{M} (\alpha_j^1 \circ \alpha_i^2) h^{j+i} \tag{6-137}$$

because (6-130) would become valid. Moreover, 1_G can be identified with

$$\underset{j \in \mathbb{N}}{\boxed{\sim}} \alpha_j h^j \in S_h[h] \tag{6-138}$$

with

$$\alpha_0 = 1_G, \qquad \alpha_j = e_{GG}, \quad j \neq 1. \tag{6-139}$$

Up to minor embellishments, then, it has been established that

$$(S_h[h], \square, e_{GG}, \circ, 1_G) \tag{6-140}$$

is a subring of the endomorphism ring.

It is appropriate at this point to illustrate an important case in which h and S_h are nontrivial.

EXAMPLE 6.5-1

Let $(G, \square, e_G)$ be a commutative group. Then it follows from the discussion of Example 4.1-3 that

$$(G^{\mathbb{Z}}, \square, e_{G^{\mathbb{Z}}}) \tag{6-141}$$

is a commutative group as well. A typical element of (6-141) can be visualized as a sequence

$$\cdots, \quad g_{-2}, \quad g_{-1}, \quad g_0, \quad g_1, \quad g_2, \quad \cdots. \tag{6-142}$$

For the basic group of the example, use (6-141). Then the endomorphism ring of interest is denoted

$$(\mathbb{EM}(G^{\mathbb{Z}}), \square, e_{(G^{\mathbb{Z}})(G^{\mathbb{Z}})}, \circ, 1_{G^{\mathbb{Z}}}), \tag{6-143}$$

where the unit for $\square$ uses notation which recalls the examples of Section 1.2. Next select a fixed endomorphism h from (6-143). For this purpose, consider the unit delay notation used in previous chapters. Define a *unit delay function*

$$\text{u.d.}: G^{\mathbb{Z}} \rightarrow G^{\mathbb{Z}} \tag{6-144}$$

by the action

$$(\text{u.d.}(f))(k) = f(k-1). \tag{6-145}$$

In this action equation, note carefully that

$$f \in G^{\mathbb{Z}} \tag{6-146}$$

is the element upon which u.d. is acting. Is u.d. a morphism? The calculation

$$\begin{aligned}
(\text{u.d.}(f_1 \square f_2))(k) &= (f_1 \square f_2)(k-1) \\
&= f_1(k-1) \square f_2(k-1) \\
&= (\text{u.d.}(f_1))(k) \square (\text{u.d.}(f_2))(k)
\end{aligned} \tag{6-147}$$

establishes this property. It now remains to determine $S_{\text{u.d.}}$. Let

$$\beta: G^{\mathbb{Z}} \rightarrow G^{\mathbb{Z}} \tag{6-148}$$

be an endomorphism. Then what are the conditions for

$$\beta \circ \text{u.d.} = \text{u.d.} \circ \beta \tag{6-149}$$

to hold? This is not a brief question. Let $f \in G^{\mathbb{Z}}$ be a given sequence. Define an example β by

$$(\beta(f))(2) = \alpha_1(f(0)) \square \alpha_2(f(1)), \tag{6-150a}$$

for

$$\alpha_i: G \to G, \qquad i = 1, 2, \tag{6-150b}$$

endomorphisms of G, and

$$(\beta(f))(j) = e_G, \qquad j \neq 2. \tag{6-150c}$$

It follows that β is a morphism because

$$\begin{aligned}
(\beta(f_1 \,\square\, f_2))(2) &= \alpha_1((f_1 \,\square\, f_2)(0)) \,\square\, \alpha_2((f_1 \,\square\, f_2)(1)) \\
&= \alpha_1(f_1(0) \,\square\, f_2(0)) \,\square\, \alpha_2(f_1(1) \,\square\, f_2(1)) \\
&= \alpha_1(f_1(0)) \,\square\, \alpha_1(f_2(0)) \,\square\, \alpha_2(f_1(1)) \,\square\, \alpha_2(f_2(1)) \\
&= \alpha_1(f_1(0)) \,\square\, \alpha_2(f_1(1)) \,\square\, \alpha_1(f_2(0)) \,\square\, \alpha_2(f_2(1)) \\
&= (\beta(f_1))(2) \,\square\, (\beta(f_2))(2), \tag{6-151}
\end{aligned}$$

with a similar calculation for (6-150c). Now

$$((\text{u.d.} \circ \beta)(f))(3) = \alpha_1(f(0)) \,\square\, \alpha_2(f(1)); \tag{6-152a}$$

but

$$((\beta \circ \text{u.d.})(f))(3) = e_G, \tag{6-152b}$$

so that

$$S_{\text{u.d.}} \neq \mathbb{EM}(G^{\mathbb{Z}}). \tag{6-153}$$

Instead of attempting to elaborate $S_{\text{u.d.}}$ completely, consider the subset

$$S \subset \mathbb{EM}(G^{\mathbb{Z}}) \tag{6-154a}$$

defined by $s \in S$ if

$$(s(f))(k) = \alpha(f(k)), \qquad k \in \mathbb{Z}, \tag{6-154b}$$

for some $\alpha \in \mathbb{EM}(G)$. In effect, such an s would take the sequence (6-142) to the sequence

$$\dots, \quad \alpha g_{-2}, \quad \alpha g_{-1}, \quad \alpha g_0, \quad \alpha g_1, \quad \alpha g_2, \quad \dots . \tag{6-155}$$

Two features of S are noteworthy. First, it is a subring of $\mathbb{EM}(G^{\mathbb{Z}})$; this feature is left as an exercise. Second,

$$S \subset S_{\text{u.d.}} . \tag{6-156}$$

This follows from

$$\begin{aligned}
((\text{u.d.} \circ s)(f))(k) &= \alpha(f(k-1)), \\
((s \circ \text{u.d.})(f))(k) &= \alpha(f(k-1)). \tag{6-157}
\end{aligned}$$

Accordingly, $S[\text{u.d.}]$ is a system class that is definitely usable in the context of this chapter. It is sometimes useful to denote the element s defined by (6-154b) by s_α, as will be seen in the following example. ∎

EXAMPLE 6.5-2

Consider the local transition equation

$$x_{k+1} = ax_k, \tag{6-158}$$

where $a: X \to X$ is an endomorphism of the commutative group $(X, \square, e_X)$. Define

$$x \in X^\mathbb{Z} \quad \text{by} \quad x(k) = x_k, \quad k \in \mathbb{Z}. \tag{6-159a,b}$$

Then a generates an element s_a of S, and

$$(\text{u.d.}^{-1}(x))(k) = x(k+1) = x_{k+1}, \tag{6-160}$$

so that the equation (6-158) can also be understood in the manner

$$\text{u.d.}^{-1}x = s_a x. \tag{6-161}$$

Adjoining $\widehat{s_a x}$ to both members brings out the fact that

$$e_{X^\mathbb{Z}} = \text{u.d.}^{-1}x \,\square\, \widehat{s_a x} = \text{u.d.}^{-1}x \,\square\, \hat{s}_a x$$
$$= (\text{u.d.}^{-1} \,\square\, \hat{s}_a)x = (\hat{s}_a \,\square\, \text{u.d.}^{-1})x. \tag{6-162}$$

A moment's reflection reveals that

$$\hat{s}_a \,\square\, \text{u.d.}^{-1} \in S[\text{u.d.}^{-1}], \tag{6-163}$$

which begins to bring out the role of this idea in GRMPSs. ∎

This section has extended the notions of Section 6.4 to the ring $S_h[h]$. Here h is a fixed endomorphism on a commutative group $(G, \square, e_G)$ to itself. S_h is the subring of $\mathbb{EM}(G)$ containing endomorphisms which commute with h. The qualitative idea is that an interconnection of the h systems can be formed in which the scalars or coefficients may be taken as endomorphisms themselves.

For illustration, the unit delay function u.d. was taken as an example of h on the group

$$(G^\mathbb{Z}, \square, e_{G^\mathbb{Z}}). \tag{6-164}$$

Though $S_{\text{u.d.}}$ was not completely explored, a ring S was established such that $S[\text{u.d.}]$ was suitable for study of certain local dynamical equations associated with GRMPSs.

The reader has probably surmised from Example 6.5-2 that there is a possibility to go further now into the notions of global transition and output.

Indeed, sequences such as those studied in Example 6.5-1 are very helpful in this regard. More will be presented along these lines in Chapter 7.

Exercises

6.5-1.   Determine that the extensions of this section to the ring

$$S_h[h]$$

are appropriate generalizations of Section 6.4 by establishing that

$$\mathbb{Z} \subset S_h$$

is a subring in a natural way.

6.5-2.   Consider the subset S defined by (6-154). Show that S is a subring of

$$\mathbb{EM}(G^{\mathbb{Z}}).$$

6.5-3.   Refer once again to the subset S defined by (6-154). Show that S is a subring of

$$S_{\text{u.d.}}.$$

6.6   A DESCRIPTION SIMPLIFICATION FOR SYSTEMS

The topic of order reduction was examined early in the book as Section 2.5. Such investigations are fundamental to most practical applications in which it is undesirable to carry more detail in the system description than is really necessary to complete the task at hand.

In the present context, a system is an endomorphism of a commutative group. If it is desired to simplify in some sense the description of a system, then it is helpful to recall that a morphism is just a special type of function, and that a function is defined by domain, codomain, and rule of action. Basically, then, the idea is to simplify the functions. Such a simplification must take due account of the three parts of the function definition.

EXAMPLE 6.6-1

Suppose that two sets S_1 and S_2 are defined by

$$S_1 = \{a, b\}, \qquad S_2 = \{c, d, e\}. \qquad (6\text{-}165\text{a,b})$$

Construct a function

$$f : S_1 \rightarrow S_2 \qquad (6\text{-}166\text{a})$$

by giving the action

$$f(a) = c, \qquad f(b) = d. \qquad (6\text{-}166\text{b,c})$$

Then f is injective, but not surjective. One way to simplify the function f is to select the subset

$$\tilde{S}_2 = \{c, d\} \subset S_2 \tag{6-167}$$

and redefine

$$f : S_1 \to \tilde{S}_2, \tag{6-168}$$

in the obvious way. This approach has been used numerous times in the volume, inasmuch as $\tilde{S}_2$ is bijectively related to the image of f. Such an adjustment in the codomain of f is compatible with the domain and action of f, and causes only the most minor reverberations.   ∎

EXAMPLE 6.6-2

   Reconsider the preceding example, but choose to change the codomain to

$$\tilde{\tilde{S}}_2 = \{c\} \subset S_2. \tag{6-169}$$

Such a choice causes a significant perturbation in the function f. Either the domain must be changed to

$$\tilde{S}_1 = \{a\} \subset S_1 \tag{6-170}$$

or the action (6-166c) must be changed to

$$f(b) = c \tag{6-171}$$

in order to accommodate (6-169).   ∎

   Now the sort of change which has been mentioned above is a simplifying type of change. With respect to the domain of a function

$$f : S \to T, \tag{6-172}$$

general guidelines for simplification have been laid down in detail in Chapter 2. See, for illustration, Figs. 2.14 and 2.15. The simplification of the domain set S is achieved by means of an equivalence relation

$$E \subset S \times S, \tag{6-173}$$

which results in a partitioning of S into a set of subsets, called equivalence classes, in such a way that each element s of S belongs to one and only one equivalence class. Assignment of an element s in S to its equivalence class $\bar{s}$ is achieved by the projection

$$p_E : S \to S/E, \tag{6-174}$$

where the quotient set S/E is just the set of equivalence classes

$$\bar{s} = p_E(s) \tag{6-175}$$

assigned to elements s in S.

Recall that the equivalence relation E simplifies the set S by effectively identifying any two elements of S which belong to the relation E. Intuitively, it is sometimes helpful to think of such a process as reducing the resolution on the set S.

Does such a resolution reduction cause any difficulties with the function f? Basically, of course, the equivalence relation suggests that the action of the function could be changed so as to assign just one result to every equivalence class $\bar{s}$. This will be impossible, however, if the original function failed to assign the same result to every element $s \in \bar{s}$. This was the essence, then, of the key triangle discussion of Section 2.6. Given a function (6-172) and a simplification (6-174) on its domain, then a unique and simpler function

$$\bar{f}: S/E \to T \tag{6-176}$$

existed under the condition that

$$s_1 \, E \, s_2 \Rightarrow f(s_1) = f(s_2) \tag{6-177}$$

for all $(s_1, s_2) \in S^2$.

These recollections describe guidelines for simplifying domains and accounting for the way in which such a simplification will interact with the action of the function.

If, instead, the equivalence relation E is placed on the codomain in the manner

$$p_E: T \to T/E, \tag{6-178}$$

then composition supplies the new function

$$\bar{f}: S \to T/E \quad \text{by} \quad \bar{f} = p_E \circ f. \tag{6-179a,b}$$

Many such constructions have been used in previous chapters. So codomain simplification by equivalence relation causes no problem in adjusting the action of the function.

EXAMPLE 6.6-3

Notice that codomain adjustment by equivalence relation takes a slightly different viewpoint than the introductory approach of Examples 6.6-1 and 6.6-2. In the first example, a partition of type

$$\{c, e\}, \{d\} \quad \text{or} \quad \{c\}, \{d, e\} \tag{6-180a,b}$$

might have been applied. Then

$$\bar{f}(a) = \{c, e\}, \qquad \bar{f}(b) = \{d\} \tag{6-181a,b}$$

and

$$\bar{f}(a) = \{c\}, \qquad \bar{f}(b) = \{d, e\} \tag{6-182a,b}$$

would have been the corresponding results. In the case of the second example, the partition

$$\{c, d, e\} \tag{6-183}$$

would be a candidate on the codomain; and the domain would be left unchanged, with

$$\bar{f}(a) = \bar{f}(b) = \{c, d, e\}. \tag{6-184}$$

∎

   Thus, not every conceivable scheme for simplification is of the relation type; but these types do represent a very effective algebraic structure in which to study simplification.

   The general case, of course, occurs when an equivalence relation E_S is placed on the domain and an equivalence relation E_T is placed on the codomain. The situation is pictured in Fig. 6.12. In turn, this figure is a generalization of Fig. 2.16. Procedures for studying the solution for $\bar{f}$ should by now be familiar. The diagram of Fig. 6.12 is converted to that of Fig. 6.13 by composing p_{E_T} with f. But Fig. 6.13 is in the standard form of Chapter 2, and a unique $\bar{f}$ exists if

$$s_1 E_S s_2 \Rightarrow (p_{E_T} \circ f)(s_1) = (p_{E_T} \circ f)(s_2) \tag{6-185}$$

for all $(s_1, s_2) \in S^2$. This means that f must take equivalent elements of S into equivalent elements of T.

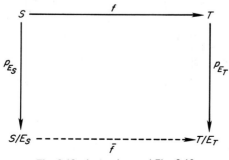

Fig. 6.12. A step beyond Fig. 2.16.

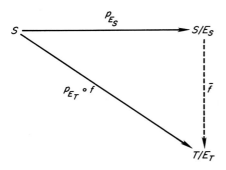

Fig. 6.13. Reorientation of Fig. 6.12.

Though not a trivial condition, (6-185) is entirely reasonable from the point of view of simplification. The statement

$$s_1 \, E_S \, s_2 \qquad (6\text{-}186)$$

implies that s_1 and s_2, though distinct elements of S, differ only in ways which are inconsequential for the study at hand. A similar statement can be made about elements t_1 and t_2 in T which satisfy

$$t_1 \, E_T \, t_2. \qquad (6\text{-}187)$$

If, therefore,

$$(p_{E_T} \circ f)(s_1) \neq (p_{E_T} \circ f)(s_2) \qquad (6\text{-}188\text{a})$$

did not imply

$$s_1 \not\equiv s_2(E_S), \qquad (6\text{-}188\text{b})$$

then the conclusion would have to be that inconsequential information, data, or details can produce significant results. If such is the situation, then an alternative simplification scheme should be taken under consideration.

The key triangle material of Section 2.6 was specialized to the case of groups in Section 4.5. The main result of that development was sketched in Fig. 4.9, where all the sets have group structure and all the functions are morphisms of groups. It is well to remember (4-127), where the nature of the equivalence relation which would permit the desired results was given. For a group $(S, \square, e_S)$, the definition was

$$s_1 \, E \, s_2 \qquad (6\text{-}189\text{a})$$

if there exists an element w in a normal subgroup W of S such that

$$s_1 = w \, \square \, s_2. \qquad (6\text{-}189\text{b})$$

The quotient set S/E was redesignated S/W and became a quotient group. The projection p_E was redesignated p_W and became a morphism of groups.

The condition for the existence of $\bar{f}$ became

$$W \subset \text{Ker } f, \tag{6-190}$$

which had the pleasing connotation that f acted to produce units on elements deemed inconsequential.

Notice that every subgroup is normal if the group is commutative. This is the assumption of the present chapter.

To finish the treatment of the general simplification of a system in the relational case, it is necessary to specialize (6-185) to the commutative group and morphism level. Let (6-185) be restated as

$$W_S \subset \text{Ker}(p_{E_T} \circ f), \tag{6-191}$$

where W_S is the subgroup of inconsequential elements which have been used to define E_S. Now

$$\text{Ker } p_{E_T} = W_T, \tag{6-192}$$

where W_T is the subgroup of inconsequential elements used to define E_T. Then the condition which specializes (6-185) to the group morphic case is

$$f(W_S) \subset W_T. \tag{6-193}$$

Again, (6-193) has a pleasing interpretation, namely, that f must take inconsequential data into inconsequential data.

In the case of a system defined by a morphism of commutative groups, then, simplification is able to proceed naturally by a process of choosing subgroups W_S and W_T of data which can be ignored in the domain and codomain. If the overall effect on the action of the system morphism is to be naturally accounted, this will have to be done so that (6-193) holds. Then

$$\bar{f} : S/W_S \rightarrow T/W_T \tag{6-194}$$

can play a basic role as a simplified system.

Exercises

6.6-1. One of the most common ways to represent physical or mathematical information about a signal is in terms of its series expansion. Let $(G, \square, e_G)$ be a group, and regard

$$G^{\mathbb{N}}$$

as a set of power series with coefficients in G. Recall that the group structure on G induces group structure on $G^{\mathbb{N}}$. Suppose that S is a normal subgroup of G. Then does it follow that $S^{\mathbb{N}}$ is a normal subgroup of $G^{\mathbb{N}}$? Prove the assertion or show a counterexample.

6.6-2. In a series expansion along the lines of Exercise 6.6-1, a common way to discard excess information about the signal is to disregard the series tail. Suppose that

$$f \in G^{\mathbb{N}},$$

and that f is represented by its image

$$f(0), \quad f(1), \quad f(2), \quad f(3), \quad f(4), \quad f(5), \quad \ldots$$

Then disregarding all but the first four coefficients can be represented by writing $\tilde{f} \in G^{\mathbb{N}}$ in the manner

$$\tilde{f}(n) = f(n), \qquad n = 0, 1, 2, 3,$$
$$\tilde{f}(n) = e_G \qquad n > 3.$$

Show that the collection of such $\tilde{f}$ is a subgroup of $G^{\mathbb{N}}$. Is it a normal subgroup?

6.6-3. Return to the series idea of Exercise 6.6-1, and suppose that two groups of series

$$G_1^{\mathbb{N}}, \qquad G_2^{\mathbb{N}}$$

are given on coefficient groups

$$(G_i, \square_i, e_{G_i}), \qquad i = 1, 2.$$

Suppose that

$$h: G_1 \rightarrow G_2$$

is a morphism of groups. Exhibit at least one way to use h so as to induce a morphism

$$h': G_1^{\mathbb{N}} \rightarrow G_2^{\mathbb{N}}.$$

6.6-4. Extend the notion of Exercise 6.6-1 to the case of a sequence

$$S(0), \quad S(1), \quad S(2), \quad \ldots$$

of subgroups of G, each of which is normal. Discuss the case in which each subgroup $S(i)$ is either G or $\{e_G\}$.

6.7 THE RING OF COMPATIBLE SYSTEMS

Return now to the details of Section 6.2. All domains and codomains are taken to be the same commutative group $(G, \square, e_G)$. Suppose that it is possible to determine a subgroup

$$W \subset G \tag{6-195a}$$

of inconsequential data. This section is concerned with studying the class of system endomorphisms which will yield natural simplification under the construction

$$p_W: G \rightarrow G/W. \tag{6-195b}$$

By the results of the preceding section, an element

$$h \in \mathbb{EM}(G) \tag{6-196}$$

will yield to such simplification if

$$h(W) \subset W; \qquad (6\text{-}197)$$

see (6-193). Denote by

$$H_W = \{h \,|\, h \in \mathbb{EM}(G) \text{ and } h(W) \subset W\}, \qquad (6\text{-}198)$$

the set of endomorphisms which satisfy (6-197). Call this set the set of endomorphisms which are compatible with the domain and codomain simplification (6-195). The intent is to show that H_W is a ring.

Inasmuch as

$$H_W \subset \mathbb{EM}(G), \qquad (6\text{-}199)$$

it is enough to show that H_W is a subring.

Clearly, H_W is nonempty. Indeed,

$$e_{GG} \in H_W \qquad \text{because} \qquad e_{GG}(W) = \{e_G\}, \qquad (6\text{-}200\text{a,b})$$

where

$$e_G \in W \qquad (6\text{-}200\text{c})$$

is a consequence of W being a subgroup. Also

$$1_G \in H_W \qquad (6\text{-}201)$$

by inspection. So proceed to showing that H_W is closed under the binary operation $\square$. If

$$h_i \in H_W, \qquad i = 1, 2, \qquad (6\text{-}202)$$

then, for arbitrary $w \in W$,

$$(h_1 \,\square\, h_2)(w) = h_1 w \,\square\, h_2 w \in W \qquad (6\text{-}203)$$

because the subgroup W is closed under $\square$ and because

$$h_i w \in W, \qquad i = 1, 2, \qquad (6\text{-}204)$$

by assumption. Again, if $h \in H_W$, then

$$\hat{h}w = \widehat{hw} \in W \qquad (6\text{-}205)$$

for each $w \in W$ because $(hw) \in W$ and because W is a subgroup which must be closed under inversion. This means that

$$(H_W, \square, e_{GG}) \qquad \text{is a subgroup of} \qquad (\mathbb{EM}(G), \square, e_{GG}). \quad (6\text{-}206\text{a,b})$$

Two steps remain. The first is to show that H_W is closed under composition; and the second is to show that H_W includes 1_G. In fact, the latter has already been observed in (6-201). Thus it only remains to complete the former. Suppose two elements (6-202); then

$$(h_1 \circ h_2)(w) = h_1(h_2(w)) \tag{6-207}$$

for every $w \in W$. Now

$$h_2(w) \in W \tag{6-208a}$$

by (6-204); then

$$h_1(h_2(w)) \in W \tag{6-208b}$$

by (6-204) also.

This completes the demonstration that

$$(H_W, \square, e_{GG}, \circ, 1_G) \tag{6-209}$$

is a subring of

$$(\mathbb{EM}(G), \square, e_{GG}, \circ, 1_G). \tag{6-210}$$

Thus the connection discussions of Section 6.2 can be applied equally well to the situation in which the decision (6-195) has been made, provided only that the systems in the interconnection belong to H_W.

EXAMPLE 6.7-1

Sections 6.3–6.5 dealt with an interconnection situation in which the basic building-block systems were all the same. Consider a fixed endomorphism $h \in \mathbb{EM}(G)$. If

$$h \in H_W, \tag{6-211}$$

then the arguments spanned by Eqs. (6-207) and (6-208) show that

$$h^2 \in H_W \tag{6-212}$$

also. By induction, it follows that

$$h^i \in H_W, \qquad i = 1, 2, 3, \ldots. \tag{6-213}$$

Inspection of (6-209) shows that (6-213) extends to $i \in \mathbb{N}$, with the usual identification of h^0 with 1_G. Also, the discussions associated with Eq. (6-205) permit $i \in \mathbb{Z}$ in the expression

$$(i)h^j \tag{6-214}$$

and for $j \in \mathbb{N}$ again. Then the subgroup nature of H_W establishes the result

$$h \in H_W \Rightarrow \mathbb{Z}[h] \subset H_W, \tag{6-215}$$

which means that the identical system studies of Section 6.4 carry over to the compatible ring case. ∎

EXAMPLE 6.7-2

Continue on to Section 6.5, and consider the possibility of generalizing the statement (6-215). Notice that

$$h \in H_W \Rightarrow \alpha \circ h^i \in H_W \qquad (6\text{-}216)$$

for every $\alpha \in H_W$. Then, in subset terms,

$$h \in H_W \Rightarrow H_W[h] \subset H_W. \qquad (6\text{-}217)$$

This is a large class of systems. An interpretation can be given as follows. From the subring H_W, select a fixed element h; then interconnections of this h with suitable coefficients from H_W will work naturally with the simplification represented by projecting group elements onto G/W. In some sense, $H_W[h]$ is the largest set which can stand in this stead, inasmuch as a polynomial

$$\boxed{\sim}_{j \in \mathbb{N}} \alpha_j h^j \qquad \text{with} \qquad \alpha_j = e_{GG}, \qquad j \neq 0, \qquad (6\text{-}218\text{a,b})$$

is always a possibility; and (6-218) would require

$$\alpha_0 h^0 = \alpha_0 1_G = \alpha_0 \in H_W. \qquad (6\text{-}219)$$

Notice that $H_W[h]$ need not be a ring.   ∎

EXAMPLE 6.7-3

Sometimes, it is convenient to make successive simplifications. Suppose that the simplification

$$p_W : G \to G/W \qquad (6\text{-}220)$$

has been carried out at each domain and codomain for a selected subgroup $W \subset G$. Later, it happens that an even larger subgroup V can be discarded, in the sense

$$W \subset V \subset G. \qquad (6\text{-}221)$$

The situation is pictured in Fig. 6.14. Because

$$\text{Ker } p_V = V \supset W, \qquad (6\text{-}222)$$

there exists a unique morphism of groups

$$\bar{p}_V : G/W \to G/V \qquad (6\text{-}223)$$

which reduces the resolution produced by p_W even further to that produced by p_V. In other words, the simplification begun by p_W can be continued by $\bar{p}_V$ to produce the same effect as if p_V had been used originally. This

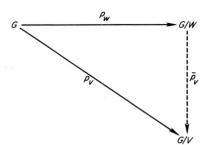

Fig. 6.14. A bigger discard.

occurs because the diagram commutes, that is,

$$p_V = \bar{p}_V \circ p_W. \tag{6-224}$$

Can the effect of the further simplification (6-223) on the ring of compatible systems be determined in general? The answer is a qualified yes. In order to have been in the original simplification step, the endomorphisms must have been elements of H_W. But in order to be in the subsequent step of simplification, (6-224) provides that they have to be in H_V. Thus the resulting ring of compatible systems is

$$H_W \cap H_V, \tag{6-225}$$

which is certainly nonempty because 1_G is an element of both. The reader may wish to verify that (6-225) represents a ring. A sketch of such a procedure would be as follows. Observe that (6-225) is a nonempty subset of the ring $\mathbb{EM}(G)$. Since 1_G is a member of (6-225), it suffices to show that (6-225) is closed under $\square$, under $\circ$, and under inverses. Refer to (4-166) for help with two of these steps. Notice that

$$(H_W \cap H_V) \subset H_W, \tag{6-226}$$

which means that the compatible ring of systems for the latter discard is contained in that for the former discard. This is as expected. But, if

$$V = G, \tag{6-227}$$

H_V becomes $\mathbb{EM}(G)$, and (6-226) is an equality. ∎

With regard to this example, it is also interesting to observe that

$$\operatorname{Ker} \bar{p}_V = V/W. \tag{6-228}$$

Also of interest, of course, is the question of choosing subgroups W in such a way that (6-197) is satisfied for some given endomorphism h. To solve (6-197), notice that it takes the specific form

$$hw_1 = w_2 \tag{6-229}$$

for

$$w_j \in W, \qquad j = 1, 2. \tag{6-230}$$

Rewrite (6-229) in the form

$$hw_1 \;\square\; \hat{1}_G w_2 = e_G, \tag{6-231}$$

and observe that the left member of (6-231) defines a morphism on the product group G^2 by the calculation associated with Fig. 4.16. Indeed, in that figure, let the function f be given by

$$h: G \to G, \tag{6-232}$$

the function g by

$$\hat{1}_G: G \to G, \tag{6-233}$$

and the function h by

$$k: G \times G \to G. \tag{6-234}$$

Then solution of (6-231) is associated with

$$\operatorname{Ker} k, \tag{6-235}$$

which is carried by the product projection morphisms of Fig. 4.14 onto the subgroups S_1 and S_2. Therefore, for each s_1 in S_1, there is an s_2 in S_2 such that

$$hs_1 = s_2; \tag{6-236}$$

moreover, for each s_2 in S_2, there is an s_1 in S_1 such that (6-236) holds again. Accordingly,

$$h(S_1) = S_2. \tag{6-237}$$

It is left as an exercise for the reader to establish that

$$S_1 = G, \qquad S_2 = \operatorname{Im} h. \tag{6-238a,b}$$

Inasmuch as

$$\operatorname{Im} h \subset G, \tag{6-239a}$$

one obvious choice of W is given by

$$W = G; \tag{6-239b}$$

however, this leads to the trivial case of (6-227). Another possibility is to choose

$$\operatorname{Ker} h \subset S_1 \tag{6-240}$$

for W, because e_G is in S_2. This choice for W, namely, $\operatorname{Ker} h$, is entirely realistic from the practical point of view. These elementary choices certainly do not exhaust the set of possibilities. However, they do make clear certain points. Thus, given h, it is not very informative to look for a greatest element in the poset of subgroups W satisfying (6-197). Indeed, (6-239b) is support for this claim. Neither is it illustrative to mount a search for a least element in the poset, because $\{e_G\}$ is immediately available and is tantamount to no simplification at all. Notice, however, that if W_1 and W_2 are two subgroups which satisfy (6-197), then the intersection of W_1 with W_2 also satisfies (6-197). Moreover, it is not a difficult exercise to see that $W_1 \square W_2$ also satisfies (6-197). It is then possible to consider the development into a lattice of the poset of W satisfying (6-197). The details are left as an exercise.

Such a lattice of subgroups may be given for each endomorphism h. Suppose an interconnection involves systems defined by endomorphisms h_1 and h_2. Denote by L_1 and L_2 the lattice of subgroups satisfying (6-197) for h_1 and h_2, respectively. Then the possibilities for subgroups W which satisfy (6-197) for both h_1 and h_2 depend upon interaction of the lattices L_1 and L_2.

Exercises

6.7-1. In Example 6.7-3, the intersection (6-225) is proposed. Consider in detail whether (6-225) is a ring.

6.7-2. Continue with the idea introduced in Exercise 6.6-2. Observe that the information discarded is actually of the form

$$e_G, \quad e_G, \quad e_G, \quad e_G, \quad f(4), \quad f(5), \quad f(6), \quad \dots.$$

Show that the collection of such series is a normal subgroup of $G^{\mathbb{N}}$.

6.7-3. Reexamine Exercise 6.7-2 and explain how it relates to Exercise 6.6-4.

6.7-4. Consider the possibility of representing the systems by elements

$$h \in \mathbb{EM}(G)^{\mathbb{N}}.$$

One way to describe the action of h on an element f in $G^{\mathbb{N}}$ is in the manner

$$(h(f))(i) = h(i)(f(i)).$$

Discuss whether such systems would be compatible with the type of subgroup of inconsequential data described in Exercise 6.7-2.

6.7-5. Continue the idea of Exercise 6.7-4, but assume G commutative and use an alternate way of defining the action of h. Think of f as a series

$$\sum_{n=0}^{\infty} f(n)x^n$$

and of h as a series

$$\sum_{i=0}^{\infty} h(i)x^i.$$

Consider the action

$$h(f) = \sum \sum h(i)(f(n))x^{i+n}.$$

Determine whether this action is a morphism. If so, repeat the discussion of Exercise 6.7-4.

6.8   THE RING OF SIMPLIFIED SYSTEMS

The diagram of Fig. 6.12 and the condition of Eq. (6-193) have spelled out the problem of simplifying a system described by

$$h:G \to G \qquad\qquad (6\text{-}241)$$

to a system described by

$$\bar{h}:G/W \to G/W. \qquad\qquad (6\text{-}242)$$

To be a candidate for such a simplification, h must satisfy

$$h(W) \subset W. \qquad\qquad (6\text{-}243)$$

For a selected subgroup W of G, the preceding section has indicated that the subset H_W of $\mathbb{EM}(G)$ with the property (6-243) is a ring on its own merits. Moreover,

$$(\mathbb{EM}(G/W), \Box, e_{(G/W)(G/W)}, \circ, 1_{G/W}) \qquad\qquad (6\text{-}244)$$

is a ring also. In this section, a function

$$P:H_W \to \mathbb{EM}(G/W) \qquad\qquad (6\text{-}245a)$$

is established by naming the action

$$P(h) = \bar{h}. \qquad\qquad (6\text{-}245b)$$

It is the initial goal to demonstrate that P is a morphism of rings.

Recall from Section 5.1 the conditions which have to be satisfied in order that P become a morphism. These conditions are stated in general terms in (5-27). For the present section, it will have to be shown that

$$P(1_G) = 1_{G/W}, \qquad\qquad (6\text{-}246a)$$
$$P(h_1 \circ h_2) = \bar{h}_1 \circ \bar{h}_2, \qquad\qquad (6\text{-}246b)$$
$$P(h_1 \Box h_2) = \bar{h}_1 \Box \bar{h}_2. \qquad\qquad (6\text{-}246c)$$

Notice carefully that the binary operations $\circ$ and $\Box$ in the left members of (6-246b) and (6-246c) are not the same functions represented by the same symbols in the right members of these equations. Indeed, in the left member of (6-246c), $\Box$ is a function

$$H_W \times H_W \to H_W, \qquad\qquad (6\text{-}247a)$$

whereas in the right member it is a function

$$\mathbb{EM}(G/W) \times \mathbb{EM}(G/W) \to \mathbb{EM}(G/W). \qquad (6\text{-}247\text{b})$$

Similar statements hold for the composition operations $\circ$ in (6-246b).

Begin with (6-246a). The construction in the left member is

$$P(1_G) = \overline{1}_G \qquad (6\text{-}248)$$

according to the diagram in Fig. 6.12. Clearly, $\overline{1}_G$ has the correct domain and codomain to be identified with $1_{G/W}$. It remains, therefore, to establish that these two functions have the same action. Indeed, for an arbitrary element $\overline{g}$ in G/W, calculate

$$\overline{1}_G \overline{g} = \overline{1}_G(p_W g) \qquad (6\text{-}249\text{a})$$
$$= (\overline{1}_G \circ p_W)(g) \qquad (6\text{-}249\text{b})$$
$$= (p_W \circ 1_G)(g) \qquad (6\text{-}249\text{c})$$
$$= p_W(1_G g) \qquad (6\text{-}249\text{d})$$
$$= p_W g \qquad (6\text{-}249\text{e})$$
$$= \overline{g} \qquad (6\text{-}249\text{f})$$

independently of the representation g selected for the coset $\overline{g}$. So (6-246a) is true.

Next, consider (6-246b). The left member performs the construction

$$P(h_1 \circ h_2) = \overline{h_1 \circ h_2}. \qquad (6\text{-}250)$$

Again, the domain and codomain are correct to identify with the right member. The action must, however, be checked. To do this, calculate for an arbitrary element $\overline{g}$ in G/W

$$\overline{h_1 \circ h_2}\,\overline{g} = \overline{h_1 \circ h_2}(p_W g) \qquad (6\text{-}251\text{a})$$
$$= ((\overline{h_1 \circ h_2}) \circ p_W)(g) \qquad (6\text{-}251\text{b})$$
$$= (p_W \circ (h_1 \circ h_2))(g) \qquad (6\text{-}251\text{c})$$
$$= ((p_W \circ h_1) \circ h_2)(g) \qquad (6\text{-}251\text{d})$$
$$= ((\overline{h}_1 \circ p_W) \circ h_2)(g) \qquad (6\text{-}251\text{e})$$
$$= (\overline{h}_1 \circ (p_W \circ h_2))(g) \qquad (6\text{-}251\text{f})$$
$$= (\overline{h}_1 \circ (\overline{h}_2 \circ p_W))(g) \qquad (6\text{-}251\text{g})$$
$$= ((\overline{h}_1 \circ \overline{h}_2) \circ p_W)(g) \qquad (6\text{-}251\text{h})$$
$$= (\overline{h}_1 \circ \overline{h}_2)(p_W g) \qquad (6\text{-}251\text{i})$$
$$= (\overline{h}_1 \circ \overline{h}_2)\overline{g}. \qquad (6\text{-}251\text{j})$$

Especially interesting here is the detailed interplay with the commutative diagram of Fig. 6.12 and the fact that composition is associative in the monoids of the rings H_W and $\mathbb{EM}(G/W)$. Because (6-251) expresses identity of action on each $\bar{g}$, it finishes the task of establishing (6-246b).

This brings the study to (6-246c). By now the approach is familiar, and so the calculation may be given directly:

$$\overline{h_1 \, \square \, h_2} \, \bar{g} = (\overline{h_1 \, \square \, h_2})(p_W g) \tag{6-252a}$$
$$= ((\overline{h_1 \, \square \, h_2}) \circ p_W)(g) \tag{6-252b}$$
$$= (p_W \circ (h_1 \, \square \, h_2))(g) \tag{6-252c}$$
$$= p_W(h_1 g \, \square \, h_2 g) \tag{6-252d}$$
$$= (p_W(h_1 g)) \, \square \, (p_W(h_2 g)) \tag{6-252e}$$
$$= ((p_W \circ h_1)g) \, \square \, ((p_W \circ h_2)g) \tag{6-252f}$$
$$= ((\bar{h}_1 \circ p_W)g) \, \square \, ((\bar{h}_2 \circ p_W)g) \tag{6-252g}$$
$$= (\bar{h}_1(p_W g)) \, \square \, (\bar{h}_2(p_W g)) \tag{6-252h}$$
$$= \bar{h}_1 \bar{g} \, \square \, \bar{h}_2 \bar{g} \tag{6-252i}$$
$$= (\bar{h}_1 \, \square \, \bar{h}_2)\bar{g}. \tag{6-252j}$$

In this calculation, notice the point that distributive rules in the ring H_W do not suffice to pass from (6-252c) to (6-252f) because $p_W \notin H_W$. The steps (6-252) for the endomorphism

$$\overline{h_1 \, \square \, h_2} = P(h_1 \, \square \, h_2) \tag{6-253}$$

complete the demonstration of (6-246c) and thus also the result that P is a morphism of rings.

Now examine

$$\operatorname{Im} P \subset \mathbb{EM}(G/W). \tag{6-254}$$

As a subset, $\operatorname{Im} P$ is also a candidate to be a subring. From the fact that P is a morphism of rings, it follows that P is a morphism of groups according to (6-246c). Thus $\operatorname{Im} P$ is a subgroup of $\mathbb{EM}(G/W)$. Because (6-246a) is already established, the only step remaining in the demonstration that $\operatorname{Im} P$ is a subring is to show that it is closed under composition. This is in effect achieved by reversing the steps of (6-251), with proviso that, given $\bar{h}_1$ and $\bar{h}_2$, the h_1 and h_2 are not unique as the argument traces backward. However, because

$$\bar{h}_i \in \operatorname{Im} P, \qquad i = 1, 2, \tag{6-255}$$

the existence of at least one h_1 and h_2 is certainly assured.

The alert reader has probably considered the possibility that the image of any morphism of rings can be shown to be a subring by the same argument. In this respect, the morphism of rings differs from the morphism of groups. For the morphism of groups has a subgroup for both its kernel and image, whereas the morphism of rings has a subring only for its image, the kernel of a ring morphism being an ideal. Though the kernel of a group morphism is of special type, being a normal subgroup, the kernel of a ring morphism, which we have called an ideal, is not generally a subring at all, because it does not contain the monoidal unit. If it did, it would be equal to the whole ring.

Because Im P is a ring in its own right, as a subring of codomain P, it is possible to redefine P. Indeed, let

$$p_H : H_W \to \text{Im } P \tag{6-256}$$

be the morphism of rings which has the same action as P, but with the result regarded as an element of Im P instead of $\mathbb{EM}(G/W)$. Then p_H is surjective as a function and can be termed an epimorphism of rings.

The observation that Im P is a ring is important from the viewpoint of this chapter. It means that the set of simplified system endomorphisms can be considered in the same way as the original set of system endomorphisms.

Moreover, the importance of the fact that p_H is a morphism of rings can scarcely be overemphasized.

These points will be presented in final form in the next section.

Exercises

6.8-1. Combine the ideas of Exercises 6.7-2 and 6.7-4. Explain how the simplified system morphisms might be represented and how they would act upon elements of the quotient group.

6.8-2. Repeat Exercise 6.8-1 for the case in which the ideas of Exercises 6.7-2 and 6.7-5 are combined.

6.9 QUOTIENTS ON INTERCONNECTED SYSTEMS

It is well now to review the ideas of Section 6.2. A system was understood to be an endomorphism of the commutative ring $(G, \square, e_G)$. A series connection of two such systems was understood as composition in the endomorphism ring $\mathbb{EM}(G)$. A parallel connection of two such systems was identified with the binary operation $\square$ on $\mathbb{EM}(G)$ induced by the corresponding operation $\square$ on G. Feedback connections, when well defined, were also explained within the $\mathbb{EM}(G)$ context.

The major idea was that, with proper use of parentheses, very many of the diagrams representing interconnected systems could be represented in turn by an algebraic expression in $\mathbb{EM}(G)$. The reader may wish to review the examples of Section 6.2.

Subsequent sections placed considerable emphasis upon showing that this ring structure was preserved in various types of situations.

In Section 6.7, a ring H_W of systems which would be compatible with a quotient construction $p_W: G \to G/W$, based upon selection of a subgroup of inconsequential data W, was established. Each element h in H_W then gave rise to a unique endomorphism $\bar{h}$ in $\mathbb{EM}(G/W)$ by a construction

$$P: H_W \to \mathbb{EM}(G/W), \tag{6-257a}$$

where P is a morphism of rings with action

$$P(h) = \bar{h}. \tag{6-257b}$$

Since $\mathrm{Im}\, P$ is a subring of $\mathbb{EM}(G/W)$, the definition

$$p_H: H_W \to \mathrm{Im}\, P \tag{6-258a}$$

was possible, where p_H is a morphism of rings with the action

$$p_H(h) = \bar{h}, \tag{6-258b}$$

but with the right member now understood as an element of $\mathrm{Im}\, P$ instead of $\mathbb{EM}(G/W)$.

In this section, it will be established that the morphism of rings defined by (6-258) is in essence setting up a quotient ring; and use will accordingly be made of the generalization of the key triangle to the ring case as discussed in Section 6.1.

Consider $\mathrm{Ker}\, p_H$. An element $h \in H_W$ which belongs to this kernel must have the property

$$P(h) = e_{(G/W)(G/W)}, \tag{6-259}$$

which transfers in a straightforward manner into the action equation

$$\bar{h}\bar{g} = e_{G/W} \tag{6-260}$$

for all $\bar{g}$ in G/W. Write

$$\bar{g} = p_W g \tag{6-261}$$

for an appropriate representative g in G. Then (6-260) becomes

$$e_{G/W} = \bar{h}(p_W g) = (\bar{h} \circ p_W)(g) \tag{6-262a,b}$$
$$= (p_W \circ h)(g) = p_W(hg). \tag{6-262c,d}$$

Moreover, (6-260) holds for all $\bar{g}$ in G/W, so that (6-262d) must hold for all g in G. Accordingly,

$$\mathrm{Im}\, h \subset \mathrm{Ker}\, p_W = W. \tag{6-263}$$

Now $\mathrm{Ker}\, p_H$ is an ideal, by the arguments explained in Section 6.1. Thus it follows that the subset of H_W with the property that each element has image contained in W is an ideal. However, it is also instructive to see this directly. Suppose

$$h_i \in \text{Ker } p_H, \qquad i = 1, 2. \tag{6-264}$$

Then

$$(h_1 \,\square\, h_2) \in \text{Ker } p_H \tag{6-265}$$

also. To see this, select an arbitrary $g \in G$ and calculate

$$(h_1 \,\square\, h_2)(g) = h_1 g \,\square\, h_2 g; \tag{6-266}$$

but

$$h_i g \in W, \qquad i = 1, 2, \tag{6-267}$$

with W being a subgroup of G. Thus the right member of (6-266) is an element of W as well. This establishes that the candidate subset of H_W is closed under $\square$. Moreover,

$$hg \in W \qquad \text{implies} \qquad \hat{h}g = \widehat{hg} \in W \tag{6-268a,b}$$

also because W is a subgroup. Thus the subset is closed under inversion. Moreover, the subset is nonempty, in view of the fact that e_{GG} is a member, itself a consequence of the subgroup property $e_G \in W$. The subset is thus a subgroup of H_W. Now consider two elements h_1 and h_2 in H_W. Suppose that

$$\text{Im } h_1 \subset W, \tag{6-269}$$

but that h_2 is arbitrary. Then

$$\text{Im}(h_1 \circ h_2) \subset \text{Im } h_1 \subset W, \tag{6-270}$$

so that the subset is closed under right ring multiplication. Moreover,

$$\begin{align*}
\text{Im}(h_2 \circ h_1) &= (h_2)_*(\text{Im } h_1) \tag{6-271a} \\
&\subset (h_2)_*(W) \tag{6-271b} \\
&\subset W, \tag{6-271c}
\end{align*}$$

the last step being a consequence of $h_2 \in H_W$. Thus the subset of H_W is closed under left ring multiplication as well. By two separate arguments, then, $\text{Ker } p_H$ is an ideal.

From Section 6.1 and the observation that

$$\text{Ker } p_H \subset \text{Ker } p_H, \tag{6-272}$$

it follows that the diagram of Fig. 6.15 can be completed uniquely with a ring morphism

$$\bar{p}_H : H_W / \text{Ker } p_H \to \text{Im } P. \tag{6-273}$$

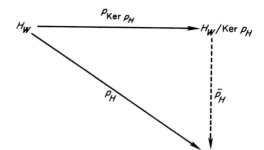

Fig. 6.15. Connection quotient idea.

Now p_H is epic, and

$$\operatorname{Im}(\bar{p}_H \circ p_{\operatorname{Ker} p_H}) \subset \operatorname{Im}\bar{p}_H, \tag{6-274a}$$

so that $\bar{p}_H$ must be epic as well. Moreover,

$$\bar{p}_H(p_{\operatorname{Ker} p_H}(h)) = (\bar{p}_H \circ p_{\operatorname{Ker} p_H})(h) = p_H h, \tag{6-274b,c}$$

so that if $\bar{p}_H$ carries $(p_{\operatorname{Ker} p_H}(h))$ into the additive unit of $\operatorname{Im} P$, then p_H must carry h to that same unit. This means that h belongs to $\operatorname{Ker} p_H$ and will thus be mapped to the additive unit by $p_{\operatorname{Ker} p_H}$. The conclusion is that $\bar{p}_H$ is monic as well.

As a morphism of rings which is epic and monic, $\bar{p}_H$ establishes an isomorphism between the ring $H_W/\operatorname{Ker} p_H$ and the ring $\operatorname{Im} P$. The construction represented by p_H may thus be visualized as isomorphically the same as a quotient ring construction.

It can now be observed that there are two quotients involved in this discussion. The first quotient is that of the domain and codomain group and is written

$$p_W : G \to G/W. \tag{6-275a}$$

The second quotient is that of the ring construction

$$p_H : H_W \to \operatorname{Im} P, \tag{6-275b}$$

which is essentially related to

$$p_{\operatorname{Ker} p_H} : H_W \to H_W/\operatorname{Ker} p_H \tag{6-276}$$

by the isomorphism (6-273).

Accordingly, since the interconnection of systems is characterized by the group and by ring elements from the endomorphism ring on the group, and since the simplification process on the interconnected system can be essentially described by two quotient calculations, one on the group and one on a subring of the endomorphism ring, it is reasonable to refer to the entire process as one of constructing quotients on interconnected systems. The simplified system may then be called a *quotient system*.

Actually, these remarks can be put on a much more substantial mathematical footing. For there is in algebraic theory a generalized type of function, called a *functor*, which can act both on sets and upon the functions which have these sets for domain and codomain. Just as functions act on sets, functors act on *categories*. It is not the purpose of this volume to explore categories and functors, however.

The remainder of the section is given over to some examples.

EXAMPLE 6.9-1

Suppose an elementary series connection is written as

$$h_1 \circ h_2, \qquad (6\text{-}277)$$

where $h_i \in H_W$, $i = 1,2$. The simplification process can be described by the calculation

$$\overline{h_1 \circ h_2} = P(h_1 \circ h_2) = P(h_1) \circ P(h_2) = \overline{h}_1 \circ \overline{h}_2. \qquad (6\text{-}278\text{a,b,c})$$

The important step is the transition from (6-278a) to (6-278b), which follows because simplification as defined in this chapter is a morphism of monoids. In words, the simplification of a series connection of systems is the series connection of the simplified systems. Qualitatively put, simplification is a morphism of series connection. ∎

EXAMPLE 6.9-2

Suppose an elementary parallel connection is written as

$$h_1 \,\square\, h_2 \qquad (6\text{-}279)$$

for $h_i \in H_W$, $i = 1,2$. Then the process of simplifying the connection can be seen by

$$\overline{h_1 \,\square\, h_2} = P(h_1 \,\square\, h_2) = P(h_1) \,\square\, P(h_2) = \overline{h}_1 \,\square\, \overline{h}_2. \qquad (6\text{-}280\text{a,b,c})$$

The step from (6-280a) to (6-280b) follows because simplification is a morphism of groups. Expressed once more in words, the simplification of a parallel connection of systems is the parallel connection of the simplified systems. Intuitively, simplification is a morphism of parallel connection. ∎

The two examples preceding are, of course, fundamental. However, the techniques readily handle more complicated cases.

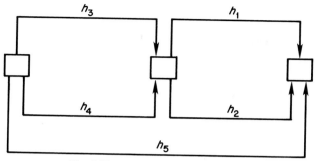

Fig. 6.16. A more complicated case.

EXAMPLE 6.9-3

Consider the connection

$$[(h_1 \,\Box\, h_2) \circ (h_3 \,\Box\, h_4)] \,\Box\, h_5. \tag{6-281}$$

This case is pictured in Fig. 6.16. On the surface, the problem of simplification seems to have become much more difficult. However, notice the ease of the calculation

$$P([(h_1 \,\Box\, h_2) \circ (h_3 \,\Box\, h_4)] \,\Box\, h_5)$$
$$= P([(h_1 \,\Box\, h_2) \circ (h_3 \,\Box\, h_4)]) \,\Box\, P(h_5) \tag{6-282a}$$
$$= [P(h_1 \,\Box\, h_2) \circ P(h_3 \,\Box\, h_4)] \,\Box\, P(h_5) \tag{6-282b}$$
$$= [(P(h_1) \,\Box\, P(h_2)) \circ (P(h_3) \,\Box\, P(h_4))] \,\Box\, P(h_5) \tag{6-282c}$$
$$= [(\bar{h}_1 \,\Box\, \bar{h}_2) \circ (\bar{h}_3 \,\Box\, \bar{h}_4)] \,\Box\, \bar{h}_5. \tag{6-282d}$$

A comparison of (6-282d) with (6-281) shows that the simplified expression is of precisely the same form as the original, except with hs replaced by $\bar{h}$s. This brings out the fact that the simplification process applied to an interconnected system produces a new interconnected system in which the connection pattern is the same as before but in which the individual systems represented by their domains, codomains, and endomorphism actions have been simplified. ∎

EXAMPLE 6.9-4

Denote by

$$f \in H_W \tag{6-283}$$

the interconnected system of (6-281), and by

$$g \tag{6-284a}$$

the interconnection

$$h_6 \circ (h_7 \,\square\, h_8) \in H_W. \tag{6-284b}$$

Then

$$\overline{f \circ g} = \overline{f} \circ \overline{g} \quad \text{and} \quad \overline{f \,\square\, g} = \overline{f} \,\square\, \overline{g}, \tag{6-285a,b}$$

which points out that the simplification idea will work just as well on interconnections of interconnected systems. ∎

EXAMPLE 6.9-5

Reconsider the case of Example 6.2-2. The governing equation is (6-53). The idea is to apply the simplification procedure to the endomorphism

$$(1_G \,\square\, (h_1 \circ h_2))^{-1} \circ h_1. \tag{6-286}$$

Notice that this will require

$$(1_G \,\square\, (h_1 \circ h_2))^{-1} \in H_W. \tag{6-287}$$

Under these conditions, it is possible to show that

$$P(1_G \,\square\, (h_1 \circ h_2)) \tag{6-288}$$

is an isomorphism as well, and that

$$P(1_G \,\square\, (h_1 \circ h_2))^{-1} = (1_{G/W} \,\square\, (\overline{h}_1 \circ \overline{h}_2))^{-1}. \tag{6-289}$$

It is straightforward to show that (6-288) is epic, because $(1_G \,\square\, (h_1 \circ h_2))$ is epic and p_W is epic. The kernel of (6-288) is a bit more subtle. It is necessary to show that there are no elements g in G but not in W with the property that

$$p_W \circ (1_G \,\square\, (h_1 \circ h_2))g = e_{G/W}. \tag{6-290}$$

But such elements satisfy

$$(1_G \,\square\, (h_1 \circ h_2))g = w \in W \tag{6-291}$$

for some w. However, (6-291) implies

$$(1_G \,\square\, (h_1 \circ h_2))^{-1}w = g \notin W, \tag{6-292}$$

which contradicts (6-287). Thus, if the original feedback system is well defined and meets the condition (6-287), then the simplified feedback system is also well defined. Establishment of (6-289) is left to the reader as an exercise. ∎

It has been stated in this section and in Section 6.2 that "very many" of the diagrams representing interconnected systems could be represented in turn by an expression in the endomorphism ring. Of course, these expressions can become complicated. In such a case, the diagram is usually replaced by a set of equations written at the connection points. If there are n such connection points, it is possible to develop the product group G^n and write the set of connection equations in the manner

$$H\tilde{g}_1 = \tilde{g}_2, \qquad \text{where} \qquad H:G^n \to G^n \qquad (6\text{-}293\text{a,b})$$

is an endomorphism of groups, $\tilde{g}_2 \in G^n$ is a given element of sources, and $\tilde{g}_1$ is an element of unknowns. Equation (6-293) defines a *generalized linear signal flow graph* (GLSFG). The ideas of this chapter can be extended to the GLSFG. If carried out, the quotient system is called a *quotient signal flow graph*.

Exercises

6.9-1. Relative to Exercise 6.8-1, assume that originally two such systems h_1 and h_2 were connected in series. Explain how the action of the morphism

$$h_1 \circ h_2$$

would be determined. Compare this action with that of

$$\bar{h}_1 \circ \bar{h}_2$$

associated with the series connection of the simplified systems.
6.9-2. Relative to Exercise 6.8-2, repeat the analyses of Exercise 6.9-1.
6.9-3. Repeat Exercise 6.9-1 for the case of parallel connection.
6.9-4. Repeat Exercise 6.9-2 for the case of parallel connection.

6.10 DISCUSSION

The chapter just concluded is intended to expand upon the use of rings in Chapter 5. In that chapter, the construction of inverse GRMPSs led to the introduction of ring-type operations on morphisms. Though the inverse GRMPS development made considerable use of diagrams of systems, with their intuitive connection patterns, it was not yet clear in those discussions that ringlike operations are intimately associated with interconnections of systems.

In this chapter, a system has been taken to be described by an endomorphism of a commutative group. The endomorphism ring of that group embodies series connections by composition and parallel connections by ring addition. A diagram of an interconnected system corresponds to an algebraic expression in the endomorphism ring.

The major idea in this regard has been to regard an interconnection in terms of a ring expression. This procedure also works for the feedback case whenever the feedback problem is well defined.

Various special ring structures carry the same connotations. Obtained as subrings of the endomorphism ring were the interconnections of identical systems, and of identical systems modulated by special endomorphisms which commuted with them.

Sections 6.6–6.9 have discussed a natural simplification procedure for interconnected systems. The concept of simplification is pursued by means of the equivalence relation. A quotient construction is placed upon the domain and codomain groups of the connection. Then a subring of endomorphisms compatible with the quotient group constructions, in the sense that it carries elements of the discarded subgroup into elements of the discarded subgroup, is identified. It then is possible to place a quotient ring construction on this subring; this corresponds to simplifying the actions of the original system endomorphisms.

A feature of this quotient simplification procedure is that the interconnection pattern is preserved, with the equivalence relations taking effect on group elements and upon endomorphism action. Thus a simplification of two systems in series or parallel consists of the two simplified systems in series or parallel, respectively. A similar statement is true for feedback situations.

When an algebraic expression in the ring gets too complicated for convenience, the ideas adapt to a generalized linear signal flow graph. In this case, the simplified system is called a quotient signal flow graph.

The ideas in this chapter touch upon the algebraic objects known as categories and functors, though these notions are not pursued here.

Finally, by now the reader should be somewhat comfortable with the notion of a ring of endomorphisms which act on a group. This leads, in the following chapter, to the question of modules.

7 ASPECTS
OF MODULE MORPHIC SYSTEMS

In the last chapter, there was an omnipresent commutative group $(G, \square, e_G)$, which served to provide inputs to systems and to receive outputs from them. A system was represented by an endomorphism $h: G \rightarrow G$ of groups; the fact that h resided in a ring

$$(\mathbb{EM}(G), \square, e_{GG}, \circ, 1_G) \tag{7-1}$$

was used to develop a way of thinking about interconnected systems. In fact, the binary operation

$$\square : \mathbb{EM}(G) \times \mathbb{EM}(G) \rightarrow \mathbb{EM}(G) \tag{7-2}$$

was used to explain the intuitive notion of parallel connection, and the binary operation

$$\circ : \mathbb{EM}(G) \times \mathbb{EM}(G) \rightarrow \mathbb{EM}(G) \tag{7-3}$$

was used to put in place a corresponding notion of series connection.

During the progress of those discussions, a great many calculations were carried out in the spirit of the two-box structure of Section 1.4. Here the reader may wish to review the intuitive material which was written there in connection with Figs. 1.20 and 1.21. There the idea was that a system would be represented by an atomic function

$$\mathbf{a} : S \rightarrow S \tag{7-4}$$

for a set S equipped with a binary operation $\square$. To construct a richer class of systems from the set of atomic functions, a set C of coefficients was introduced in such a way that $c \in C$ would imply that another system

$$ca : S \to S \tag{7-5}$$

could be determined. It was pointed out at that time that the two-box structure was a particularly convenient one from the algebraic point of view. In fact, this has already been borne out by the discussions of Chapter 6. In more concrete terms, Section 6.4 has elaborated the ring

$$\mathbb{Z}[h] = \left\{ \boxed{\sim}_{j \in \mathbb{N}} (k_j) h^j \mid k_j \in \mathbb{Z} \right\} \tag{7-6}$$

of polynomials in a fixed endomorphism h. Here the atomic function (7-4) is identified with $h : G \to G$, and the set C of coefficients is taken to be the integers $\mathbb{Z}$. In turn, these concepts were extended in Section 6.5 to the ring

$$(S_h[h], \square, e_{GG}, \circ, 1_G) \tag{7-7}$$

of polynomials in h but with coefficients taken from S_h, the subring of $\mathbb{EM}(G)$ consisting of endomorphisms which commute with h in the sense of (6-131).

Later sections of Chapter 6 then expanded to the cases in which the atomic morphism h need not be the only one available.

The reader, therefore, has probably begun already to surmise a close relationship between the two-box idea of Chapter 1 and the ring structures above.

This chapter formalizes the two-box idea, in such a way that every ring becomes an example. Consider an arbitrary ring

$$(R, *, e_R, \triangle, 1_R). \tag{7-8}$$

Notice that the structure

$$(R, *, e_R) \tag{7-9}$$

is a commutative group. Then the ring (7-8) can act on the commutative group (7-9) in the sense that there is a natural function

$$f : R \times R \to R \qquad \text{given by} \qquad f(r_1, r_2) = r_1 \triangle r_2. \tag{7-10a,b}$$

The key is to recognize that the domain of (7-10a) can be regarded as the product of a ring with a commutative group, while the codomain of (7-10a) can be regarded as a commutative group.

In this sense, every ring represents a two-box structure. However, many of the most interesting of such structures obtain their commutative groups *outside* the ring. This idea is clarified in the following section.

7.1 MODULES

This section uses as its building blocks a commutative group

$$(G, \square, e_G) \tag{7-11}$$

and a ring

$$(R, *, e_R, \triangle, 1_R). \tag{7-12}$$

If (7-11) and (7-12) are given, then an *R-module G* is defined to be the commutative group (7-11) together with a function

$$f : R \times G \to G \tag{7-13}$$

satisfying the four axioms

$$f(r, g_1 \square g_2) = f(r, g_1) \square f(r, g_2), \tag{7-14a}$$
$$f(r_1 * r_2, g) = f(r_1, g) \square f(r_2, g), \tag{7-14b}$$
$$f(r_1 \triangle r_2, g) = f(r_1, f(r_2, g)), \tag{7-14c}$$
$$f(1_R, g) = g \tag{7-14d}$$

for all r, r_1, and r_2 in the ring and for all g, g_1, and g_2 in the commutative group.

EXAMPLE 7.1-1

As in Chapter 6, select (7-11) as the commutative group and

$$(\mathbb{EM}(G), \square, e_{GG}, \circ, 1_G) \tag{7-15}$$

as the ring. Define the function (7-13) by an action

$$f(h, g) = h(g), \tag{7-16}$$

the result of applying the endomorphism h to the element g in G. Then an $\mathbb{EM}(G)$-module G has been established. It is instructive to verify the axioms. First,

$$f(h, g_1 \square g_2) = h(g_1 \square g_2) = (hg_1) \square (hg_2) \tag{7-17a,b}$$
$$= f(h, g_1) \square f(h, g_2), \tag{7-17c}$$

with the key step from (7-17a) to (7-17b) a consequence of the fact that h is a morphism. Second,

$$f(h_1 \square h_2, g) = (h_1 \square h_2)(g) = (h_1 g) \square (h_2 g) \tag{7-17d,e}$$
$$= f(h_1, g) \square f(h_2, g), \tag{7-17f}$$

by definition of $\square$ on $\mathbb{EM}(G)$. Third,

$$f(h_1 \circ h_2, g) = (h_1 \circ h_2)(g) = h_1(h_2 g) \qquad (7\text{-}17\text{g,h})$$
$$= h_1(f(h_2, g)) \qquad (7\text{-}17\text{i})$$
$$= f(h_1, f(h_2, g)), \qquad (7\text{-}17\text{j})$$

by definition of $\circ$ on $\mathbb{EM}(G)$. Fourth,

$$f(1_G, g) = 1_G(g) = g, \qquad (7\text{-}17\text{k})$$

as required. ∎

The function f of (7-13) is often called *scalar multiplication* of the commutative group element. Moreover, the explicit notation

$$f(r, g) \qquad \text{is typically replaced by} \qquad rg. \qquad (7\text{-}18\text{a,b})$$

For Example 7.1-1, the reader can observe that (7-18b) corresponds to

$$hg. \qquad (7\text{-}19)$$

It is possible, therefore, to regard the action of a system on its input as scalar multiplication in a module. This occurs, however, only when the input and output groups of the system morphism are the same, as in Chapter 6.

Now let two commutative groups

$$(G_i, \square_i, e_{G_i}), \qquad i = 1, 2, \qquad (7\text{-}20)$$

be given. Develop them into R-modules G_1 and G_2, respectively, by appropriate scalar multiplications

$$f_i : R \times G_i \to G_i, \qquad i = 1, 2 \qquad (7\text{-}21)$$

with reference to the ring (7-12). Suppose that a morphism

$$h : G_1 \to G_2 \qquad (7\text{-}22)$$

of groups is given. Then h becomes a *morphism of R-modules* if

$$h(f_1(r, g)) = f_2(r, h(g)) \qquad (7\text{-}23)$$

for all r in R and for all g in G_1. In terms of (7-18b), (7-23) appears in the form

$$h(rg) = rh(g). \qquad (7\text{-}24)$$

EXAMPLE 7.1-2

Suppose that h is a morphism of R-modules. What can be said of the morphism of groups $\hat{h}$? The calculation

$$\hat{h}(f_1(r, g)) = \widehat{h(f_1(r, g))} = h(\widehat{f_1(r, g)}) \qquad (7\text{-}25\text{a,b})$$

suggests that some knowledge of

$$f_1(r, g) \tag{7-26}$$

would be helpful. To examine (7-26), begin by observing that, for a *fixed r* in R, the axiom (7-14a) defines an endomorphism of groups. It then follows that

$$\widehat{f_1(r, g)} = f_1(r, \hat{g}), \tag{7-27}$$

which means that (7-25) may be developed further:

$$h(\widehat{f_1(r, g)}) = h(f_1(r, \hat{g})) = f_2(r, h(\hat{g})) \tag{7-28a,b}$$
$$= f_2(r, \widehat{h(g)}) = f_2(r, \hat{h}(g)). \tag{7-28c,d}$$

But (7-25) and (7-28) imply that $\hat{h}$ is also a morphism of R-modules.   ∎

EXAMPLE 7.1-3

Consider the set of endomorphisms of a commutative group (7-11). It was determined in the previous chapter that this set is a ring, namely, (7-15). Now develop $(G, \square, e_G)$ into an R-module G by an appropriate scalar multiplication $f: R \times G \to G$. Some of the endomorphisms in $\mathbb{EM}(G)$ are also endomorphisms of R-modules. For example,

$$1_G(f(r, g)) = f(r, g) = f(r, 1_G(g)). \tag{7-29}$$

Moreover, if h_1 and h_2 in $\mathbb{EM}(G)$ are morphisms of R-modules, then

$$(h_1 \square h_2)(f(r, g)) = h_1(f(r, g)) \square h_2(f(r, g)) \tag{7-30a}$$
$$= f(r, h_1(g)) \square f(r, h_2(g)) \tag{7-30b}$$
$$= f(r, h_1(g) \square h_2(g)) \tag{7-30c}$$
$$= f(r, (h_1 \square h_2)(g)), \tag{7-30d}$$

where the step from (7-30b) to (7-30c) follows by the axiom (7-14a). Together with Example 7.1-2, these calculations establish that the subset of $\mathbb{EM}(G)$ containing endomorphisms which are also endomorphisms of R-modules is a subgroup of $\mathbb{EM}(G)$. In view of (7-29), subring status for this subset is established by the one additional calculation

$$(h_1 \circ h_2)(f(r, g)) = h_1(h_2(f(r, g))) \tag{7-31a}$$
$$= h_1(f(r, h_2(g))) \tag{7-31b}$$
$$= f(r, h_1(h_2(g))) \tag{7-31c}$$
$$= f(r, (h_1 \circ h_2)(g)). \tag{7-31d}$$

∎

The arguments above can be generalized somewhat to the case in which the module morphism does not have the same domain and codomain modules. Thus, when defined, the composition of two morphisms of R-modules is another morphism of R-modules. Of course, the ring structure of Example 7.1-3 is not attained in these circumstances.

Consider next an R-module G, where the commutative group and the ring are described by (7-11) and (7-12), respectively, and the scalar multiplication function is given by (7-13). Then a *submodule* of the R-module G is a subgroup S of G which has the property

$$f_*(R, S) \subset S. \tag{7-32}$$

This permits a restriction

$$f \,|\, R \times S : R \times S \to S \tag{7-33}$$

of the function (7-13), which already satisfies the axioms (7-14).

EXAMPLE 7.1-4

The kernel of a morphism of R-modules is a submodule. In the notation of (7-20) to (7-23), h must be a morphism of groups; thus Ker h must be a subgroup. Moreover, if

$$hg_1 = e_{G_2}, \tag{7-34a}$$

then

$$h(f_1(r, g_1)) = f_2(r, hg_1) = f_2(r, e_{G_2}) = e_{G_2}, \tag{7-34b,c,d}$$

which completes the submodule demonstration. ∎

EXAMPLE 7.1-5

The image of a morphism of R-modules is a submodule. Indeed, Section 4.1 already established that it is a subgroup. Moreover, if

$$hg_1 = g_2 \tag{7-35a}$$

for some $g_2 \in \text{Im}\, h$, then

$$f_2(r, g_2) = f_2(r, hg_1) = h(f_1(r, g_1)), \tag{7-35b,c}$$

so that

$$f_2(r, g_2) \in \text{Im}\, h \tag{7-36}$$

as well. ∎

The reader may have noted that the kernel and image of a morphism of *R*-modules are both the same type of subsets. In this respect, they differ from the kernels and images of morphisms of groups and rings. Recall that the image of a morphism of groups was a subgroup whereas its kernel was a normal subgroup. Similarly, the image of a morphism of rings was a subring, whereas its kernel was an ideal.

Example 7.1-1 has indicated that much of what was said in Chapter 6 may be understood also in module theoretic terms. Actually, however, considerably more is true. For the discussion on GRMPSs and their inverses also has such an interpretation. This idea is a consequence of the fact that every commutative group is a $\mathbb{Z}$-module. This is the content of the next example.

EXAMPLE 7.1-6

Let $(G, \square, e_G)$ be a commutative group. Then a $\mathbb{Z}$-module G can be constructed by defining a scalar multiplication function

$$f. \mathbb{Z} \times G \to G \tag{7-37a}$$

having the action

$$f(0, g) = e_G, \tag{7-37b}$$

$$f(i, g) = \coprod_{k=1}^{i} g, \qquad i = 1, 2, 3, \ldots, \tag{7-37c}$$

$$f(-i, g) = \coprod_{k=1}^{i} \hat{g}, \qquad i = 1, 2, 3, \ldots. \tag{7-37d}$$

Here, of course, the symbol

$$\coprod_{k=1}^{i} g \qquad \text{stands for} \qquad g \,\square\, g \,\square \cdots \square\, g, \tag{7-38a,b}$$

where g appears i times. The axioms may be checked. Thus

$$f(0, g_1 \,\square\, g_2) = e_G = e_G \,\square\, e_G = f(0, g_1) \,\square\, f(0, g_2); \tag{7-39}$$

$$f(i, g_1 \,\square\, g_2) = \coprod_{k=1}^{i} (g_1 \,\square\, g_2)$$

$$= (g_1 \,\square\, g_2) \,\square\, (g_1 \,\square\, g_2) \,\square \cdots \square\, (g_1 \,\square\, g_2)$$

$$= \left(\coprod_{k=1}^{i} g_1 \right) \square \left(\coprod_{k=1}^{i} g_2 \right)$$

$$= f(i, g_1) \,\square\, f(i, g_2); \tag{7-40}$$

a similar calculation on (7-37d) completes axiom (7-14a). Axiom (7-14b) is a straightforward exercise, as is axiom (7-14d). Axiom (7-14c) requires manipulation of expressions of the type

$$\prod_{k_1=1}^{i}\left(\prod_{k_2=1}^{j} g\right), \quad \prod_{k_1=1}^{i}\left(\prod_{k_2=1}^{j} \hat{g}\right), \quad \prod_{k_1=1}^{i}\left(\widehat{\prod_{k_2=1}^{j} g}\right), \quad \prod_{k_1=1}^{i}\left(\widehat{\prod_{k_2=1}^{j} \hat{g}}\right),$$

(7-41a,b,c,d)

all of which can be worked out by standard commutative group properties. These details are left as exercises.　∎

It should be observed that the definition (7-37b) is not arbitrary. Indeed, axiom (7-14b) shows that, for a *fixed* element of the commutative group G, f becomes a morphism of commutative groups in the manner

$$(R, *, e_R) \rightarrow (G, \Box, e_G). \tag{7-42}$$

Because such a morphism must carry units to units, the definition (7-37b) is *necessary*.

The idea of Example 7.1-6 has been prefigured in Section 4.7. See, for example, (4-193). In fact, (4-196) introduced a notation

$$(\mathbb{Z})g = \{(z)g \,|\, z \in \mathbb{Z}\} \tag{7-43}$$

for the set of all elements in G which could be generated by a process of scalar multiplication with elements from $\mathbb{Z}$. It is not hard to see that $(\mathbb{Z})g$ is a subgroup of G. Axiom (7-14b) establishes closure under the binary operation $\Box$ on G, and closure under inversion follows from

$$\widehat{f(r,g)} = f(\hat{r}, g), \tag{7-44}$$

which is itself a consequence of the calculation

$$e_G = f(0, g) = f(r * \hat{r}, g) = f(r, g) \,\Box\, f(\hat{r}, g). \tag{7-45}$$

Since closure under scalar multiplication is a consequence of (7-14c), it follows that (7-43) is a submodule of the $\mathbb{Z}$-module G.

The fact that every commutative group is a $\mathbb{Z}$-module certainly adds to the evidence of utility for this type of algebraic structure. Moreover, submodules of type (7-43) and their generalizations

$$Rg = \{rg \,|\, r \in R\} \tag{7-46}$$

are very important in system theory. This topic is pursued further in the next section.

Exercises

7.1-1. With the aid of the commutative group $(G, \square, e_G)$ of (7-11) and the ring $\mathbb{Z}[h]$ of (7-6), define a scalar multiplication function

$$\mathbb{Z}[h] \times G \to G$$

and thus establish a $\mathbb{Z}[h]$-module G. Verify explicitly the four scalar multiplication axioms (7-14).

7.1-2. Repeat Exercise 7.1-1, but this time with the ring $S_h[h]$ of (7-7).

7.1-3. In Example 7.1-6, establish axiom (7-14b) for the scalar multiplication (7-37). Notice carefully that there are multiple cases to be considered, because the actions (7-37b)–(7-37d) are defined on various subsets of the domain of the scalar multiplication function. Also show axiom (7-14d).

7.1-4. In Example 7.1-6, establish axiom (7-14c).

7.1-5. Explain the step from (7-34c) to (7-34d).

7.1-6. Refer to (7-43). Show that $(\mathbb{Z})g$ is a subgroup of G.

7.1-7. Explain in detail why $(\mathbb{Z})g$ of (7-43) is closed under scalar multiplication. Use (7-14c).

7.1-8. Consider two R-modules G_1 and G_2 as introduced in (7-20) and (7-21). Determine whether or not

$$h_1 \square_2 h_2$$

is a morphism of R-modules when

$$h_i : G_1 \to G_2, \qquad i = 1, 2,$$

are morphisms of R-modules.

7.1-9. Develop R-modules G_i, $i = 1, 2, 3$, and define morphisms

$$h_1 : G_1 \to G_2, \qquad h_2 : G_2 \to G_3$$

of R-modules. Show that

$$h_2 \circ h_1 : G_1 \to G_3$$

is a morphism of R-modules.

7.2 SIGNAL REPRESENTATION

Consider once more the R-module G with R and G denoted again by (7-11) and (7-12) in the preceding section. It has been pointed out at the end of Section 7.1 that there are submodules Rg defined for each fixed element g in G. Rg is called a *cyclic module* and is said to be *spanned* by g or *generated* by g.

By means of the notation developed in Section 4.6, it is then possible to develop subgroups

$$\underset{i=1}{\overset{n}{\square}} \, Rg_i \tag{7-47a}$$

consisting of all *R-linear combinations*

$$\left\{ \bigsqcup_{i=1}^{n} r_i g_i \,\middle|\, r_i \in R \right\} \tag{7-47b}$$

of the *n* group elements g_i in *G*. That (7-47) also defines a submodule follows from the calculation

$$r \left\{ \bigsqcup_{i=1}^{n} r_i g_i \right\} = \bigsqcup_{i=1}^{n} r(r_i g_i) = \bigsqcup_{i=1}^{n} (r \triangle r_i) g_i, \tag{7-48a,b}$$

where the first step is a consequence of (7-14a) and the second step is a consequence of (7-14c). Notice that the convention (7-18b) is applied here, for simplicity.

The submodule (7-47a) is said to be *spanned* by the elements $g_1, g_2, \ldots, g_n$ in *G*. The *R*-module *G* is said to be of *finite type* if there is a natural number *n* in $\mathbb{N}$ and *n* elements g_i in *G* such that

$$G = \bigsqcup_{i=1}^{n} R g_i. \tag{7-49}$$

If an *R*-module *G* is of finite type, then it follows that every element *g* in *G* can be written in the manner

$$g = \bigsqcup_{i=1}^{n} r_i g_i \tag{7-50}$$

for an appropriate *R*-linear combination of the spanning elements g_i, $i = 1$, $2, \ldots, n$. This suggests at once the possibility of representing a signal *g* by the element

$$(r_1, r_2, \ldots, r_n) \tag{7-51}$$

in the product R^n. Pitfalls, however, can await the unwary in pursuing this notion.

EXAMPLE 7.2-1

Let the ring be the integers $(\mathbb{Z}, +, 0, \cdot, 1)$, and consider the submodule $\mathbb{Z}g$ for a nonunit element *g* in *G*. Then *g* spans $\mathbb{Z}g$, and every element in $\mathbb{Z}g$ can be written

$$zg \tag{7-52}$$

for an appropriate *z* in $\mathbb{Z}$. Thus such elements are candidates to be represented by the simplest possible version of (7-51), namely,

$$(z) \tag{7-53}$$

with n equal to 1. However, consider the statement

$$\mathbb{Z}g = \mathbb{Z}(2g) \,\square\, \mathbb{Z}(3g). \tag{7-54}$$

Is (7-54) a verity? The answer is yes, because

$$(-4)(2g) \,\square\, (3)(3g) = (-8)g \,\square\, (9)g = (-8 + 9)g = 1g = g. \tag{7-55a,b,c,d}$$

The steps in (7-55) follow from (7-14c), (7-14b), and (7-14d), respectively. Therefore the generator g of $\mathbb{Z}g$ can itself be generated from $2g$ and $3g$; and so $2g$ and $3g$ span $\mathbb{Z}g$. From this viewpoint, a pair

$$(z_1, z_2) \tag{7-56}$$

might be proposed to represent elements in $\mathbb{Z}g$ with n equal to 2. Moreover, it is not possible to write either

$$2g = (z)3g \qquad \text{or} \qquad 3g = (\bar{z})2g \tag{7-57a,b}$$

for any choices of z or $\bar{z}$ in $\mathbb{Z}$. Neither $2g$ nor $3g$ can be generated from the other. Notice, however, that

$$(-3)(2g) \,\square\, (2)(3g) = (-6)g \,\square\, (6)g = 0g = e_G. \tag{7-58a,b,c}$$

∎

Example 7.2-1 shows that a set of spanning elements for an R-module G may be greater in number than necessary. Moreover, such a spanning set may not always be easy to replace with a spanning set containing fewer elements. The possibility for doing so may be dependent upon the ring R.

Now suppose that an R-module G of finite type is given. A set of n non-trivial elements g_i, $i = 1, 2, \ldots, n$ in G, is said to be R-*linearly independent* if

$$\square_{i=1}^{n} \, r_i g_i = e_G \qquad \text{implies} \qquad r_i = e_R, \qquad i = 1, 2, \ldots, n. \tag{7-59a,b}$$

It may be helpful to recall from the remarks leading up to (7-42) that

$$e_R g = e_G \tag{7-60}$$

for all g in G.

If an R-module G satisfies

$$G = \square_{i=1}^{n} \, Rg_i \tag{7-61}$$

for R-linearly independent elements $g_1, g_2, \ldots, g_n$ in G, then these elements are said to be a *basis* for the module; and the R-module G is said to be *free* on this basis.

If an R-module G has a basis consisting of the n elements $g_1, g_2, \ldots,$ g_n, then the representation (7-50) is unique. Indeed, assume an alternative representation

$$g = \bigsqcup_{i=1}^{n} \tilde{r}_i g_i. \tag{7-62}$$

Then (7-50) and (7-62) imply

$$\bigsqcup_{i=1}^{n} r_i g_i = \bigsqcup_{i=1}^{n} \tilde{r}_i g_i \tag{7-63}$$

and consequently

$$e_G = \left(\bigsqcup_{i=1}^{n} r_i g_i \right) \square \left(\overparen{\bigsqcup_{i=1}^{n} \tilde{r}_i g_i} \right) \tag{7-64a}$$

$$= \left(\bigsqcup_{i=1}^{n} r_i g_i \right) \square \left(\bigsqcup_{i=1}^{n} \widehat{\tilde{r}_i g_i} \right) \tag{7-64b}$$

$$= \left(\bigsqcup_{i=1}^{n} r_i g_i \right) \square \left(\bigsqcup_{i=1}^{n} \hat{\tilde{r}}_i g_i \right) \tag{7-64c}$$

$$= \bigsqcup_{i=1}^{n} (r_i * \hat{\tilde{r}}_i) g_i. \tag{7-64d}$$

But R-linear independence then implies

$$r_i * \hat{\tilde{r}}_i = e_R, \qquad i = 1, 2, \ldots, n, \tag{7-65a}$$

which gives

$$r_i = \tilde{r}_i, \qquad i = 1, 2, \ldots, n. \tag{7-65b}$$

EXAMPLE 7.2-2

Let $(R, *, e_R, \triangle, 1_R)$ be a ring. Then $(R, *, e_R)$ is a commutative group, and it is possible to form the product group

$$R^n = \{(r_1, r_2, \ldots, r_n) \mid r_i \in R\} \tag{7-66}$$

with the elementwise binary operation

$$(r_1, r_2, \ldots, r_n) ** (\tilde{r}_1, \tilde{r}_2, \ldots, \tilde{r}_n) = (r_1 * \tilde{r}_1, r_2 * \tilde{r}_2, \ldots, r_n * \tilde{r}_n) \tag{7-67a}$$

having unit

$$(e_R, e_R, \ldots, e_R) \tag{7-67b}$$

and an inverse

$$(\hat{r}_1, \hat{r}_2, \ldots, \hat{r}_n) \qquad \text{for an element} \qquad (r_1, r_2, \ldots, r_n) \qquad \text{(7-67c,d)}$$

in R^n. Then R^n can be made into an R-module by defining the scalar multiplication

$$r(r_1, r_2, \ldots, r_n) = (r \triangle r_1, r \triangle r_2, \ldots, r \triangle r_n). \qquad \text{(7-68)}$$

So equipped, the R-module R^n is free on the n basis elements

$$(1_R, e_R, e_R, \ldots, e_R, e_R),$$
$$(e_R, 1_R, e_R, \ldots, e_R, e_R),$$
$$(e_R, e_R, 1_R, \ldots, e_R, e_R),$$
$$\vdots$$
$$(e_R, e_R, e_R, \ldots, 1_R, e_R),$$
$$(e_R, e_R, e_R, \ldots, e_R, 1_R), \qquad \text{(7-69)}$$

as the reader may verify. ∎

With the aid of the R-module R^n, the idea of uniquely representing elements of an R-module G of finite type can be formalized. Suppose that $g_1, g_2, \ldots, g_n$ are a basis for such a module. Then each element g in G has the unique representation (7-50). Define a function

$$\alpha : G \to R^n \qquad \text{by the action} \qquad \alpha(g) = (r_1, r_2, \ldots, r_n). \qquad \text{(7-70a,b)}$$

Then α is a morphism of groups by means of the demonstration

$$\alpha(g \square \tilde{g}) = (r_1 * \tilde{r}_1, r_2 * \tilde{r}_2, \ldots, r_n * \tilde{r}_n)$$
$$= (r_1, r_2, \ldots, r_n) ** (\tilde{r}_1, \tilde{r}_2, \ldots, \tilde{r}_n)$$
$$= \alpha(g) ** \alpha(\tilde{g}), \qquad \text{(7-71)}$$

where

$$g = \underset{i=1}{\overset{n}{\square}} r_i g_i, \qquad \tilde{g} = \underset{i=1}{\overset{n}{\square}} \tilde{r}_i g_i. \qquad \text{(7-72)}$$

Moreover,

$$\alpha(rg) = \alpha \left(r \underset{i=1}{\overset{n}{\square}} r_i g_i \right) = \alpha \left(\underset{i=1}{\overset{n}{\square}} (r \triangle r_i) g_i \right)$$
$$= (r \triangle r_1, r \triangle r_2, \ldots, r \triangle r_n)$$
$$= r(r_1, r_2, \ldots, r_n) = r\alpha(g), \qquad \text{(7-73)}$$

so that α is a morphism of R-modules as well. If

$$\alpha(g) = (r_1, r_2, \ldots, r_n) = (e_R, e_R, \ldots, e_R), \qquad \text{(7-74a)}$$

then

$$r_i = e_R, \quad i = 1, 2, \ldots, n, \quad \text{and} \quad g = \coprod_{i=1}^{n} e_R g_i = e_G. \tag{7-74b,c}$$

Thus

$$\text{Ker } \alpha = \{e_G\}, \tag{7-74d}$$

the unit submodule. Moreover, it follows easily that

$$\text{Im } \alpha = R^n. \tag{7-75}$$

Thus α is an isomorphism

$$G \approx R^n \tag{7-76}$$

of R-modules.

The isomorphism (7-70) makes it possible to set up matrices to describe morphisms of free R-modules. Even though that is true, however, the reader should not be lulled into complacency. And for this reason, another example is offered.

EXAMPLE 7.2-3

Consider the set

$$\mathbb{Z}_4 = \{0, 1, 2, 3\} \tag{7-77a}$$

equipped with the addition operation

+	0	1	2	3
0	0	1	2	3
1	1	2	3	0
2	2	3	0	1
3	3	0	1	2

$$\tag{7-77b}$$

and the multiplication operation

·	0	1	2	3
0	0	0	0	0
1	0	1	2	3
2	0	2	0	2
3	0	3	2	1

$$\tag{7-77c}$$

These operations develop $\mathbb{Z}_4$ into a ring

$$(\mathbb{Z}_4, +, 0, \cdot, 1), \tag{7-78}$$

and the product $(\mathbb{Z}_4)^2$ develops into a $\mathbb{Z}_4$-module. The interesting thing to note is that an element of $(\mathbb{Z}_4)^2$ which is not a unit, as for example

$$(2, 2), \tag{7-79a}$$

can be scalar multiplied by an element of $\mathbb{Z}_4$ which is not a unit, as for example

$$2, \tag{7-79b}$$

to give the unit in $(\mathbb{Z}_4)^2$, as illustrated by

$$2(2, 2) = (2 \cdot 2, 2 \cdot 2) = (0, 0). \tag{7-79c}$$

Such a phenomenon is called *torsion*, and will be considered further in a later section. ∎

The reader should now be in a position to see that the GRMPS theory of Chapter 4 can be generalized in several ways. First, commutative input, state, and output groups can be recognized as $\mathbb{Z}$-modules, and the group morphisms (a, b, c, d) as morphisms of $\mathbb{Z}$-modules. Second, more general R-modules and morphisms thereof can be developed. Third, if the R-modules describing the inputs, states, and outputs are free on given bases, then matrix descriptions can be developed. The next section begins to explore these points.

Exercises

7.2-1. What is the algebraic explanation for a step of the type which passes from (7-64b) to (7-64c)?

7.2-2. Demonstrate in complete detail how to pass from (7-64c) to (7-64d).

7.2-3. In Example 7.2-2, verify the scalar multiplication axioms (7-14) for the definition (7-68).

7.2-4. In Example 7.2-2, verify that the R-module R^n is free on the n basis elements (7-69).

7.2-5. What justifies the first step in (7-71)? Explain in detail.

7.2-6. Carefully detail the steps which lead from (7-74a) to (7-74b).

7.2-7. Establish (7-75).

7.2-8. Let $g_1, g_2, \ldots, g_n$ be a basis for the R-module G. Show that, if $r \neq e_R$,

$$rg_i = e_G$$

is not possible for $i = 1, 2, 3, \ldots, n$.

7.2-9. Consider commutative groups

$$(G_i, \square_i, e_{G_i}), \qquad i = 1, 2,$$

and a morphism of groups

$$h: G_1 \to G_2.$$

If the groups G_i are regarded as $\mathbb{Z}$-modules, show that h also becomes a morphism of $\mathbb{Z}$-modules.

7.3   MODULE MORPHIC SYSTEMS

Section 4.2 has introduced the notion of a group morphic system (GRMPS). The ingredients were an input group

$$(U, \triangle, e_U), \qquad (7\text{-}80\text{a})$$

a state group

$$(X, \square, e_X), \qquad (7\text{-}80\text{b})$$

and an output group

$$(Y, *, e_Y), \qquad (7\text{-}80\text{c})$$

together with four morphisms

$$a: X \to X \qquad b: U \to X, \qquad c: X \to Y \qquad d: U \to Y \quad (7\text{-}81\text{a,b,c,d})$$

of groups. The local dynamical equations for a GRMPS were given by

$$x_{k+1} = ax_k \,\square\, bu_k, \qquad y_k = cx_k * du_k. \qquad (7\text{-}82\text{a,b})$$

In the present section, it is assumed that the GRMPS (a, b, c, d) is commutative, in the sense of Section 5.6, which is the same as requiring that each of the three groups in (7-80) be commutative.

Under these assumptions, (7-80a) can be regarded as a $\mathbb{Z}$-module U; (7-80b) can be viewed as a $\mathbb{Z}$-module X; and (7-80c) has interpretation as a $\mathbb{Z}$-module Y. Actually, more is true, as is brought out in the following example.

EXAMPLE 7.3-1

Suppose that

$$\beta: G_1 \to G_2 \qquad (7\text{-}83\text{a})$$

is a morphism of the commutative groups

$$(G_i, \square_i, e_{G_i}), \qquad i = 1, 2. \qquad (7\text{-}83\text{b})$$

Next develop (7-83b) into a pair of $\mathbb{Z}$-modules by the process described in Example 7.1-6. Denote the resulting scalar multiplication functions by

$$f_i: \mathbb{Z} \times G_i \to G_i, \qquad i = 1, 2. \qquad (7\text{-}84)$$

Now consider the computations

$$\beta(f_1(i, g)) = \beta\left(\overset{i}{\underset{k=1}{\square}} g\right) = \overset{i}{\underset{k=1}{\square}} \beta g = f_2(i, \beta g), \qquad (7\text{-}85)$$

and

$$\beta(f_1(-i,g)) = \beta\left(\coprod_{k=1}^{i} \hat{g}\right) = \coprod_{k=1}^{i} \beta\hat{g} = \coprod_{k=1}^{i} \widehat{\beta g} = f_2(-i, \beta g), \qquad (7\text{-}86)$$

and

$$\beta(f_1(0,g)) = \beta e_{G_1} = e_{G_2} = f_2(0, \beta g). \qquad (7\text{-}87)$$

This shows that morphisms of commutative groups develop into morphisms of $\mathbb{Z}$-modules when the commutative groups develop into $\mathbb{Z}$-modules. ∎

From these deliberations, it follows that a commutative group morphic system is also a module morphic system in a clear sense. To make this notion precise, let the three groups of (7-80) be commutative. Define three scalar multiplication functions

$$f_U: R \times U \to U, \qquad f_X: R \times X \to X, \qquad f_Y: R \times Y \to Y \quad (7\text{-}88\text{a,b,c})$$

for the three groups, and require them to satisfy the axioms (7-14), for a ring

$$(R, *, e_R, \triangle, 1_R). \qquad (7\text{-}89)$$

Then (7-88) develops (7-80) into R-modules U, X, and Y of inputs, states, and outputs respectively. Finally, require the morphisms of groups (7-81) to be morphisms also of R-modules. Then the dynamical system

$$(U, X, Y; a, b, c, d) \qquad (7\text{-}90)$$

is a *module morphic system* (MOMPS).

EXAMPLE 7.3-2

The commutative group morphic systems (GRMPSs) are also module morphic systems (MOMPSs) in the current sense, with the ring (7-89) chosen to be $(\mathbb{Z}, +, 0, \cdot, 1)$. ∎

The discussion now returns to the topic of reachability, as examined in Sections 4.6 and 4.7. For a MOMPS, the inputs, states, and outputs are elements of commutative groups. Aside from the scalar multiplication functions (7-88), then, a MOMPS is a commutative GRMPS. This means that the sets

$$X_i, \qquad i = 0, 1, 2, \ldots, \qquad (7\text{-}91)$$

of states which can be reached from the unit state e_X at index i are subgroups of X. The purpose here is to establish that in the case of a MOMPS the X_i are submodules as well. To accomplish this end, reconsider (4-164). In that equation a function

$$\prod_{j=0}^{i} a^{i-j}b: \tilde{U}_{i+1} \to X \tag{7-92a}$$

was set up with action

$$\left(\prod_{j=0}^{i} a^{i-j}b\right)(u_0, u_1, u_2, \ldots, u_i) = \prod_{j=0}^{i} a^{i-j}bu_j. \tag{7-92b}$$

At that point, the morphisms $a: X \to X$ and $b: U \to X$ were of groups; at this point, they are regarded as morphisms of R-modules. It will be shown that (7-92a) has the interpretation of a morphism of R-modules. Once that fact is established, it follows from Example 7.1-5 that the image of that morphism is a submodule of the state module of the MOMPS.

As a preliminary fact, it is convenient to show that the composition of two morphisms of R-modules is another morphism of R-modules. This conclusion has been hinted in Section 7.1, but will be detailed here. Consider commutative groups

$$(G_i, \square_i, e_{G_i}), \qquad i = 1, 2, 3, \tag{7-93a}$$

equipped with scalar multiplication functions

$$f_i: R \times G_i \to G_i, \qquad i = 1, 2, 3, \tag{7-93b}$$

with the ring denoted by (7-89). Define two R-module morphisms

$$\alpha: G_1 \to G_2 \quad \text{and} \quad \beta: G_2 \to G_3. \tag{7-94a,b}$$

Now examine

$$\beta \circ \alpha: G_1 \to G_3 \tag{7-95}$$

to verify that it is a morphism of R-modules. Begin by examining

$$(\beta \circ \alpha)(f_1(r, g)) = \beta(\alpha(f_1(r, g))) = \beta(f_2(r, \alpha g)) \tag{7-96a,b}$$
$$= f_3(r, \beta(\alpha g)) = f_3(r, (\beta \circ \alpha)(g)); \tag{7-96c,d}$$

then recall from Section 4.3 that $\beta \circ \alpha$ is already a morphism of groups because α and β must be morphisms of groups in order to be morphisms of R-modules. Thus a composition (7-95) of two R-module morphisms (7-94) is also an R-module morphism.

Next establish an R-module structure on $\tilde{U}_{i+1}$. To do this, construct the product group

$$U^{i+1} = \{(u_0, u_1, u_2, \ldots, u_i)\} \qquad (7\text{-}97)$$

in the usual manner, as introduced in Section 4.2. This product group develops naturally into an R-module by a definition of scalar multiplication

$$r(u_0, u_1, u_2, \ldots, u_i) = (f_U(r, u_0), f_U(r, u_1), \ldots, f_U(r, u_i)). \qquad (7\text{-}98)$$

Finally, then, verification that (7-92) is an R-module morphism rests upon two calculations. The first of these, namely,

$$\left(\bigsqcup_{j=0}^{i} a^{i-j} b\right)(u_0 \triangle \tilde{u}_0, u_1 \triangle \tilde{u}_1, \ldots, u_i \triangle \tilde{u}_i)$$

$$= \bigsqcup_{j=0}^{i} a^{i-j} b(u_j \triangle \tilde{u}_j)$$

$$= \bigsqcup_{j=0}^{i} \{(a^{i-j} b u_j) \,\square\, (a^{i-j} b \tilde{u}_j)\}$$

$$= \left\{\bigsqcup_{j=0}^{i} a^{i-j} b u_j\right\} \,\square\, \left\{\bigsqcup_{j=0}^{i} a^{i-j} b \tilde{u}_j\right\}$$

$$= \left(\bigsqcup_{j=0}^{i} a^{i-j} b\right)(u_0, u_1, \ldots, u_i)$$

$$\square \left(\bigsqcup_{j=0}^{i} a^{i-j} b\right)(\tilde{u}_0, \tilde{u}_1, \ldots, \tilde{u}_i), \qquad (7\text{-}99)$$

shows that (7-92) is a morphism of groups. The second, which is

$$\left(\bigsqcup_{j=0}^{i} a^{i-j} b\right)(f_U(r, u_0), f_U(r, u_1), \ldots, f_U(r, u_i)) = \bigsqcup_{j=0}^{i} a^{i-j} b(f_U(r, u_j))$$

$$= \bigsqcup_{j=0}^{i} f_X(r, a^{i-j} b u_j)$$

$$= f_X\left(r, \bigsqcup_{j=0}^{i} a^{i-j} b u_j\right), \qquad (7\text{-}100)$$

with the last step a consequence of (7-14a), completes the argument that (7-92) is a morphism of R-modules.

Thus the subgroups (7-91) become submodules of the R-module X. Now the explanations leading up to Fig. 4.10 have established

$$X_0 \subset X_1 \subset X_2 \subset \cdots \subset X, \tag{7-101}$$

which is an ascending chain of submodules. In Chapter 4, it was shown by means of arguments based upon the complete lattice

$$(\mathbb{P}(X), \subset, \cap, \cup) \tag{7-102}$$

that the chain has a least upper bound

$$X^* = \sup\{X_i \,|\, i \in \mathbb{N}\}. \tag{7-103}$$

The MOMPS is reachable only if

$$X^* = X; \tag{7-104}$$

but the complete discussion of the question whether

$$X^* = X_i \tag{7-105}$$

for some finite i in $\mathbb{N}$ was postponed. The answer to this question is carried out in part in the next example.

EXAMPLE 7.3-3

Consider an R-module G, and an arbitrary ascending chain

$$G_0 \subset G_1 \subset G_2 \subset \cdots \subset G \tag{7-106}$$

of submodules. If there exists for every such chain a natural number i in $\mathbb{N}$ such that

$$G_i = G_{i+1} = G_{i+2} = \cdots, \tag{7-107}$$

then it must follow that every submodule of the R-module G is of finite type. For suppose that $S \subset G$ is a submodule which is not of finite type. Then no finite number of elements in S can span S. Select s_0 in S, and define

$$S_0 = Rs_0. \tag{7-108a}$$

Then there must be an element s_1 in S such that $s_1 \notin S_0$. Define

$$S_1 = S_0 \,\square\, Rs_1 \tag{7-108b}$$

and observe that

$$S_0 \subset S_1. \tag{7-108c}$$

Now suppose that

$$S_p = \coprod_{k=0}^{p} Rs_k \qquad (7\text{-}109a)$$

for $s_k \in S$, $k = 0, 1, 2, \ldots, p$. There must be an element s_{p+1} in S which is not in S_p. So

$$S_{p+1} = S_p \square Rs_{p+1} \quad \text{and} \quad S_p \subset S_{p+1}. \qquad (7\text{-}109b,c)$$

In this way an ascending chain

$$S_0 \subset S_1 \subset S_2 \subset \cdots \subset G \qquad (7\text{-}110)$$

which does not satisfy (7-107) is determined, in contradiction to assumption. Thus, a necessary condition for (7-107) to hold for every (7-106) is that each submodule of the R-module G is of finite type. To show that this condition is sufficient as well, consider the union of the submodules (7-106). An element g in the union is an element in at least one of the submodules G_k. Therefore $rg \in G_k$ and $\hat{g} \in G_k$ because G_k is a submodule. Suppose next that elements g and $\tilde{g}$ in the union are given. Then $g \in G_k$ for some k and $\tilde{g} \in G_j$ for some j. Let m be the greater of k and j. Then g and $\tilde{g}$ are both elements of G_m, and so is $g \square \tilde{g}$. This means that the union is a submodule and is consequently of finite type, being spanned by some finite number of elements. Each of these elements must be a member of at least one of the submodules in (7-106). For each of the elements, select one of those submodules. If G_i is any of the submodules in (7-106) which contains all those so chosen, then (7-107) is satisfied. ∎

This example shows that R-modules X in which every submodule is of finite type will have the property (7-105) for an appropriate index i. Thus, a MOMPS with such a state module will have a *reachable submodule*.

Now one possible submodule of the state module is, of course, itself, so that the state module itself must be of finite type. When the state module is assumed to be of finite type, then it can be shown that (7-105) does hold for certain types of rings. One of these types of rings is $(\mathbb{Z}, +, 0, \cdot, 1)$, and so the conclusions of Section 4.7 are reinforced. In particular, since any commutative GRMPS can be understood as a MOMPS over the ring of integers, it follows that the fact that $(X, \square, e_X)$ is of finite type is a viable assumption to establish (4-192) for a commutative GRMPS.

The feature of the integer ring which brings about the existence of the reachable subgroup is a very important and well-known ring feature. Such features will be discussed shortly.

In the meantime, it is interesting to point out the possibility of a matrix theory for certain MOMPSs. This is the topic of the next section.

Exercises

7.3-1. Suppose that two MOMPSs are connected in series, one with the other. Study the problem of whether or not the series interconnection can also be interpreted as a MOMPS. If conditions are necessary to make this result a reality, state them as assumptions. Then determine the morphisms of modules which characterize the resulting interconnected system.

7.3-2. Repeat the exercise above for the case of two MOMPSs connected in parallel, one with the other.

7.3-3. Explain in detail why the definition (7-98) makes (7-97) into an R-module. Verify axioms (7-14).

7.4 MATRIX MODULE MORPHIC SYSTEMS

A great deal of special attention is attached to the case in which the R-modules of inputs, states, and outputs are of finite type and are free on bases

$$\{u_1, u_2, \ldots, u_m\}, \qquad \{x_1, x_2, \ldots, x_n\}, \qquad \text{and} \qquad \{y_1, y_2, \ldots, y_p\}.$$

$$(7\text{-}111\text{a},\text{b},\text{c})$$

In such a situation, the R-module morphisms a, b, c, and d have matrix representations, which are denoted $[a]$, $[b]$, $[c]$, and $[d]$, respectively.

It is not difficult to see how this comes about. Consider, say, the morphism

$$b: U \to X \qquad (7\text{-}112)$$

of R-modules. Let b act upon the ith basis element of U, which is u_i from (7-111a). The result is an element bu_i in X, which is free on the basis (7-111b). Thus the result bu_i can be written

$$bu_i = \bigsqcup_{k=1}^{n} b_{ik} x_k \qquad (7\text{-}113)$$

for unique elements b_{ik} in R. Moreover, the expansion (7-113) can be repeated for $i = 1, 2, \ldots, m$. The result is a matrix

$$[b] = \begin{bmatrix} b_{11} & b_{12} & \cdots & b_{1n} \\ b_{21} & b_{22} & \cdots & b_{2n} \\ \vdots & \vdots & & \vdots \\ b_{m1} & b_{m2} & \cdots & b_{mn} \end{bmatrix}. \qquad (7\text{-}114)$$

Consider, then, an arbitrary element u in U. Express u uniquely by

$$u = \bigtriangleup_{i=1}^{m} r_i u_i. \qquad (7\text{-}11?)$$

Can $[b]$ be used to calculate bu? The answer is yes, if care is exercised. From (7-115) and the fact that b is a morphism of R-modules, it follows that

$$bu = b\left(\bigwedge_{i=1}^{m} r_i u_i\right) \tag{7-116a}$$

$$= \prod_{i=1}^{m} b(r_i u_i) \tag{7-116b}$$

$$= \prod_{i=1}^{m} r_i(bu_i) \tag{7-116c}$$

$$= \prod_{i=1}^{m} r_i \left\{\prod_{k=1}^{n} b_{ik} x_k\right\} \tag{7-116d}$$

$$= \prod_{i=1}^{m} \prod_{k=1}^{n} r_i(b_{ik} x_k) \tag{7-116e}$$

$$= \prod_{i=1}^{m} \prod_{k=1}^{n} (r_i \triangle b_{ik}) x_k \tag{7-116f}$$

$$= \prod_{k=1}^{n} \prod_{i=1}^{m} (r_i \triangle b_{ik}) x_k \tag{7-116g}$$

$$= \prod_{k=1}^{n} \left\{ \mathop{*}_{i=1}^{m} (r_i \triangle b_{ik}) \right\} x_k. \tag{7-116h}$$

If the definition

$$[u] = [r_1 \quad r_2 \quad \cdots \quad r_m] \tag{7-117}$$

is made, then the quantity

$$\left\{ \mathop{*}_{i=1}^{m} (r_i \triangle b_{ik}) \right\} \tag{7-118}$$

in (7-116h) can be recognized as the kth column element in the ordinary matrix product

$$[u][b]. \tag{7-119}$$

Moreover, because bu is in X, with X free on the basis (7-111b), then

$$bu = \prod_{k=1}^{n} \tilde{r}_k x_k \tag{7-120}$$

as well, for unique $\tilde{r}_k$ in R. Thus the column elements produced by the matrix multiplication (7-119) are just the unique elements $\tilde{r}_k$. Now represent

$$[bu] = [\tilde{r}_1 \quad \tilde{r}_2 \quad \cdots \quad \tilde{r}_n]; \qquad (7\text{-}121)$$

then it has been shown that

$$[bu] = [u][b]. \qquad (7\text{-}122)$$

In a similar way, the actions

$$ax_i = \coprod_{k=1}^{n} a_{ik}x_k, \qquad i = 1, 2, \ldots, n, \qquad (7\text{-}123\text{a})$$

$$cx_i = \underset{k=1}{\overset{p}{\text{\Large*}}} c_{ik}y_k, \qquad i = 1, 2, \ldots, n, \qquad (7\text{-}123\text{b})$$

$$du_i = \underset{k=1}{\overset{p}{\text{\Large*}}} d_{ik}y_k, \qquad i = 1, 2, \ldots, m \qquad (7\text{-}123\text{c})$$

can be used to develop the matrices

$$[a] = \begin{bmatrix} a_{11} & a_{12} & \cdots & a_{1n} \\ a_{21} & a_{22} & \cdots & a_{2n} \\ \vdots & \vdots & & \vdots \\ a_{n1} & a_{n2} & \cdots & a_{nn} \end{bmatrix}, \qquad (7\text{-}124\text{a})$$

$$[c] = \begin{bmatrix} c_{11} & c_{12} & \cdots & c_{1p} \\ c_{21} & c_{22} & \cdots & c_{2p} \\ \vdots & \vdots & & \vdots \\ c_{n1} & c_{n2} & \cdots & c_{np} \end{bmatrix}, \qquad (7\text{-}124\text{b})$$

$$[d] = \begin{bmatrix} d_{11} & d_{12} & \cdots & d_{1p} \\ d_{21} & d_{22} & \cdots & d_{2p} \\ \vdots & \vdots & & \vdots \\ d_{m1} & d_{m2} & \cdots & d_{mp} \end{bmatrix}. \qquad (7\text{-}124\text{c})$$

With the aid of these matrix representations of the quadruple of R-module morphisms (a, b, c, d), the local dynamical equations of a MOMPS can be rewritten in matrix form. Before doing this, however, it is useful to have one more observation.

EXAMPLE 7.4-1

Let $g_1, g_2, g_3, \ldots, g_n$ be a basis for the R-module G of finite type, and let g and $\tilde{g}$ be two elements in G with the unique representations

$$g = \coprod_{k=1}^{n} r_k g_k, \qquad \tilde{g} = \coprod_{k=1}^{n} \tilde{r}_k g_k. \qquad (7\text{-}125\text{a,b})$$

Then

$$g \,\square\, \tilde{g} = \left\{ \bigsqcup_{k=1}^{n} r_k g_k \right\} \,\square\, \left\{ \bigsqcup_{k=1}^{n} \tilde{r}_k g_k \right\}$$

$$= \bigsqcup_{k=1}^{n} (r_k g_k \,\square\, \tilde{r}_k g_k) = \bigsqcup_{k=1}^{n} (r_k * \tilde{r}_k) g_k. \tag{7-126}$$

As a consequence,

$$[g \,\square\, \tilde{g}] = [r_1 * \tilde{r}_1 \quad r_2 * \tilde{r}_2 \quad \cdots \quad r_n * \tilde{r}_n]$$

$$= [r_1 \quad r_2 \quad \cdots \quad r_n] \,**\, [\tilde{r}_1 \quad \tilde{r}_2 \quad \cdots \quad \tilde{r}_n]$$

$$= [g] \,**\, [\tilde{g}], \tag{7-127}$$

where the binary operation $**$ in the right member of (7-127) takes place in the R-module R^n. ∎

Return at this point to the consideration of the local dynamical equations. The local transition equation

$$x_{k+1} = ax_k \,\square\, bu_k \tag{7-128}$$

is easily converted to matrix form. To avoid any possible confusion with the notations for the basis elements in the various R-modules, rewrite (7-128) in the alternative, earlier form

$$x(k + 1) = ax(k) \,\square\, bu(k). \tag{7-129}$$

The conversion begins by writing the matrix for both members of (7-129) and then applying the properties just deduced. Thus

$$[x(k + 1)] = [ax(k) \,\square\, bu(k)] = [ax(k)] \,**\, [bu(k)] \tag{7-130a,b}$$

$$= [x(k)][a] \,**\, [u(k)][b]; \tag{7-130c}$$

the convention

$$[x(k)] = [x]_k, \tag{7-131}$$

together with a similar agreement on inputs, permits the matrix local transition equation

$$[x]_{k+1} = [x]_k[a] \,**\, [u]_k[b]. \tag{7-132}$$

In like manner, the local output equation

$$y_k = cx_k * du_k \tag{7-133}$$

undergoes the conversions

$$y(k) = cx(k) * du(k), \tag{7-134a}$$

and

$$[y(k)] = [cx(k) * du(k)] = [cx(k)] ** [du(k)] \qquad \text{(7-134b,c)}$$
$$= [x(k)][c] ** [u(k)][d], \qquad \text{(7-134d)}$$

and finally

$$[y]_k = [x]_k[c] ** [u]_k[d]. \qquad \text{(7-135)}$$

Together, (7-132) and (7-135) define a module morphic system with R-modules of inputs, states, and outputs given by R^m, R^n, and R^p, respectively. Such a system might be described by

$$([a], [b], [c], [d]; R^m, R^n, R^p) \qquad \text{(7-136)}$$

to convey the fact that the matrices of morphisms in (7-132) and (7-135) appear on the right of matrices of module elements, whereas the morphism symbols themselves appear to the left of such elements in (7-128) and (7-133).

EXAMPLE 7.4-2

Let $(G_i, \square_i, e_{G_i})$, $i = 1,2,3$, be commutative groups, and let

$$\alpha: G_1 \to G_2, \qquad \beta: G_2 \to G_3 \qquad \text{(7-137a,b)}$$

be morphisms of free R-modules of finite type. Assume that suitable scalar multiplications are defined. Then

$$[(\beta \circ \alpha)g] = [\beta(\alpha g)] = [\alpha g][\beta] = [g][\alpha][\beta]; \qquad \text{(7-138a,b,c)}$$

but

$$[(\beta \circ \alpha)g] = [g][\beta \circ \alpha] \qquad \text{(7-139)}$$

as well, and so

$$[g][\beta \circ \alpha] = [g][\alpha][\beta]. \qquad \text{(7-140a)}$$

By successive selection of

$$[g] = [e_R \quad \cdots \quad e_R \quad 1_R \quad e_R \quad \cdots \quad e_R] \qquad \text{(7-140b)}$$

with the 1_R in the ith position as i runs from 1 to the number of basis elements of G_1, it follows that

$$[\beta \circ \alpha] = [\alpha][\beta]. \qquad \text{(7-141)}$$

∎

The observations of the preceding example, when combined with those foregoing, permit straightforward development of counterparts for many of the results of preceding chapters.

EXAMPLE 7.4-3

Consider the calculation of (4-161), namely,

$$\coprod_{j=0}^{i} a^{i-j}bu_j. \tag{7-142}$$

This expression can be changed into matrix form by

$$\left[\coprod_{j=0}^{i} a^{i-j}bu_j\right] = \divideontimes\divideontimes[a^{i-j}bu_j] = \divideontimes\divideontimes[u]_j[b][a]^{i-j}. \tag{7-143a,b}$$

■

An interesting question arises in connection with the matrix module morphic system. This question is related to the issue of why it is that

$$[bu] = [u][b], \tag{7-144}$$

with the right member having a reverse order: many readers may have suspected that the right member would be $[b][u]$ instead.

This revealing point turns out to be a consequence of the way in which scalar multiplication was defined as a function

$$f : R \times G \rightarrow G \tag{7-145}$$

in Section 7.1. There is an alternative, namely,

$$f : G \times R \rightarrow G. \tag{7-146}$$

In fact, both types of modules can be studied. Equipped with (7-145), the structure becomes a *left* R-module. Equipped with (7-146), it becomes a *right* R-module. The discussions preceding have thus been in focus upon left modules.

Suppose that this focus were exchanged with that of the right module. Then (7-113) would become

$$bu_i = \coprod_{k=1}^{n} x_k b_{ki}, \qquad i = 1, 2, \ldots, m, \tag{7-147}$$

and the matrix $[b]$ could be written down as

$$[b] = \begin{bmatrix} b_{11} & b_{12} & \cdots & b_{1m} \\ b_{21} & b_{22} & \cdots & b_{2m} \\ \vdots & \vdots & & \vdots \\ b_{n1} & b_{n2} & \cdots & b_{nm} \end{bmatrix}. \tag{7-148}$$

Moreover, the counterpart of (7-115) would be

$$u = \bigwedge_{i=1}^{m} u_i r_i, \tag{7-149}$$

and the calculation (7-116) would proceed in the manner

$$bu = b\left(\bigwedge_{i=1}^{m} u_i r_i\right) \tag{7-150a}$$

$$= \coprod_{i=1}^{m} b(u_i r_i) \tag{7-150b}$$

$$= \coprod_{i=1}^{m} (bu_i) r_i \tag{7-150c}$$

$$= \coprod_{i=1}^{m} \left\{ \coprod_{k=1}^{n} x_k b_{ki} \right\} r_i \tag{7-150d}$$

$$= \coprod_{i=1}^{m} \coprod_{k=1}^{n} (x_k b_{ki}) r_i \tag{7-150e}$$

$$= \coprod_{i=1}^{m} \coprod_{k=1}^{n} x_k (b_{ki} \bigtriangleup r_i) \tag{7-150f}$$

$$= \coprod_{k=1}^{n} \coprod_{i=1}^{m} x_k (b_{ki} \bigtriangleup r_i) \tag{7-150g}$$

$$= \coprod_{k=1}^{m} x_k \left\{ \underset{i=1}{\overset{m}{*}} (b_{ki} \bigtriangleup r_i) \right\}. \tag{7-150h}$$

In this context, it follows that

$$[bu] = [b][u], \tag{7-151}$$

with $[u]$ having a column interpretation.

Some remarks are in order. First, the reader should take note of the subscript difference between (7-113) and (7-147). This is not accidental. Indeed, if (7-113) had been written

$$\coprod_{k=1}^{n} b_{ki} x_k, \tag{7-152a}$$

then (7-118) would have become

$$\underset{i=1}{\overset{m}{*}} (r_i \bigtriangleup b_{ki}), \tag{7-152b}$$

which is not a natural way to write matrix multiplication. Second, note that these distinctions tend to disappear rapidly if the ring R is commutative. In such a case, (7-152b) becomes

$$\underset{i=1}{\overset{m}{*}} (b_{ki} \triangle r_i) \qquad (7\text{-}152\text{c})$$

and (7-151) would follow again. Third, it should also be recognized that the definition of a scalar multiplication $R \times G \to G$ as opposed to $G \times R \to G$ can be taken, in some cases, as a matter of taste. Fourth, to emphasize the latter point, consider that to write the result of a function $f : S \to T$ at a point s in the manner $f(s)$ is also arbitrary and could be changed to $(s)f$.

Nonetheless, when the usual ideas of matrix multiplication are taken into account, together with the usual conventions for functions and the possibility that the ring R may not be commutative, there is a reasonable case for the conclusion (7-144) being associated with (7-145).

In conclusion, a MOMPS in this section, where finite type and free behavior on bases is assumed, might be denoted by [MOMPS], with the brackets connoting matrices.

Finally, the existence of a basis for U permits the introduction of impulse sequences

$$u_i, \quad e_U, \quad e_U, \quad \ldots \qquad (7\text{-}153)$$

for $i = 1, 2, \ldots, m$ to drive a dynamical system. These in turn permit a concept of impulse response, which will be encountered again at a point later in the chapter.

Exercises

7.4-1.   Suppose that $\alpha : G \to G$ is an automorphism of the finitely generated free R-module G onto itself. Select a basis in G, and show that

$$[\alpha^{-1}] = [\alpha]^{-1}.$$

7.4-2.   What is the explanation for the step from (7-116g) to (7-116h)?

7.4-3.   Find the matrix representation for a series connection of two free [MOMPS]s of finite type.

7.4-4.   Find the matrix representation for a parallel connection of two free [MOMPS]s of finite type.

7.4-5.   Adapt the axioms (7-14) for scalar multiplication in a left module to a corresponding list appropriate for a right module. Find illustrations of the use of the new axioms in (7-150).

7.5 COMMUTATIVE RINGS AND SPECIAL MODULES

The early sections in this chapter have for the most part examined general R-modules without any particular assumptions on the ring R. This section assumes R to be a commutative ring, and then explores various special cases. To emphasize the commutative nature of the ring, the general ring symbol R is replaced by CR, with straightforward connotation. Thus, the ring of record here is denoted by

$$(CR, *, e_{CR}, \triangle, 1_{CR}). \tag{7-154}$$

EXAMPLE 7.5-1

One remarkable feature of a commutative ring CR is that

$$r_1 \triangle r_2 = e_{CR} \tag{7-155a}$$

is possible, even when

$$r_i \neq e_{CR}, \qquad i = 1, 2. \tag{7-155b}$$

By way of an example, select the ring

$$(\mathbb{Z}_4, +, 0, \cdot, 1) \tag{7-156}$$

with the pair of binary operations

+	0	1	2	3		·	0	1	2	3
0	0	1	2	3		0	0	0	0	0
1	1	2	3	0		1	0	1	2	3
2	2	3	0	1		2	0	2	0	2
3	3	0	1	2		3	0	3	2	1

$$\tag{7-157}$$

indicated. Then

$$2 \cdot 2 = 0. \tag{7-158}$$

This type of behavior is not limited to rings which are commutative, but is reiterated here because the assumption of its absence is a common one in the commutative ring case. ∎

Suppose that a ring (7-154) has the property that

$$r_1 \triangle r_2 = e_{CR} \tag{7-159}$$

can occur only when at least one of the elements r_1 and r_2 are equal to e_{CR}. Suppose further that the ring is nontrivial, which means that

$$e_{CR} \neq 1_{CR}. \tag{7-160}$$

Then the ring is said to be an *integral domain*. Often the adjective integral is dropped, and the ring is simply called a *domain*.

EXAMPLE 7.5-2

If the ring (7-154) is a domain, then there is a familiar step which can be carried out. When $r_1 \neq e_{CR}$ and when

$$r_1 \triangle r_2 = r_1 \triangle r_3, \qquad (7\text{-}161)$$

the r_1 can be cancelled. The reasoning is as follows. Perform the calculation

$$e_{CR} = (r_1 \triangle r_3) * \overline{(r_1 \triangle r_3)} = (r_1 \triangle r_2) * (r_1 \triangle \hat{r}_3) \qquad (7\text{-}162\text{a,b})$$
$$= r_1 \triangle (r_2 * \hat{r}_3). \qquad (7\text{-}162\text{c})$$

Because $r_1 \neq e_{CR}$, (7-162c) implies

$$r_2 * \hat{r}_3 = e_{CR}, \qquad \text{from which} \qquad r_2 = r_3. \qquad (7\text{-}163\text{a,b})$$

∎

Now recall from Section 6.1 that an ideal I in a ring CR is a subgroup with the property that i in I and r in CR implies $i \triangle r$ is in I. A *principal ideal* is an ideal which is spanned in CR by one element. Thus a principal ideal I can be written

$$I = (CR)i \qquad (7\text{-}164)$$

for some element i in I. There are some rather fascinating rings in which every ideal is of the type (7-164). A *principal ideal domain* ring is an integral domain ring in which every ideal is a principal ideal.

EXAMPLE 7.5-3

$(\mathbb{Z}, +, 0, \cdot, 1)$ is a principal ideal domain ring. ∎

Among the most familiar of rings is the *field*, which is an integral domain with the property that each element

$$r \neq e_{CR} \qquad (7\text{-}165)$$

has an inverse $\tilde{r}$ under $\triangle$ such that

$$r \triangle \tilde{r} = 1_{CR}. \qquad (7\text{-}166)$$

A field will be denoted

$$(F, +, 0, \cdot, 1), \qquad (7\text{-}167)$$

and there are many well-known illustrations.

EXAMPLE 7.5-4

The rings

$$(\mathbb{R}, +, 0, \cdot, 1), \qquad (\mathbb{C}, +, 0, \cdot, 1), \qquad (\mathbb{Q}, +, 0, \cdot, 1) \qquad \text{(7-168a,b,c)}$$

are fields. ∎

If a ring CR is also a field F, then the terminology CR-module G is replaced by F-*vector space* V. In other words, the group symbol G is replaced by V. It is traditional in such cases to write

$$(V, +, 0) \qquad \text{(7-169)}$$

for the commutative group, in place of $(G, \square, e_G)$. Moreover, the inverse $\hat{v}$ of an element v in V under $+$ is usually written as

$$-v, \qquad \text{(7-170)}$$

as will be no surprise to the reader. Inverses under $+$ in the field (7-167) are also written in the same spirit, namely,

$$-f; \qquad \text{(7-171)}$$

inverses under $\cdot$ in (7-167) are written

$$f^{-1}. \qquad \text{(7-172)}$$

One of the most dominant features of an F-vector space V is the fact that if it is of finite type then it is always free on a suitable basis. The number of elements in the basis is called, of course, the *dimension* of V. The reason for this simplicity is not hard to see. If two nonzero elements v_1 and v_2 in V are F-linearly dependent in the manner

$$f_1 v_1 + f_2 v_2 = 0, \qquad \text{(7-173a)}$$

with $f_i \neq 0$, $i = 1, 2$, then the steps

$$f_2 v_2 = -f_1 v_1 \qquad \text{and} \qquad v_2 = f_2^{-1}(-f_1 v_1) \qquad \text{(7-173b,c)}$$

permit v_2 to be directly expressed in terms of v_1. For CR-modules in general, however, such inverses in the ring of scalars may not be available.

One of the most important rings encountered in system theory is that of polynomials having coefficients in a field F. The construction is analogous to that used in Section 6.4; but there are differences. Begin with the function set

$$F^{\mathbb{N}}. \qquad \text{(7-174)}$$

It is a consequence of Section 4.1 that (7-174) admits commutative group structure. Consider now the subgroup S of (7-174) with the property that

$f \in S$ implies

$$f(n) = 0 \qquad (7\text{-}175\text{a})$$

for all but a finite number of natural numbers n in $\mathbb{N}$. This subgroup can be made into a ring; and a prevalent way to do this is to regard an element

$$f(0), \quad f(1), \quad f(2), \quad \dots, \quad f(p), \quad 0, \quad 0, \quad \dots \qquad (7\text{-}175\text{b})$$

in S as a polynomial

$$\sum_{n=0}^{p} f(n)x^n. \qquad (7\text{-}176)$$

In considering (7-176), it is necessary to realize that x is not a variable taking values in some set. Rather, it is a placemarker of sorts which goes with an element in F to determine its place in the sequence. Accordingly, x is often called an *indeterminate*. Inasmuch as

$$f(n) \in F, \qquad \text{write} \qquad f(n) = f_n; \qquad (7\text{-}177\text{a,b})$$

(7-176) becomes

$$\sum_{n=0}^{p} f_n x^n. \qquad (7\text{-}177\text{c})$$

If

$$\sum_{k=0}^{m} g_k x^k \qquad (7\text{-}178)$$

represents another element in S, then the ring structure is completed by defining

$$\left(\sum_{n=0}^{p} f_n x^n \right) \cdot \left(\sum_{k=0}^{m} g_k x^k \right) = \sum_{n=0}^{p} \sum_{k=0}^{m} (f_n \cdot g_k) x^{n+k}, \qquad (7\text{-}179)$$

where the operation $(f_n \cdot g_k)$ takes place in F. In fact, the right member of (7-179) can be rewritten in the manner

$$\sum_{q=0}^{p+m} \left\{ \sum_{i=0}^{q} (f_i \cdot g_{q-i}) \right\} x^q, \qquad (7\text{-}180)$$

where the operations

$$\left\{ \sum_{i=0}^{q} (f_i \cdot g_{q-i}) \right\} \qquad (7\text{-}181)$$

are all carried out in F. So equipped, the subgroup S of $F^{\mathbb{N}}$ is denoted $F[x]$.

The polynomial ring $F[x]$ shares with the integer ring $\mathbb{Z}$ the property of being a principal ideal domain.

In turn, the standard proof for $F[x]$ being a principal ideal domain depends upon the *division algorithm*. If

$$p(x) = \sum_{i=0}^{m} f_i x^i \in F[x],$$ (7-182)

then its degree is the greatest natural number i in $\mathbb{N}$ for which

$$f_i \neq 0.$$ (7-183)

Traditionally, this degree is m in (7-182); but this need not be the case, because f_m may be equal to 0.

Suppose that $p_1(x) \in F[x]$ is a nonzero polynomial, then the division algorithm states that for every other polynomial $p_2(x) \in F[x]$ there exist unique polynomials $v(x)$ and $w(x)$ in $F[x]$ with the property that

$$p_2(x) = v(x)p_1(x) + w(x),$$ (7-184)

where the degree of $w(x)$ is less than the degree of $p_1(x)$. Notice first the special cases. If the degree of $p_2(x)$ is less than that of $p_1(x)$, the choice $v(x)$ equal to the zero polynomial and $w(x)$ equal to $p_2(x)$ is clear. Also, by convention, the degree of the zero polynomial is taken to be negative, to cover the case in which $p_1(x)$ has degree zero. The division algorithm can be proved by induction. At this point, however, it is more convenient simply to recall its familiar features with an example.

EXAMPLE 7.5-5

Choose

$$p_1(x) = 5 + 2x, \qquad p_2(x) = x^2,$$ (7-185a,b)

with F the field $\mathbb{R}$ of real numbers. Then perform classical long division

$$
\begin{array}{r}
0.5x - 1.25 \\
2x + 5 \overline{)x^2} \\
x^2 + 2.5x \\
\overline{ - 2.5x} \\
- 2.5x - 6.25 \\
\overline{ 6.25}
\end{array}
$$ (7-186)

to determine that

$$v(x) = 0.5x - 1.25, \qquad w(x) = 6.25.$$ (7-187a,b)

∎

The division algorithm can be used to show that $F[x]$ is a principal ideal domain. To show that it is a domain is left as an exercise. Consider an arbitrary ideal in $F[x]$. If the ideal contains only the zero polynomial, then it is trivial to argue that the ideal is principal. If the ideal contains a nonzero polynomial, choose an element $p_1(x)$ of least degree in the ideal. Now let $p_2(x)$ be any other nonzero polynomial in the ideal. By the division algorithm,

$$w(x) = p_2(x) - v(x)p_1(x) \qquad (7\text{-}188)$$

for a polynomial $w(x)$ of degree less than that of $p_1(x)$. But the right member of (7-188) is also an element of the ideal, so that $w(x)$ has to be an ideal element as well. The only way this can happen without contradiction is in the case of $w(x)$ the zero polynomial. But then

$$p_2(x) = v(x)p_1(x), \qquad (7\text{-}189)$$

and the ideal can be written

$$(F[x])p_1(x), \qquad (7\text{-}190)$$

as asserted.

The fact that $F[x]$ is a principal ideal domain can be used to show that any sequence

$$I_0 \subset I_1 \subset I_2 \subset \cdots \subset F[x] \qquad (7\text{-}191)$$

of ideals has the property

$$I_m = I_{m+1} = I_{m+2} = \cdots \qquad (7\text{-}192)$$

for some natural number m. By an argument along the same lines as that used in the latter part of Section 7.3, the union of all the ideals in (7-191) is also an ideal. Indeed, $F[x]$ is a ring and thus admits the structure of an $F[x]$-module. Then the ideals are just submodules. As an ideal in $F[x]$, the union is spanned by some polynomial $p(x)$ in $F[x]$. But to be an element of the union, $p(x)$ must be an element of one of the ideals, say I_m. But this means that the union

$$(F[x])p(x) \subset I_m, \qquad (7\text{-}193)$$

so that the inclusions

$$I_m \subset I_{m+1} \subset I_{m+2} \cdots \qquad (7\text{-}194)$$

must become equalities as asserted.

The time is still not quite right to complete the reachable subgroup argument of Chapter 4, but there is one preliminary technicality which will make it possible to do so in a later section.

Let the commutative groups $(G_i, \square_i, e_{G_i})$ be CR-modules over the ring $(CR, *, e_{CR}, \triangle, 1_{CR})$ for $i = 1, 2, 3$. In like vein, let $\alpha : G_1 \to G_2$ and $\beta : G_2 \to G_3$ be morphisms of CR-modules. Express this pictorially by

$$G_1 \overset{\alpha}{\to} G_2 \overset{\beta}{\to} G_3. \qquad (7\text{-}195)$$

The diagram (7-195) is called an *exact sequence* of morphisms if

$$\text{Im } \alpha = \text{Ker } \beta. \tag{7-196}$$

Now denote by 0 the CR-module developed on a trivial group consisting of just the single element $\{e_G\}$ through the straightforward process of stating scalar multiplication

$$re_G = e_G \tag{7-197}$$

for all r in CR. Then the sequence

$$0 \to G_1 \xrightarrow{\alpha} G_2 \tag{7-198a}$$

is exact if α is a monomorphism; the sequence

$$G_2 \xrightarrow{\beta} G_3 \to 0 \tag{7-198b}$$

is exact if β is an epimorphism; and the sequence

$$0 \to G_1 \xrightarrow{\alpha} G_2 \xrightarrow{\beta} G_3 \to 0 \tag{7-199}$$

is a *short exact sequence* if the sequences (7-195) and (7-198) are exact. Note that a morphism $h:0 \to G_1$ in (7-198a) is easily defined by the action

$$h(e_G) = e_{G_1}. \tag{7-200a}$$

Similarly, a morphism $h:G_3 \to 0$ is defined by

$$h(g_3) = e_G \tag{7-200b}$$

for all g_3 in G_3.

Now consider a short exact sequence (7-199) of CR-modules. If every submodule of G_1 is of finite type, and if every submodule of G_3 is of finite type, then it can be shown that every submodule of G_2 is of finite type. Suppose

$$S_0 \subset S_1 \subset S_2 \subset \cdots \subset G_2 \tag{7-201}$$

is an ascending sequence of submodules. Then construct

$$\beta_* S_0 \subset \beta_* S_1 \subset \beta_* S_2 \subset \cdots \subset G_3 \tag{7-202a}$$

and

$$\alpha^* S_0 \subset \alpha^* S_1 \subset \alpha^* S_2 \subset \cdots \subset G_1. \tag{7-202b}$$

Here the reader may wish to review Section 2.1 for the notation α^*. By assumption, and by the results of Section 7.3, there is a natural number m in $\mathbb{N}$ such that

$$\beta_* S_m = \beta_* S_{m+1} = \cdots \qquad \text{and} \qquad \alpha^* S_m = \alpha^* S_{m+1} = \cdots. \tag{7-203a,b}$$

Examine a natural number $p > m$, and let $s \in S_p$. Then βs is in $\beta_* S_p$; by (7-203a) it follows that

$$\beta s = \beta \tilde{s} \tag{7-204}$$

for some $\tilde{s}$ in S_m. But (7-204) implies that

$$(s \,\square_2\, \hat{\tilde{s}}) \in \operatorname{Ker} \beta = \operatorname{Im} \alpha, \tag{7-205}$$

so that

$$(s \,\square_2\, \hat{\tilde{s}}) = \alpha g_1 \tag{7-206}$$

for some g_1 in G_1. Thus g_1 must be an element of $\alpha^* S_p$, because $s \in S_p$ and $\tilde{s} \in S_m \subset S_p$. Now apply (7-203b) to see that g_1 is also an element of $\alpha^* S_m$, which means that the right member of (7-206) is also in S_m. Because $\tilde{s} \in S_m$, it then follows that $s \in S_m$, so that $S_p = S_m$. Then by the results of Section 7.3, it follows that every submodule of G_2 must be of finite type, as desired.

In this section, certain special commutative rings and modules have been discussed; of interest here was the familiar topic of rings which are also fields. The modules were then called vector spaces. A major feature of vector spaces is that they are free on a basis whenever they are of finite type. This is quite a bit different from the general case for modules. Another emphasized ring was $F[x]$, the polynomials having coefficients in a field F. This ring was developed on a subgroup of $F^{\mathbb{N}}$, with the indeterminate x introduced for ease in understanding the ring multiplication. $F[x]$ turned out to be a principal ideal domain and to have special properties relative to nested sequences of ideals (7-191). Showing that $F[x]$ was a principal ideal domain used the classical division algorithm. Since the integer ring $\mathbb{Z}$ also has a well-known division algorithm, these arguments need be modified only slightly to cover that case. Finally, a little more machinery was put into place in order to resolve the issue of the existence of a reachable submodule for a MOMPS. The results of this line of thought must await the introduction of the key triangle for the module case.

Exercises

7.5-1.   It will be illustrated in the section following how a domain D can be imbedded naturally into a field F called the *quotient field of D*. Accordingly, any square matrix containing elements in D can have associated with it a determinant in the usual way. Let

$$\alpha : D^n \to D^n$$

be an automorphism of D-modules, and let $\triangle$ be the multiplication operation in D. Show that

$$\det[\alpha]$$

must have an inverse under $\triangle$ in D.

7.5-2. The ring

$$(\mathbb{Z}, +, 0, \cdot, 1)$$

is a domain. Consider automorphisms

$$\alpha: \mathbb{Z}^2 \to \mathbb{Z}^2.$$

Show that

$$\begin{bmatrix} 0 & 1 \\ 1 & 0 \end{bmatrix}, \quad \begin{bmatrix} 1 & z \\ 0 & 1 \end{bmatrix}, \quad \begin{bmatrix} 1 & 0 \\ z & 1 \end{bmatrix}$$

are possible matrices for such automorphisms. Are there any interesting cases in which

$$\begin{bmatrix} z & 0 \\ 0 & 1 \end{bmatrix}, \quad \begin{bmatrix} 1 & 0 \\ 0 & z \end{bmatrix}$$

can represent automorphisms?

7.5-3. Consider the set of 2×2 matrices whose elements are real numbers. Establish ring structure on this set using the usual definitions of matrix multiplication and addition. Refer to Example 7.5-1. Show that the same type of phenomenon can occur in this matrix ring.

7.5-4. Explain the step from (7-162a) to (7-162b).

7.5-5. Why can an element in the subgroup S of $F^{\mathbb{N}}$ be expressed in the manner (7-175b)? Refer to (7-175a).

7.5-6. Carry out the complete verification that (7-179) makes $F[x]$ a ring.

7.5-7. Prove the division algorithm for $F[x]$.

7.5-8. Fix a polynomial $p_1(x)$ in $F[x]$, and consider two other arbitrary polynomials

$$p(x) = v_p(x)p_1(x) + w_p(x), \qquad q(x) = v_q(x)p_1(x) + w_q(x)$$

in $F[x]$. Establish a binary relation B by

$$(p(x), q(x)) \in B \qquad \text{if} \qquad w_p(x) = w_q(x).$$

Study the r-s-t properties of B.

7.5-9. Show that $F[x]$ is an integral domain.

7.6 THE TRANSFER FUNCTION: A SIMPLE CASE

Section 7.5 introduced the principal ideal domain ring $F[x]$ of polynomials in an indeterminate x with coefficients from a field F. Earlier chapters, moreover, have established a considerable utility for the ubiquitous equivalence relation. In this section, a well-known equivalence relation is placed upon the product of the ring $F[x]$ with its subset of all nonzero polynomials.

Recall from Section 2.3 that a relation on a set S to a set T is just a subset of the product set $S \times T$. Define

$$D[x] = \{p(x) \mid 0 \neq p(x) \in F[x]\} \tag{7-207}$$

to be the subset of nonzero polynomials in $F[x]$. Next form the product

$$F[x] \times D[x]. \tag{7-208}$$

The equivalence relation E to be formed is just a subset

$$E \subset (F[x] \times D[x])^2. \tag{7-209}$$

Let

$$(n_i(x), d_i(x)) \in F[x] \times D[x] \tag{7-210}$$

for $i = 1, 2$. Then

$$((n_1(x), d_1(x)), (n_2(x), d_2(x))) \in E \tag{7-211a}$$

if

$$n_1(x)d_2(x) = n_2(x)d_1(x), \tag{7-211b}$$

with the calculation (7-211b) being made in the ring $F[x]$. For convenience, (7-211a) can be written

$$(n_1(x), d_1(x)) \equiv (n_2(x), d_2(x)). \tag{7-212}$$

Because

$$n(x)d(x) = n(x)d(x), \tag{7-213a}$$

it follows immediately that

$$(n(x), d(x)) \equiv (n(x), d(x)), \tag{7-213b}$$

so that the relation E is reflexive. Moreover,

$$n_1(x)d_2(x) = n_2(x)d_1(x) \qquad \text{implies} \qquad n_2(x)d_1(x) = n_1(x)d_2(x), \tag{7-214a,b}$$

which means that

$$(n_1(x), d_1(x)) \equiv (n_2(x), d_2(x)) \qquad \text{gives} \qquad (n_2(x), d_2(x)) \equiv (n_1(x), d_1(x)); \tag{7-214c,d}$$

thus, E is symmetric. Finally,

$$n_1(x)d_2(x) = n_2(x)d_1(x) \qquad \text{and} \qquad n_2(x)d_3(x) = n_3(x)d_2(x) \tag{7-215a,b}$$

can be used to establish

$$n_1(x)d_3(x) = n_3(x)d_1(x). \tag{7-215c}$$

To achieve this, multiply both members of (7-215a) by $d_3(x)$ to give

$$(n_1(x)d_2(x))d_3(x) = (n_2(x)d_1(x))d_3(x). \tag{7-215d}$$

But multiplication in $F[x]$ is associative and commutative, so that

$$(n_2(x)d_1(x))d_3(x) = n_2(x)(d_1(x)d_3(x)) \qquad (7\text{-}216\text{a})$$
$$= n_2(x)(d_3(x)d_1(x)) \qquad (7\text{-}216\text{b})$$
$$= (n_2(x)d_3(x))d_1(x) \qquad (7\text{-}216\text{c})$$
$$= (n_3(x)d_2(x))d_1(x) \qquad (7\text{-}216\text{d})$$
$$= n_3(x)(d_2(x)d_1(x)) \qquad (7\text{-}216\text{e})$$
$$= n_3(x)(d_1(x)d_2(x)) \qquad (7\text{-}216\text{f})$$
$$= (n_3(x)d_1(x))d_2(x), \qquad (7\text{-}216\text{g})$$

with the step (7-216d) following from (7-215b). Moreover, the left member of (7-215d) can be written

$$(n_1(x)d_3(x))d_2(x) \qquad (7\text{-}217)$$

by a similar reasoning. So

$$(n_1(x)d_3(x))d_2(x) = (n_3(x)d_1(x))d_2(x) \qquad (7\text{-}218)$$

with $d_2(x)$ nonzero in $F[x]$. From Example 7.5-2, then, the desired conclusion (7-215c) follows. Thus,

$$(n_1(x), d_1(x)) \equiv (n_2(x), d_2(x)) \qquad (7\text{-}219\text{a})$$

and

$$(n_2(x), d_2(x)) \equiv (n_3(x), d_3(x)) \qquad (7\text{-}219\text{b})$$

imply

$$(n_1(x), d_1(x)) \equiv (n_3(x), d_3(x)), \qquad (7\text{-}219\text{c})$$

so that E is transitive. As an r-s-t binary relation, E is therefore a legitimate equivalence relation.

Denote by $[n(x), d(x)]$ the equivalence class containing $(n(x), d(x))$. It turns out that the quotient set $F[x] \times D[x]/E$ of these equivalence classes admits commutative group structure. Define

$$+ : (F[x] \times D[x]/E) \times (F[x] \times D[x]/E) \to (F[x] \times D[x]/E) \quad (7\text{-}220\text{a})$$

by the action

$$[n_1(x), d_1(x)] + [n_2(x), d_2(x)] = [n_1(x)d_2(x) + n_2(x)d_1(x), d_1(x)d_2(x)].$$
$$(7\text{-}220\text{b})$$

It is left as an exercise for the reader to show that (7-220) is associative and commutative. The unit for (7-220) is just $[0, d(x)]$, and an inverse for $[n(x), d(x)]$ under (7-220) is $[-n(x), d(x)]$. Next define a second binary operation

$$\cdot : (F[x] \times D[x]/E) \times (F[x] \times D[x]/E) \to (F[x] \times D[x]/E) \qquad (7\text{-}221\text{a})$$

by the action

$$[n_1(x), d_1(x)] \cdot [n_2(x), d_2(x)] = [n_1(x)n_2(x), d_1(x)d_2(x)]. \quad (7\text{-}221b)$$

The binary operation (7-221) is commutative and associative, and has a unit $[1, 1]$ with 1 the multiplicative unit in F. To show that

$$(F[x] \times D[x]/E, +, [0, d(x)], \cdot, [1, 1]) \quad (7\text{-}222)$$

is a commutative ring, it remains to verify a distributive law. Suppress the indeterminate x, and write

$$[n_1, d_1] \cdot \{[n_2, d_2] + [n_3, d_3]\} = [n_1, d_1] \cdot [n_2 d_3 + n_3 d_2, d_2 d_3] \quad (7\text{-}223a)$$
$$= [n_1(n_2 d_3 + n_3 d_2), d_1(d_2 d_3)] \quad (7\text{-}223b)$$
$$= [n_1 n_2 d_3 + n_1 n_3 d_2, d_1 d_2 d_3]; \quad (7\text{-}223c)$$

on the other hand,

$$[n_1, d_1] \cdot [n_2, d_2] = [n_1 n_2, d_1 d_2], \quad (7\text{-}224a)$$
$$[n_1, d_1] \cdot [n_3, d_3] = [n_1 n_3, d_1 d_3], \quad (7\text{-}224b)$$

and the sum of (7-224a) and (7-224b) provides

$$[n_1 n_2 d_1 d_3 + n_1 n_3 d_1 d_2, d_1 d_1 d_2 d_3], \quad (7\text{-}224c)$$

which is equivalent to (7-223c). From (7-221) and the fact that $F[x]$ is an integral domain, it follows that (7-222) is an integral domain. In fact, if

$$[n(x), d(x)] \neq [0, d(x)], \quad (7\text{-}225a)$$

then

$$[d(x), n(x)] \cdot [n(x), d(x)] = [1, 1], \quad (7\text{-}225b)$$

so that (7-222) is also a field, called the *quotient field of* $F[x]$ and denoted by $F(x)$.

The development above can be completed for any integral domain ring.

EXAMPLE 7.6-1

The field $(\mathbb{Q}, +, 0, \cdot, 1)$ is the quotient field of the ring $(\mathbb{Z}, +, 0, \cdot, 1)$.   ∎

It is very much the common practice in applications to replace the revealing notation

$$[n(x), d(x)] \quad (7\text{-}226)$$

by

$$\frac{n(x)}{d(x)}, \quad (7\text{-}227)$$

with (7-227) being called a *transfer function*. This type of usage has already been pointed out in (1-5) of Section 1.1. The quotient field $F(x)$ clarifies several potentially confusing points in transfer function calculations, by making clear that (7-227) is but a particular representative of an equivalence class.

As an element of a field $F(x)$, (7-227) can serve to define a matrix

$$\left[\frac{n(x)}{d(x)}\right] \tag{7-228}$$

having one row and one column and arising from a morphism of a first 1-dimensional $F(x)$-vector space into a second 1-dimensional $F(x)$-vector space. The task of relating this transfer function viewpoint to the module morphic system viewpoint of Sections 7.3 and 7.4 represents one of the most heavily studied topics in algebraic system theory. In the remainder of this section, some of these ideas are brought forward.

Consider a MOMPS

$$x_{k+1} = ax_k + bu_k, \qquad y_k = cx_k + du_k, \tag{7-229a,b}$$

where all the modules are F-vector spaces of dimension 1.

Example 7.6-2

From Section 7.4, it follows that a MOMPS such as (7-229) has a matrix or [MOMPS] version

$$[x]_{k+1} = [a][x]_k + [b][u]_k, \qquad [y]_k = [c][x]_k + [d][u]_k \tag{7-230a,b}$$

for appropriate choice of one basis vector each in U, X, and Y. Because fields are commutative rings, the form (7-230) can be chosen over the forms (7-132) and (7-135). Typical illustrations might be $(F = \mathbb{R})$

$$[a] = [-1], \qquad [b] = [1], \qquad [c] = [2], \qquad [d] = [0]. \tag{7-231}$$

In fact, these cases often see the matrix bracket symbol suppressed in the manner

$$x_{k+1} = -x_k + u_k, \qquad y_k = 2x_k. \tag{7-232a,b}$$

∎

Input sequences to the system (7-229) have been visualized in previous chapters as elements

$$u_0, \quad u_1, \quad u_2, \quad \ldots \qquad \text{of} \qquad U^{\mathbb{N}}, \tag{7-233a,b}$$

where U is in this case a 1-dimensional $\mathbb{R}$-vector space. However, there is no real need to have the inputs associated with natural numbers. Thus,

(7-233b) might be replaced by

$$U^{\mathbb{Z}}, \tag{7-234a}$$

provided that (7-233a) is prefixed by zero vectors in the manner

$$\ldots, \quad 0, \quad 0, \quad 0, \quad u_0, \quad u_1, \quad u_2, \quad \ldots . \tag{7-234b}$$

Once the step (7-234) has been taken, it is soon seen that there is nothing special about having the first nonzero u_i in (7-234b) occur at i equal to zero. An input sequence

$$\ldots, \quad 0, \quad 0, \quad u_{-2}, \quad u_{-1}, \quad u_0, \quad u_1, \quad \ldots \tag{7-235}$$

serves much the same purpose. Elements of type (7-234b) and (7-235) in $U^{\mathbb{Z}}$ are said to be of *finite left support*, because there exists an integer p such that u_i vanishes for i less than p. By arguments similar to those used in Section 7.5 and in Chapter 4, it follows that (7-234a) is a commutative group. It is an easy exercise to show that the subset of elements of finite left support is a subgroup. Denote this subgroup by

$$\mathrm{FLS}(U^{\mathbb{Z}}). \tag{7-236}$$

There is a useful way to convert (7-236) into a vector space. To do this, it is very helpful to use a procedure similar to that used in constructing the ring $F[x]$ in Section 7.5. Identify sequences of finite left support, such as (7-235), with series

$$\sum u_i z^{-i}, \tag{7-237}$$

where i takes values in $\mathbb{Z}$. Here again, z is not a variable taking values in a set but is simply an indeterminate placemarker in the same spirit as x in $F[x]$. Denote the set of series (7-237) corresponding to (7-236) by

$$U((z^{-1})), \tag{7-238}$$

and observe that (7-238) inherits directly the commutative group structure of (7-236). $U((z^{-1}))$ is called the set of *formal Laurent series* in the indeterminate z^{-1} and with coefficients in the F-vector space U. $U((z^{-1}))$ can be made into an $F(z)$-vector space. In seeing how this is done, it is revealing to consider the next example.

Example 7.6-3

Let $\alpha(z)$ be a typical element of $\mathbb{R}(z)$. Then $\alpha(z)$ can be identified with an element

$$\sum r_i z^{-i} \in \mathbb{R}((z^{-1})) \tag{7-239}$$

by classical long division. As an illustration, suppose

$$\alpha(z) = \frac{z^2}{1+z};$$
(7-240)

then

$$
z + 1 \,\overline{)z^2} \quad
\begin{array}{l}
z - 1 + z^{-1} - z^{-2} + \cdots \\
\hline
z^2 + z \\
\hline
\quad -z \\
\quad -z - 1 \\
\hline
\qquad 1 \\
\qquad 1 + z^{-1} \\
\hline
\qquad\quad -z^{-1} \\
\qquad\quad -z^{-1} - z^{-2} \\
\hline
\qquad\qquad \vdots
\end{array}
$$
(7-241)

shows that

$$
\begin{aligned}
r_i &= 0, & i &< -1, \\
r_i &= (-1)^{i+1}, & i &\geq -1.
\end{aligned}
$$
(7-242)

In using the symbol $\mathbb{R}((z^{-1}))$, it is useful to notice that the field of real numbers $\mathbb{R}$ can also be regarded as an $\mathbb{R}$-vector space. ∎

With the intuition afforded by this example, $U((z^{-1}))$ becomes an $F(z)$-vector space by means of the following two-step scalar multiplication. Identify an element

$$f(z) \in F(z) \quad \text{with an element} \quad \sum f_j z^{-j} \in F((z^{-1})). \quad (7\text{-}243\text{a,b})$$

Multiply according to

$$\left(\sum f_j z^{-j}\right)\left(\sum u_i z^{-i}\right) = \sum\sum (f_j u_i) z^{-(i+j)},$$
(7-243c)

where $(f_j u_i)$ is the scalar multiplication in the F-vector space U.

As an $F(z)$-vector space, however, $U((z^{-1}))$ is not 1-dimensional. However, there is a natural subspace with this property. Define by

$$U[z]$$
(7-244a)

the subgroup of $U((z^{-1}))$ consisting of elements

$$\sum u_i z^{-i}$$
(7-244b)

satisfying $u_i = 0$ for $i > 0$. Next define the subspace

$$U(z)$$
(7-244c)

of the $F(z)$-vector space $U((z^{-1}))$ by the property that

$$u(z) \in U(z) \qquad\qquad (7\text{-}244\text{d})$$

if there exists a nonzero element $f(z)$ in $F(z)$ such that

$$f(z)u(z) \in U[z]. \qquad\qquad (7\text{-}244\text{e})$$

The $F(z)$-vector space $U(z)$ has dimension 1 for the transfer function of this section.

Return now to (7-229), and focus upon the local output equation. Regard (7-229b) as an identity on $Y(z)$, namely,

$$\sum y_k z^{-k} = \sum (cx_k + du_k)z^{-k} = \sum (cx_k)z^{-k} + \sum (du_k)z^{-k}. \quad (7\text{-}245\text{a})$$

Define morphisms

$$\tilde{c}: X(z) \rightarrow Y(z) \qquad \text{and} \qquad \tilde{d}: U(z) \rightarrow Y(z) \qquad (7\text{-}245\text{b,c})$$

of $F(z)$-vector spaces by

$$\tilde{c}\sum x_k z^{-k} = \sum (cx_k)z^{-k} \qquad \text{and} \qquad \tilde{d}\sum u_k z^{-k} = \sum (du_k)z^{-k}, \quad (7\text{-}246\text{a,b})$$

respectively. Then (7-245a) becomes

$$\sum y_k z^{-k} = \tilde{c}\sum x_k z^{-k} + \tilde{d}\sum u_k z^{-k}. \qquad (7\text{-}247)$$

Along similar lines, the right member of (7-229a) becomes

$$\tilde{a}\sum x_k z^{-k} + \tilde{b}\sum u_k z^{-k}, \qquad\qquad (7\text{-}248\text{a})$$

with the left member following

$$\sum_k x_{k+1} z^{-k} = \sum_j x_j z^{-(j-1)} = \sum_j x_j z z^{-j} = z\sum_j x_j z^{-j}. \quad (7\text{-}248\text{b,c,d})$$

Thus the local transition equation on $X(z)$ gives

$$z\sum x_k z^{-k} = \tilde{a}\sum x_k z^{-k} + \tilde{b}\sum u_k z^{-k}, \qquad (7\text{-}249\text{a})$$

which can be written

$$(z - \tilde{a})\sum x_k z^{-k} = \tilde{b}\sum u_k z^{-k}, \qquad (7\text{-}249\text{b})$$

if $z: X(z) \rightarrow X(z)$ is regarded as a morphism of $F(z)$-vector spaces, and solved to give

$$\sum x_k z^{-k} = (z - \tilde{a})^{-1}\tilde{b}\sum u_k z^{-k}. \qquad (7\text{-}249\text{c})$$

The existence of the inverse is treated in the next section. Finally, substitute (7-249c) into (7-247) to give

$$\begin{aligned}\sum y_k z^{-k} &= \tilde{c}(z - \tilde{a})^{-1}\tilde{b}\sum u_k z^{-k} + \tilde{d}\sum u_k z^{-k} \\ &= \{\tilde{c}(z - \tilde{a})^{-1}\tilde{b} + \tilde{d}\}\sum u_k z^{-k}.\end{aligned} \qquad (7\text{-}250)$$

The morphism

$$\tilde{c}(z - \tilde{a})^{-1}\tilde{b} + \tilde{d} \qquad (7\text{-}251)$$

of 1-dimensional $F(z)$-vector spaces has a matrix

$$[\tilde{c}(z - \tilde{a})^{-1}\tilde{b} + \tilde{d}] \qquad (7\text{-}252)$$

with one row and one column and containing a transfer function (7-227). Now

$$[\tilde{a}] = [a], \qquad (7\text{-}253)$$

in the sense that each matrix contains the same element of the field F, provided that the chosen basis element in the F-vector space X is used to span the $F(z)$-vector space $X(z)$. Let x be that basis element. Then

$$\sum x_k z^{-k} = \sum (f_k x) z^{-k} = (\sum f_k z^{-k}) x, \qquad (7\text{-}254)$$

where x is now regarded as an element of $X(z)$; the fact that there exists $f(z)$ in $F(z)$ such that

$$f(z) \sum x_k z^{-k} \in X[z] \qquad \text{means that} \qquad f(z) \sum f_k z^{-k} x = p(z)x \quad (7\text{-}255\text{a,b})$$

for $p(z)$ in $F[z]$. Accordingly,

$$\sum f_k z^{-k} = (f(z))^{-1} p(z) \in F(z); \qquad (7\text{-}256)$$

and $X(z)$ is also spanned by x. Of course, the element of F in the matrix of the left member of (7-253) is to be regarded as an element of $F(z)$. Similar statements

$$[\tilde{b}] = [b], \qquad [\tilde{c}] = [c], \qquad [\tilde{d}] = [d] \qquad (7\text{-}257)$$

follow by analogous arguments.

EXAMPLE 7.6-4

For (7-232) of Example 7.6-2, the matrix (7-252) becomes

$$[\tilde{c}][(z - \tilde{a})^{-1}][\tilde{b}] = [\tilde{c}][z - \tilde{a}]^{-1}[\tilde{b}] \qquad (7\text{-}258\text{a})$$
$$= [\tilde{c}]([z] - [\tilde{a}])^{-1}[\tilde{b}] \qquad (7\text{-}258\text{b})$$
$$= [2]([z] - [-1])^{-1}[1] \qquad (7\text{-}258\text{c})$$
$$= [2][z + 1]^{-1}[1] \qquad (7\text{-}258\text{d})$$
$$= \left[\frac{2}{z + 1}\right]. \qquad (7\text{-}258\text{e})$$

∎

It should be remarked in passing that the development of the morphism (7-251) depends upon the MOMPS being in the zero state when the first nonzero input arrives. If such is not the case, adjustments can be made by assuming a finite sequence of inputs which transitions the system from the zero state to the state in question. Such a finite sequence can be prefixed to any sequence of finite left support in $U(z)$ to produce another sequence of finite left support in $U(z)$. Clearly, the notion of reachability pertains here.

Exercises

7.6-1. The construction of this section for the field $F(x)$ of transfer functions from the principal ideal domain ring $F[x]$ can take place in a more general setting. The important fact is that $F[x]$ is a domain. So, let D be an arbitrary domain. Select the subset of D consisting of elements

$$d \neq e_D$$

and repeat the steps used in constructing $F(x)$ from $F[x]$ to obtain the *quotient field* $Q(D)$ of D. Show that this type of construction elaborates Example 7.6-1.

7.6-2. Refer to the binary operation $+$ defined in (7-220). Show that $+$ is associative and commutative.

7.6-3. Refer to the binary operation $\cdot$ defined in (7-221). Show that $\cdot$ is associative and commutative.

7.6-4. Explain in detail why the right members of (7-230) may be written as shown instead of being written

$$[x]_k[a] + [u]_k[b], \qquad [x]_k[c] + [u]_k[d].$$

7.6-5. Establish that the subset of elements of finite left support in U^Z is a subgroup.

7.6-6. Carry out the details of showing that an element (7-243a) can be identified with an element (7-243b). Establish that the scalar multiplication (7-243c) satisfies axioms (7-14).

7.7 TRANSFER FUNCTIONS AND IMPULSE RESPONSE

The preceding section dealt with the morphism

$$z - \tilde{a} : X(z) \rightarrow X(z) \tag{7-259}$$

in the case for which $a : X \rightarrow X$ was a morphism of 1-dimensional F-vector spaces. This situation was especially simple, because the action of a is then essentially the same as a scalar multiplication.

In this section, the same ideas are presented in a more general framework with greater attention to detail.

The MOMPS (a, b, c, d) is assumed to be defined for F-vector spaces of inputs, states, and outputs having dimensions m, n, and p, respectively. The morphisms

$$a : X \rightarrow X, \qquad b : U \rightarrow X, \qquad c : X \rightarrow Y, \qquad d : U \rightarrow Y \tag{7-260a,b,c,d}$$

of F-vector spaces induce morphisms

$$\tilde{a}: X((z^{-1})) \to X((z^{-1})), \qquad \tilde{b}: U((z^{-1})) \to X((z^{-1})), \quad \text{(7-261a,b)}$$
$$\tilde{c}: X((z^{-1})) \to Y((z^{-1})), \qquad \tilde{d}: U((z^{-1})) \to Y((z^{-1})) \quad \text{(7-261c,d)}$$

of $F(z)$-vector spaces by the actions

$$\tilde{a}\sum x_k z^{-k} = \sum (ax_k)z^{-k}, \qquad \tilde{b}\sum u_k z^{-k} = \sum (bu_k)z^{-k}, \quad \text{(7-262a,b)}$$
$$\tilde{c}\sum x_k z^{-k} = \sum (cx_k)z^{-k}, \qquad \tilde{d}\sum u_k z^{-k} = \sum (du_k)z^{-k}. \quad \text{(7-262c,d)}$$

The local transition equation

$$x_{k+1} = ax_k + bu_k \tag{7-263a}$$

has the interpretation

$$\sum_k x_{k+1} z^{-k} = \sum_k (ax_k + bu_k)z^{-k} \tag{7-263b}$$

on $X((z^{-1}))$ if it is agreed that the state of the system is zero when the initial nonzero input arrives. This point was also discussed at the end of Section 7.6. By the group operation on $X((z^{-1}))$, and by (7-262a) and (7-262b), the right member of (7-263b) becomes

$$\tilde{a}\sum x_k z^{-k} + \tilde{b}\sum u_k z^{-k}. \tag{7-263c}$$

Through a change of index

$$q = k + 1, \tag{7-264}$$

the left member can be written

$$\sum_k x_{k+1} z^{-k} = \sum_q x_q z^{-(q-1)} = z\sum_q x_q z^{-q}. \tag{7-265a,b}$$

Axioms (7-14a) and (7-14c), together with the fact that the field $F(z)$ is a commutative ring, lead to the conclusion that the scalar multiplication by z in (7-265b) can be regarded as a morphism of $F(z)$-vector spaces. Then (7-263b) becomes

$$(z - \tilde{a})\sum x_k z^{-k} = \tilde{b}\sum u_k z^{-k} \tag{7-266}$$

with the aid of (7-263c) and (7-265b).

The morphism

$$(z - \tilde{a}): X((z^{-1})) \to X((z^{-1})) \tag{7-267}$$

is different in a rather subtle way from (7-259), even if the dimension of the F-vector spaces is ignored. To obtain an extension of (7-259), it will be necessary to show that (7-267) restricts in the manner

$$(z - \tilde{a})\,|\,X(z): X(z) \to X(z). \tag{7-268}$$

By definition of restriction, and by the fact that $X(z)$ is a subspace of $X((z^{-1}))$, it follows immediately that the restriction

$$(z - \tilde{a})\,|\,X(z) : X(z) \rightarrow X((z^{-1})) \tag{7-269}$$

can be made. To reach (7-268), it remains to show that the image of (7-269) is contained in $X(z)$. Let

$$\sum x_k z^{-k} \in X(z). \tag{7-270a}$$

Then there is a nonzero $f(z)$ in $F(z)$ such that

$$f(z) \sum x_k z^{-k} \in X[z]. \tag{7-270b}$$

But

$$z_*(X[z]) \subset X[z] \qquad \text{and} \qquad \tilde{a}_*(X[z]) \subset X[z], \tag{7-271a,b}$$

so that

$$(z - \tilde{a})_*(X[z]) \subset X[z]. \tag{7-271c}$$

Now (7-271c) combines with (7-270b) to give

$$f(z)(z - \tilde{a}) \sum x_k z^{-k} = (z - \tilde{a})f(z) \sum x_k z^{-k} \in X[z] \tag{7-272}$$

because $(z - \tilde{a})$ is a morphism of $F(z)$-vector spaces. Thus

$$(z - \tilde{a})_*(X(z)) \subset X(z), \tag{7-273}$$

and (7-268) is established. For convenience, the restriction sign in (7-268) is now suppressed.

It is not difficult to see that (7-268) is a monomorphism. In fact, begin by making the observation that the restrictions

$$\tilde{a} : X(z) \rightarrow X(z), \qquad \tilde{b} : U(z) \rightarrow X(z),$$
$$\tilde{c} : X(z) \rightarrow Y(z), \qquad \tilde{d} : U(z) \rightarrow Y(z) \tag{7-274a,b,c,d}$$

are possible by the same type of reasoning. Then calculate

$$0 = (z - \tilde{a}) \sum x_k z^{-k} = z \sum x_k z^{-k} + \sum (-ax_k)z^{-k}$$
$$= \sum_k (x_{k+1} - ax_k)z^{-k}. \tag{7-275}$$

But (7-275) implies

$$x_{k+1} = ax_k, \qquad k \in \mathbb{Z}. \tag{7-276}$$

Now suppose that there is a nonzero series (7-270a) such that (7-275) is satisfied. Then there must exist a p in $\mathbb{Z}$ such that

$$x_{p-1} = 0, \qquad x_p \neq 0. \tag{7-277}$$

However, (7-277) contradicts (7-276). Thus the only series to satisfy (7-275) is the zero series.

Is (7-268) an epimorphism as well? Assume an arbitrary (7-270a). It is necessary to exhibit another such series which is carried by (7-268) into the assumed series. To exhibit such a series, write

$$z^{-1} \sum x_k z^{-k} + z^{-2} \tilde{a} \sum x_k z^{-k} + z^{-3} \tilde{a}^2 \sum x_k z^{-k} + z^{-4} \tilde{a}^3 \sum x_k z^{-k} + \cdots.$$

$$(7\text{-}278)$$

Each of the terms in (7-278) is in $X(z)$, and the overall sum can be calculated in each coefficient by a finite number of steps. Indeed, (7-278) can be rewritten as

$$\sum x_{k-1} z^{-k} + \sum (a x_{k-2}) z^{-k} + \sum (a^2 x_{k-3}) z^{-k} + \sum (a^3 x_{k-4}) z^{-k} + \cdots. \quad (7\text{-}279)$$

Because of the assumption (7-270a), the sum

$$x_{k-1} + a x_{k-2} + a^2 x_{k-3} + a^3 x_{k-4} + \cdots \qquad (7\text{-}280)$$

cannot proceed beyond a finite number of terms for each k. It is an easy exercise to apply $(z - \tilde{a})$ to the series (7-278) and obtain (7-270a) back.

A technicality remains, because it has not yet been established that (7-278) is actually in $X(z)$. An alternative approach here is revealing.

To complete the argument, recall that the F-vector space X has a basis (7-111b). Understood as elements of $X(z)$, these vectors become a basis for the $F(z)$-vector space $X(z)$ as well. To see this, write

$$\sum \tilde{x}_k z^{-k} = \sum \left(\sum_{j=1}^{n} f_{kj} x_j \right) z^{-k} = \sum_{j=1}^{n} \left(\sum f_{kj} x_j z^{-k} \right) \qquad (7\text{-}281\text{a,b})$$

$$= \sum_{j=1}^{n} \left(\sum f_{kj} z^{-k} \right) x_j = \sum_{j=1}^{n} f_j(z) x_j, \qquad (7\text{-}281\text{c,d})$$

where the demonstration

$$\sum f_{kj} z^{-k} = f_j(z) \in F(z), \qquad j = 1, 2, \dots, n, \qquad (7\text{-}281\text{e})$$

is left as an exercise. Thus the $F(z)$- vector space $X(z)$ is n-dimensional.

Now the images

$$(z - \tilde{a}) x_i, \qquad i = 1, 2, \dots, n, \qquad (7\text{-}282)$$

of the basis vectors cannot be $F(z)$-linearly dependent, because $(z - \tilde{a})$ is a monomorphism. It follows that (7-282) must span $X(z)$ as well. This finishes the discussion of (7-268) as an isomorphism of $F(z)$-vector spaces.

Return now to the local transition interpretation (7-266). If

$$\sum u_k z^{-k} \in U(z), \qquad (7\text{-}283)$$

then the right member of (7-266) is in $X(z)$; and the isomorphism (7-268) can be used to solve (7-266) in the manner

$$\sum x_k z^{-k} = (z - \tilde{a})^{-1} \tilde{b} \sum u_k z^{-k}, \qquad (7\text{-}284)$$

where the composition symbol has been suppressed. The local output equation

$$y_k = cx_k + du_k \qquad (7\text{-}285\text{a})$$

has the interpretation

$$\sum y_k z^{-k} = \tilde{c} \sum x_k z^{-k} + \tilde{d} \sum u_k z^{-k}; \qquad (7\text{-}285\text{b})$$

and (7-283) together with the isomorphism (7-268) then imply that

$$\sum y_k z^{-k} \in Y(z). \qquad (7\text{-}286)$$

Accordingly, (7-284) and (7-285b) combine to yield

$$\sum y_k z^{-k} = \{\tilde{c}(z - \tilde{a})^{-1} \tilde{b} + \tilde{d}\} \sum u_k z^{-k}. \qquad (7\text{-}287)$$

The morphism

$$\tilde{c}(z - \tilde{a})^{-1} \tilde{b} + \tilde{d} : U(z) \to Y(z) \qquad (7\text{-}288)$$

of $F(z)$-vector spaces is called the *transfer function* of the MOMPS as defined in this section. For simplicity, the symbol

$$\tilde{h} = \tilde{c}(z - \tilde{a})^{-1} \tilde{b} + \tilde{d} \qquad (7\text{-}289)$$

is assigned to the transfer function.

Now assume the bases of (7-111) for the F-vector spaces U, X, and Y. Follow the procedure of (7-281) to develop these into bases for the $F(z)$-vector spaces $U(z)$, $X(z)$, and $Y(z)$ respectively. Then there is a *transfer function matrix*

$$[\tilde{h}] = [\tilde{c}(z - \tilde{a})^{-1} \tilde{b} + \tilde{d}] = [\tilde{c}(z - \tilde{a})^{-1} \tilde{b}] + [\tilde{d}] \qquad (7\text{-}290\text{a,b})$$
$$= [\tilde{c}][(z - \tilde{a})^{-1}][\tilde{b}] + [\tilde{d}] \qquad (7\text{-}290\text{c})$$

by the results of Section 7.4. Moreover, since

$$(z - \tilde{a})^{-1} \circ (z - \tilde{a}) = 1_{X(z)}, \qquad (7\text{-}291\text{a})$$

Example 7.4-2 implies that

$$[(z - \tilde{a})^{-1}][z - \tilde{a}] = [1_{X(z)}], \qquad (7\text{-}291\text{b})$$

which in turn implies by the usual matrix theory that

$$[(z - \tilde{a})^{-1}] = [z - \tilde{a}]^{-1}. \qquad (7\text{-}291\text{c})$$

So (7-290c) becomes

$$[\tilde{h}] = [\tilde{c}][z - \tilde{a}]^{-1}[\tilde{b}] + [\tilde{d}].\qquad(7\text{-}292)$$

As has already been pointed out in the section preceding, the way in which the bases were developed for the $F(z)$-vector spaces $U(z)$, $X(z)$, and $Y(z)$ from the bases for the F-vector spaces U, X, and Y assures that

$$[\tilde{a}] = [a],\qquad [\tilde{b}] = [b],\qquad [\tilde{c}] = [c],\qquad [\tilde{d}] = [d],\quad(7\text{-}293a,b,c,d)$$

in the sense that left members and right members of the four equations (7-293) contain the same elements of the field F. In the left members, however, these elements of F are to be regarded as elements of $F(z)$. Thus, an element $f \in F$ appearing in a left member of (7-293) should be understood as standing for an equivalence class $[f, 1]$ in $F(z)$. By convention, of course, the equivalence class symbol $[\ ,1]$ is omitted; and only $f/1$ is written, with the notation $/1$ rarely included.

The transfer function

$$\tilde{h}: U(z) \rightarrow Y(z)\qquad(7\text{-}294)$$

can be used in a natural way to define a typical response. Let

$$\sum u_k z^{-k} \in U(z)\qquad(7\text{-}295a)$$

be a sequence of inputs having the property that

$$k \neq 0 \Rightarrow u_k = 0.\qquad(7\text{-}295b)$$

If u_0 is not zero, then (7-295) results in an *impulse sequence*. There are m $F(z)$-linearly independent impulse sequences, corresponding to the m inputs (7-111a). Denote an input impulse sequence (7-295) by the input u_0 which occurs at index zero. Then (7-295) becomes

$$\sum u_k z^{-k} = u_0,\qquad(7\text{-}296a)$$

and the response of the system to (7-296a) can be written $\tilde{h}(u_0)$. With the notation of (7-111a), then, the m *impulse responses* are

$$\tilde{h}(u_i),\qquad i = 1, 2, \ldots, m.\qquad(7\text{-}296b)$$

It is customary in the usage (7-296b) to regard (7-294) as a morphism of F-vector spaces.

Because the notion of impulse response in (7-296b) depends upon the basis chosen in (7-111a), it is convenient also to define an *abstract impulse response*

$$\tilde{h}(U),\qquad(7\text{-}296c)$$

where (7-296c) is to be thought of as a subspace of $Y(z)$, regarded as an F-vector space.

Exercises

7.7-1.   Establish that the actions (7-262) lead to the functions (7-261) being morphisms of $F(z)$-vector spaces.

7.7-2.   In the $F(z)$-vector space $X((z^{-1}))$, show that scalar multiplication by z can be regarded as a morphism of $F(z)$-vector spaces.

7.7-3.   In (7-280), why cannot the sum proceed beyond a finite number of terms for each k?

7.7-4.   Give another argument to show that (7-278) is an element of $X(z)$.

7.7-5.   Supply the details which will establish (7-281e).

7.8   THE KEY TRIANGLE—ONE MORE TIME

To complete this chapter's brief glimpse into some aspects of the theory of module morphic systems, two more steps will be taken. The first step is to finish the discussion on reachable submodules. That will be accomplished in this section. The second step is to take a short look at the reverse question associated with Section 7.7, namely, what can be said about such a system from its transfer function. This will be done in the following section. For both of these steps, it is necessary to put in place the concept of a quotient module. Thus, for the last time in this volume, a key triangle is again studied.

The last discussion of a key triangle occurred in Section 6.1, for the case of rings. Because a ring is a special case of a module, it may be expected that the present discussion will bear some similarity to that of Section 6.1. This is indeed the case. However, there are some differences, and so the ideas are sketched out here.

Just as in the case of a ring, a module is constructed upon a commutative group. So, once again, the development can proceed from the key triangle for commutative groups. Here the reader may wish to review the early part of Section 6.1, or perhaps the complete details of Section 2.6. Either way, the basic observation is the same: a module is a commutative group together with a scalar multiplication, and a morphism of modules must necessarily be a morphism of groups.

Let $(S, \square, e_S)$ and $(T, *, e_T)$ be commutative groups, and develop on them by appropriate scalar multiplication functions

$$f_S: R \times S \to S \quad \text{and} \quad f_T: R \times T \to T \qquad \text{(7-297a,b)}$$

the structure of R-modules over the ring

$$(R, *, e_R, \triangle, 1_R). \qquad \text{(7-298)}$$

Further, let

$$f: S \to T \qquad \text{(7-299)}$$

be a morphism of R-modules.

It is then known that the results of Fig. 6.1 apply. In fact, if $W \subset S$ is any subgroup, and if

$$W \subset \operatorname{Ker} f, \tag{7-300a}$$

then there exists a unique morphism

$$\bar{f} : S/W \to T \tag{7-300b}$$

of groups such that

$$f = \bar{f} \circ p_W, \qquad \text{where} \qquad p_W : S \to S/W \tag{7-300c,d}$$

is the projection morphism onto the quotient group S/W. Along the lines of these studies earlier in the volume, the idea is to develop the quotient group $(S/W, \square, e_{S/W})$ into an R-module in such a way that (7-300d) becomes a morphism of R-modules.

What conditions on the subgroup W are appropriate so that such developments will be feasible? Once again, Ker f serves as a guide. Example 7.1-4 has shown that Ker f is a submodule of S. Therefore, it will be assumed that the subgroup W is also a submodule of S.

As a scalar multiplication

$$f_{S/W} : R \times S/W \to S/W, \tag{7-301a}$$

consider the function whose action is

$$f_{S/W}(r, \bar{s}) = p_W(f_S(r, s)) \tag{7-301b}$$

for some $s \in S$ satisfying

$$p_W(s) = \bar{s}. \tag{7-301c}$$

It is necessary to show that (7-301) is well defined in the sense that the result does not depend upon which representation s of the equivalence class $\bar{s}$ is chosen. If

$$w \mathbin{\square} s \tag{7-301d}$$

is any other representative, then

$$f_S(r, w \mathbin{\square} s) = f_S(r, w) \mathbin{\square} f_S(r, s) \tag{7-301e}$$

by (7-14a). By the fact that p_W is a morphism of groups, it then follows that

$$p_W(f_S(r, w \mathbin{\square} s)) = p_W(f_S(r, w)) \mathbin{\square} p_W(f_S(r, s))$$
$$= p_W(f_S(r, w)) \mathbin{\square} f_{S/W}(r, \bar{s}). \tag{7-301f}$$

Now W is a submodule of S, by assumption. Thus

$$f_S(r, w) \in W, \tag{7-301g}$$

which as the kernel of p_W means

$$p_W(f_S(r, w)) = e_{S/W}. \tag{7-301h}$$

Accordingly,

$$p_W(f_S(r, w \square s)) = f_{S/W}(r, \bar{s}) \tag{7-301i}$$

for all $s \in S$ satisfying (7-301c), and (7-301b) is a well-defined function.

It then becomes appropriate to verify that (7-301b) satisfies the axioms for scalar multiplication. For (7-14a), calculate

$$f_{S/W}(r, \bar{s}_1 \square \bar{s}_2) = p_W(f_S(r, s_1 \square s_2)) \tag{7-302a}$$
$$= p_W(f_S(r, s_1) \square f_S(r, s_2)) \tag{7-302b}$$
$$= p_W(f_S(r, s_1)) \square p_W(f_S(r, s_2)) \tag{7-302c}$$
$$= f_{S/W}(r, \bar{s}_1) \square f_{S/W}(r, \bar{s}_2), \tag{7-302d}$$

and notice that (7-14a) applied to f_S induces (7-14a) for $f_{S/W}$. A similar technique establishes (7-14b). For (7-14c), construct

$$f_{S/W}(r_1 \triangle r_2, \bar{s}) = p_W(f_S(r_1 \triangle r_2, s)) \tag{7-303a}$$
$$= p_W(f_S(r_1, f_S(r_2, s))) \tag{7-303b}$$
$$= f_{S/W}(r_1, \overline{f_S(r_2, s)}) \tag{7-303c}$$
$$= f_{S/W}(r_1, p_W(f_S(r_2, s))) \tag{7-303d}$$
$$= f_{S/W}(r_1, f_{S/W}(r_2, \bar{s})). \tag{7-303e}$$

Finally, for (7-14d),

$$f_{S/W}(1_R, \bar{s}) = p_W(f_S(1_R, s)) = p_W(s) = \bar{s}, \tag{7-304a,b,c}$$

as desired.

Thus $(S/W, \square, e_{S/W})$ is made into an R-module by (7-301b); and p_W becomes a morphism of R-modules.

It only remains to show that (7-300b) becomes a morphism of R-modules. It is already a morphism of groups, and so it suffices to check

$$\bar{f}(f_{S/W}(r, \bar{s})) = \bar{f}(p_W(f_S(r, s))) \tag{7-305a}$$
$$= (\bar{f} \circ p_W)(f_S(r, s)) \tag{7-305b}$$
$$= f(f_S(r, s)) \tag{7-305c}$$
$$= f_T(r, f(s)) \tag{7-305d}$$
$$= f_T(r, (\bar{f} \circ p_W)(s)) \tag{7-305e}$$
$$= f_T(r, \bar{f}(p_W(s))) \tag{7-305f}$$
$$= f_T(r, \bar{f}(\bar{s})). \tag{7-305g}$$

Notice that step (7-305d) is a consequence of the fact that (7-299) is a morphism of R-modules.

These remarks show that the diagram of Fig. 6.1 carries over to the R-module case, provided that W has the added feature of being a submodule of S.

To conclude the discussion of reachable submodules in earlier sections, the use of quotient modules will now be considered. The idea of a reachable submodule arose initially in Sections 4.6 and 4.7, in the group context. Section 7.3 proceeded to establish that the subgroup in question is actually a submodule for commutative MOMPSs when the integer ring $\mathbb{Z}$ is deployed in the usual way. This brought the discussion to the study of ascending chains of submodules (7-101). Example 7.3-3 showed that the existence of a reachable submodule depended upon whether or not every submodule of the state module X was of finite type. Though this meant that X itself had to be of finite type, there remained a question: What sorts of rings lead to modules having the property that their submodules are of finite type when they themselves are of finite type? In Section 7.5, the concept of principal ideal domain ring was used to establish that the ascending chain of ideals (7-191) in $F[x]$ had the property given by (7-192). A similar argument could have been used in the ring of integers. Section 7.5 concluded with some technicalities about short exact sequences and modules having the property that all their submodules are of finite type. Those technicalities are now employed.

Let R be a ring. As pointed out earlier, such a ring is an R-module. An ideal in the ring is then just a submodule. Thus a ring can have the property that each of its ideals is of finite type. If this is so, it is possible to establish that R-modules of finite type built upon the ring have the property that each of their submodules is of finite type.

Consider the R-module R^n. Let

$$p : R^n \to R \tag{7-306a}$$

be a projection of R^n upon one of its factors, with action

$$p(r_1, r_2, \ldots, r_n) = r_n. \tag{7-306b}$$

Then p is trivially a morphism of R-modules. Also, define the product group insertion

$$i : R^{n-1} \to R^n \tag{7-307a}$$

by

$$i(r_1, r_2, \ldots, r_{n-1}) = (r_1, r_2, \ldots, r_{n-1}, e_R). \tag{7-307b}$$

Notice that (7-307) is also trivially a morphism of R-modules. Moreover,

$$\text{Im}\, i = \text{Ker}\, p, \tag{7-308}$$

so that the short exact sequence

$$0 \to R^{n-1} \overset{i}{\to} R^n \overset{p}{\to} R \to 0 \qquad (7\text{-}309)$$

is created. When n is 2, it follows from the discussion at the end of Section 7.5 that every submodule of the R-module R^2 is of finite type. An induction gives that property on R^n for n greater than 2.

If G is an arbitrary R-module of finite type, then

$$G = \coprod_{i=1}^{n} Rg_i \qquad (7\text{-}310)$$

for some natural number n in $\mathbb{N}$. Establish an epimorphism

$$\alpha : R^n \to G \qquad (7\text{-}311)$$

of R-modules by assigning the ith basis element of (7-69) to g_i. This morphism is uniquely determined by such an assignment. Define

$$p_{\mathrm{Ker}\,\alpha} : R^n \to R^n/\mathrm{Ker}\,\alpha \qquad (7\text{-}312)$$

as a quotient module construction according to the ideas of this section. By the properties of the key triangle, there exists a unique morphism

$$\bar{\alpha} : R^n/\mathrm{Ker}\,\alpha \to G \qquad (7\text{-}313)$$

of R-modules such that

$$\alpha = \bar{\alpha} \circ p_{\mathrm{Ker}\,\alpha}. \qquad (7\text{-}314)$$

But (7-313) is epic because (7-311) is epic. Now suppose that

$$\bar{\alpha}(p_{\mathrm{Ker}\,\alpha} x) = e_G \qquad (7\text{-}315)$$

for some $x \in R^n$. Then $x \in \mathrm{Ker}\,\alpha$ by (7-314), and so $\bar{\alpha}$ is a monomorphism as well. So $\bar{\alpha}$ is an isomorphism of R-modules from $R^n/\mathrm{Ker}\,\alpha$ onto G.

The main interest centers upon the fact that (7-311) is epic, as a result of which

$$\mathrm{Im}\,\alpha = G. \qquad (7\text{-}316)$$

There is now a natural way to relate submodules of G to submodules of R^n. Indeed, if

$$S \subset G \qquad (7\text{-}317)$$

is a submodule of G, then this establishes that

$$\alpha^*(S) \subset R^n \qquad (7\text{-}318)$$

is a submodule of R^n. In fact, the construction

$$S \to \alpha^*(S) \qquad (7\text{-}319)$$

is a bijection from the set of all submodules of G to the set of all submodules of R^n which contain $\text{Ker}\,\alpha$. As a result of this bijection, an ascending chain of submodules in G relates directly to an ascending chain of submodules in R^n, for which the desired property has already been established.

Finally, therefore, the discussion of reachable submodules has been completed. When the state module X is of finite type, and when it draws its scalar multiples from a ring with the property that each of its ideals is of finite type, then it can be argued that there is a natural concept of reachable submodule.

As principal ideal domains, the rings $F[x]$ and $\mathbb{Z}$ have the requisite property. In fact, because any ideal in these rings is a principal ideal, $F[x]$ and $\mathbb{Z}$ need only one generator for each of their ideals. Accordingly, though these two rings are very useful, other rings could be visualized as well.

The availability of $\mathbb{Z}$ in this context, of course, reflects the entire set of conclusions back to commutative GRMPSs.

Exercises

7.8-1. Discussion subsequent to (7-300) shows that W a submodule of S is a sufficient condition to develop the key triangle for groups into a corresponding triangle for R-modules. Can you show that this condition is also necessary?

7.8-2. Establish axiom (7-14b) for the scalar multiplication function (7-301).

7.8-3. Explain in detail why it follows that (7-301) makes p_W into a morphism of R-modules.

7.8-4. Show that (7-306) is indeed a morphism of R-modules.

7.8-5. Show that (7-307) is also a morphism of R-modules.

7.8-6. Show that (7-311) is uniquely determined as a morphism of R-modules by its action on the basis elements (7-69).

7.8-7. Show that (7-318) is in fact a submodule.

7.8-8. Provide the details to back up the assertion that (7-319) establishes a bijection from the set of all submodules of G to the set of all submodules of R^n which contain $\text{Ker}\,\alpha$.

7.9 REMARKS ON REALIZATION THEORY

Section 7.7 introduced the idea of an abstract transfer function

$$\tilde{h}:U(z) \to Y(z) \tag{7-320}$$

for a MOMPS (a,b,c,d), defined on F-vector spaces of inputs, states, and outputs, having dimensions m,n, and p, respectively. The basic idea was to begin with local dynamical equations

$$x_{k+1} = ax_k + bu_k, \qquad y_k = cx_k + du_k \tag{7-321a,b}$$

associated with morphisms

$$a:X \to X, \qquad b:U \to X, \qquad c:X \to Y, \qquad d:U \to Y \quad (7\text{-}322\text{a,b,c,d})$$

of F-vector spaces. Each of the morphisms (7-322) in turn induces a morphism of $F(z)$-vector spaces in the manner

$$\tilde{a}:X((z^{-1})) \to X((z^{-1})), \tag{7-323a}$$
$$\tilde{b}:U((z^{-1})) \to X((z^{-1})), \tag{7-323b}$$
$$\tilde{c}:X((z^{-1})) \to Y((z^{-1})), \tag{7-323c}$$
$$\tilde{d}:U((z^{-1})) \to Y((z^{-1})), \tag{7-323d}$$

by the natural actions

$$\tilde{a}\sum x_k z^{-k} = \sum (ax_k)z^{-k}, \tag{7-324a}$$
$$\tilde{b}\sum u_k z^{-k} = \sum (bu_k)z^{-k}, \tag{7-324b}$$
$$\tilde{c}\sum x_k z^{-k} = \sum (cx_k)z^{-k}, \tag{7-324c}$$
$$\tilde{d}\sum u_k z^{-k} = \sum (du_k)z^{-k}. \tag{7-324d}$$

In turn, the morphisms (7-323) are restricted—this was one of the primary efforts of Section 7.7—to morphisms

$$\begin{aligned} \tilde{a}:X(z) \to X(z), \qquad \tilde{b}:U(z) \to X(z), \\ \tilde{c}:X(z) \to Y(z), \qquad \tilde{d}:U(z) \to Y(z), \end{aligned} \qquad (7\text{-}325\text{a,b,c,d})$$

of $F(z)$-vector spaces. Then

$$\tilde{c}(z - \tilde{a})^{-1}\tilde{b} + \tilde{d} \tag{7-326}$$

was the explicit representation of $\tilde{h}$ in (7-320). Pictorially, the situation might be regarded in the manner of Fig. 7.1.

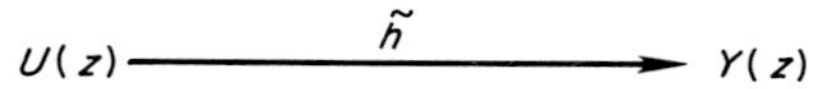

Fig. 7.1. The MOMPS transfer function. $\qquad U(z) \xrightarrow{\quad\tilde{h}\quad} Y(z)$

In this section, the goal is to give a brief introduction to the basic idea of reversing the construction of the transfer function $\tilde{h}$ with the purpose of learning something about the behavior of the underlying MOMPS. Many systems investigators would vote for this topic as the beginning of the development of algebraic system theory in general. Even if they did not ascribe to this as the starting point, they would probably agree that it consumed the lion's share of the early efforts in algebraic system theory.

As a descriptive term for the section the phrase "realization theory" suggests itself rather naturally. Think of the MOMPS as a black box, the specifics of whose innards are not yet known. By the methods appropriate to the case at hand, various input sequences are supplied to the box, with careful measurement of the resulting output sequences. In this way, suppose that $\tilde{h}$ has been inferred. The next task would be to determine what is inside the box. In the situation of MOMPSs, this would amount to exhibiting (a, b, c, d) that would return the $\tilde{h}$ found by measurement. Once (a, b, c, d) is available, it is in principle possible to build, construct, or "realize" the insides of the box. Needless to say, the techniques for solving black-box problems were in full-scale development long before investigators considered the issue's relation to module theory. However, the insights afforded by module theory have produced new viewpoints and techniques which now offer promise of resolving problems not well understood in the classical lore.

Some investigators tend to use the phrase "algebraic system theory" almost interchangeably with "realization theory." Though such usage is consonant with the origins of the subject a decade and more ago, it fails to reflect adequately the rapid development of algebraic methods in system theory at the present time.

Recall now from Section 7.6 the introduction of the subgroup

$$U[z] \tag{7-327a}$$

of $U((z^{-1}))$ consisting of elements

$$\sum u_i z^{-i} \tag{7-327b}$$

with the property that

$$u_i = 0, \qquad i = 1, 2, 3, \ldots . \tag{7-327c}$$

$U[z]$ might be visualized as a set of polynomials in the indeterminate z having coefficients in the F-vector space U of dimension m. Though $U((z^{-1}))$ is an $F(z)$-vector space, $U[z]$ is not a subspace. Indeed, the point is demonstrated simply by selecting an element

$$\sum u_i z^i \in U[z] \qquad \text{with the assumption} \qquad u_i = 0, \quad i \neq 0. \tag{7-328a,b}$$

Then a scalar multiplication by

$$z^{-1} \in F(z) \tag{7-329}$$

gives

$$u_0 z^{-1} \notin U[z], \tag{7-330}$$

.o that $U[z]$ is not closed under scalar multiplication within the $F(z)$-vector space structure of $U((z^{-1}))$.

However, notice that

$$F[z] \subset F(z), \tag{7-331}$$

so that the same scalar multiplication which made $U((z^{-1}))$ an $F(z)$-vector space also makes $U((z^{-1}))$ into an $F[z]$-module.

Now consider $U[z]$ as a subgroup of the $F[z]$-module $U((z^{-1}))$. In this situation, it is a straightforward exercise to show that $U[z]$ is closed under scalar multiplication from the ring $F[z]$. Thus $U[z]$ is a submodule of $U((z^{-1}))$. In fact, with reference once more to Section 7.6,

$$U[z] \subset U(z), \tag{7-332}$$

which is also a submodule of the $F[z]$-module $U((z^{-1}))$.

Accordingly, there is an insertion morphism

$$i: U[z] \rightarrow U(z) \tag{7-333}$$

of $F[z]$-modules. This morphism is now added to the diagram of Fig. 7.1 in order to display that of Fig. 7.2. A comment is in order with regard to Fig. 7.2. Notice that $U(z)$ and $Y(z)$ can also be regarded as $F[z]$-modules, and that $\tilde{h}$ can then be understood as a morphism of $F[z]$-modules. In that case, the diagram represents an interconnection of $F[z]$-module morphisms. On the other hand, unlike $\tilde{h}$, i does not admit the interpretation of a morphism of $F(z)$-vector spaces; on that level, there is essentially no interconnection.

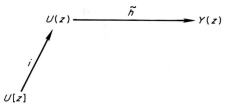

Fig. 7.2. Add the insertion morphism.

A word is in order on the nature of the input sequences contained in $U[z]$. These can be visualized in terms of inputs which begin at a finite time "in the past" and end "now." Indeed, if

$$\sum_{i=0}^{p} u_i z^i \in U[z] \tag{7-334a}$$

is such an element, then it corresponds from the discussions of Section 7.7 to a sequence of inputs

$$\dots, \quad 0, \quad 0, \quad u_p, \quad u_{p-1}, \quad \dots, \quad u_1, \quad u_0, \quad 0, \quad 0, \quad \dots. \tag{7-334b}$$

This interpretation can be seen by writing (7-334a) in the more revealing form

$$\sum_{j=-p}^{0} \tilde{u}_j z^{-j}, \quad \text{where} \quad \tilde{u}_j = u_{-j}. \tag{7-335a,b}$$

Notice that when the sequence of inputs (7-334b) shuts off with the last input at index zero, the result will be to place the **MOMPS** in some state x_1, which it reaches from the zero state under the action of the sequence. The next step in realization discussions is to examine the output sequences produced by x_1 in the presence of no further inputs.

Clearly, these output sequences will be of the general form

$$y_1, \quad y_2, \quad y_3, \quad \ldots, \tag{7-336}$$

with

$$y_j, \quad j \leq 0, \tag{7-337}$$

being disregarded.

The process of disregarding (7-337), as probably suspected by the reader, can be made more explicit. The method is to observe that the information (7-337) is contained in $Y[z]$, which is a submodule of the $F[z]$-module $Y(z)$. To formalize, simply set up the quotient projection

$$p_{Y[z]} : Y(z) \rightarrow Y(z)/Y[z] \tag{7-338}$$

as a morphism of $F[z]$-modules. For simplicity, write

$$p = p_{Y[z]}. \tag{7-339}$$

Then the diagram of Fig. 7.2 can be expanded to that of Fig. 7.3. Inasmuch as i, $\tilde{h}$, and p are all morphisms of $F[z]$-modules, it is possible to define the composition

$$\tilde{h}\# : U[z] \rightarrow Y(z)/Y[z] \quad \text{by} \quad \tilde{h}\# = p \circ \tilde{h} \circ i, \tag{7-340a,b}$$

so that the diagram commutes. By discussions in this chapter, (7-340) is a morphism of $F[z]$-modules.

Now construct a quotient projection

$$\tilde{p} : U[z] \rightarrow U[z]/\mathrm{Ker}\,\tilde{h}\#, \tag{7-341}$$

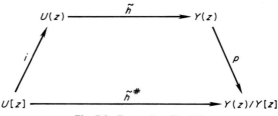

Fig. 7.3. Expanding Fig. 7.2.

and recall that $\tilde{p}$ is an epimorphism of $F[z]$-modules. From the key triangle discussion of Section 7.8, there then exists a unique monomorphism

$$\overline{\tilde{h}\#} : U[z]/\operatorname{Ker}\tilde{h}\# \;\to\; Y(z)/Y[z] \tag{7-342}$$

of $F[z]$-modules such that

$$\overline{\tilde{h}\#} \circ \tilde{p} = \tilde{h}\#. \tag{7-343}$$

This situation is reflected in Fig. 7.4.

Just why Fig. 7.4 might be called a realization diagram is an interesting question. Before remarks are made on this point, however, it is appropriate to comment upon the algebraic nature of the $F[z]$-module

$$U[z]/\operatorname{Ker}\tilde{h}\#. \tag{7-344}$$

Consider a nonzero element

$$\overline{u}(z) \in U[z]/\operatorname{Ker}\tilde{h}\#. \tag{7-345}$$

Inasmuch as (7-342) is a monomorphism of $F[z]$-modules, it follows that

$$\overline{\tilde{h}\#}(\overline{u}(z)) \neq 0 \in Y(z)/Y[z]. \tag{7-346}$$

Now (7-341) is an epimorphism of $F[z]$-modules. Thus there is a

$$u(z) \in U[z] \tag{7-347}$$

such that

$$\tilde{p}(u(z)) = \overline{u}(z). \tag{7-348}$$

Then commutativity of the diagram in Fig. 7.4 provides that

$$\begin{aligned}
\overline{\tilde{h}\#}(\overline{u}(z)) &= \overline{\tilde{h}\#}(\tilde{p}(u(z))) && \text{(7-349a)}\\
&= (\overline{\tilde{h}\#} \circ \tilde{p})(u(z)) && \text{(7-349b)}\\
&= \tilde{h}\#(u(z)) && \text{(7-349c)}\\
&= (p \circ \tilde{h} \circ i)(u(z)) && \text{(7-349d)}\\
&= p(y(z)) && \text{(7-349e)}
\end{aligned}$$

for

$$y(z) = (\tilde{h} \circ i)(u(z)) \in Y(z). \tag{7-350}$$

By definition of $Y(z)$, there then exists an element

$$0 \neq p(z) \in F[z] \tag{7-351}$$

with the property

$$p(z)y(z) \in Y[z]. \tag{7-352}$$

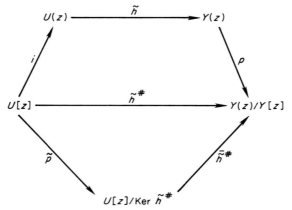

Fig. 7.4. Realization diagram.

However, (7-352) would imply that

$$p(p(z)y(z)) = 0 \in Y(z)/Y[z]. \tag{7-353}$$

If it is now recalled that p is a morphism of $F[z]$-modules, it is straightforward to see that (7-353) implies

$$p(z)p(y(z)) = 0 \in Y(z)/Y[z]. \tag{7-354}$$

But (7-354), together with (7-349), gives

$$p(z)\overline{\tilde{h}} \# (\overline{u}(z)) = 0 \in Y(z)/Y[z], \tag{7-355}$$

and the fact that $\overline{\tilde{h}} \#$ is a morphism leads to

$$\overline{\tilde{h}} \# (p(z)\overline{u}(z)) = 0 \in Y(z)/Y[z]. \tag{7-356}$$

Because $\overline{\tilde{h}} \#$ is a monomorphism, (7-356) can be true only if

$$p(z)\overline{u}(z) = 0 \in U[z]/\mathrm{Ker}\,\tilde{h} \# . \tag{7-357}$$

These steps of reasoning show that every nonzero element (7-345) is a *torsion element*, that is, a module element which can be scalar multiplied into the zero element of the module group by a nonzero scalar. The existence of such behavior was illustrated earlier in this chapter. However, in this case,

$$U[z]/\mathrm{Ker}\,\tilde{h} \# \tag{7-358}$$

is a *torsion module*, which is a module with the property that all of its elements are torsion elements.

Notice that the torsion module (7-358) is finitely generated. Indeed, select a basis $u_1, u_2, \ldots, u_m$ for the F-vector space U. Next, regard each of these u_i as elements of $U[z]$. Then the u_i generate $U[z]$. As a consequence, they map into a set of generators

$$\tilde{p}(u_i), \qquad i = 1, 2, \ldots, m \tag{7-359}$$

for (7-358).

Finitely generated $F[z]$-modules possess many classic algebraic properties, which are usually discussed in texts on rings and module theory.

Most important here, however, is the fact that the action of a morphism

$$a : \tilde{X} \to \tilde{X} \tag{7-360}$$

of F-vector spaces essentially can be identified with that of scalar multiplication by z in the $F[z]$-module (7-358). To be more precise, rename

$$U[z]/\mathrm{Ker}\,\tilde{h}\# = \tilde{X} \tag{7-361}$$

as a commutative group. Then for a given

$$\tilde{x} \in \tilde{X}, \tag{7-362}$$

define the action of (7-360) by

$$a\tilde{x} = z\tilde{x}, \tag{7-363}$$

where the right member is found by scalar multiplication in the $F[z]$-module (7-358).

It turns out that the dimension of the F-vector space $\tilde{X}$ may be less than n, if the original MOMPS which led to $\tilde{h}$ fails to be both reachable and observable.

The action of a suitable morphism

$$b : U \to \tilde{X} \tag{7-364}$$

of F-vector spaces can then be deduced from $\tilde{p}$ in (7-341). The idea here is to express the generators (7-359) as linear combinations of a suitable basis for $\tilde{X}$.

Similarly, the action of another morphism

$$c : \tilde{X} \to Y \tag{7-365}$$

of F-vector spaces can be inferred from $\bar{\tilde{h}}\#$. Here the idea is to represent elements in $Y(z)/Y[z]$ by series

$$\sum_{j=1}^{\infty} y_j z^{-j}. \tag{7-366}$$

With the aid of a basis $\{x_i\}$ for $\tilde{X}$, the constructions

$$\bar{\tilde{h}}\#(x_i) \tag{7-367}$$

can be carried out and expressed in the manner (7-366). Let

$$y_1(i) \in Y \tag{7-368}$$

be the first series coefficient in a representation (7-366) for the ith construction (7-367). Each coefficient (7-368) can be expressed as a linear combination of basis vectors in Y. In this way, the action of (7-365) can be determined.

As the reader is no doubt aware, there are many ways to select the basis $\{x_i\}$ in $\tilde{X}$. Moreover, there are competing methods for calculating $\operatorname{Ker} \tilde{h} \#$. In fact, the choice (7-361) is but one way to name $\tilde{X}$. Another interesting procedure is to choose

$$\tilde{\tilde{X}} = \operatorname{Im} \tilde{h} \# . \tag{7-369}$$

All of these have specific details associated with them.

The fundamental idea, however, is the conceptual approach to Fig. 7.4. As an algebraic tool, such an approach makes it possible to consider realization problems from a vantage point not foreseen by classical black-box procedures; among the possibilities are the use of novel types of rings and modules to describe the MOMPS.

Exercises

7.9-1. Consider that an F-vector space V is given, and let R be a subring of F. Show that the same scalar multiplication used to make V into an F-vector space now makes V into an R-module. Be careful to check all the axioms (7-14).

7.9-2. Show that $U[z]$ is a submodule of the $F[z]$-module $U((z^{-1}))$.

7.9-3. Use the definition given in Section 7.6 to show that $U[z] \subset U(z)$.

7.9-4. Demonstrate that the insertion function defined in Section 2.1 becomes a morphism of $F[z]$-modules in the case (7-333).

7.9-5. Recall the definition of $Y(z)$ from an earlier section. How do the assertions (7-351) and (7-352) differ from that definition? Show that the use of $F[z]$ in place of $F(z)$ can be deduced from the earlier definition.

7.9-6. Explain how an element u in U can be regarded as an element of $U[z]$. Can an insertion morphism be used for this purpose?

7.9-7. Explain how a basis for U can generate $U[z]$.

7.9-8. Why do the elements (7-359) generate (7-358)?

7.9-9. Establish that (7-361) is an F-vector space.

7.10   DISCUSSION

The concept of groups as a fundamental building block in system theory was introduced and developed in Chapter 4. In Chapter 5, the concept of rings was laid out as a backdrop for the study of interconnected group morphic systems. An initial interweaving of the notions of ring and commutative group was accomplished unobtrusively in Chapter 6, where the group elements played the role of signals and the ring elements played the role of systems.

The present chapter has made a formal introduction to the notion of a module of commutative group elements equipped with a scalar multiplication making use of ring elements. Clearly, the number of possibilities for discussion expands quite rapidly in this context. Either of the notions of group and ring can fill chapters without even a consideration of the possible interplays between them.

Accordingly, careful selections have had to be made.

A final elaboration of ideas on the reachable submodule has involved consideration of certain special types of rings and modules. The emphasis has been placed upon commutative rings in which every ideal is of finite type, as a submodule; special cases are $F[z]$ and $\mathbb{Z}$, which have the property that each of their ideals is generated by one element. The classical division algorithms for polynomials and integers underlie such a feature. Modules of finite type over such rings have a corresponding property, which was used to complete the reachable submodule idea. The results here have implications all the way back to Chapter 4, because every commutative group is a $\mathbb{Z}$-module in a natural way.

The reachable submodule thread, which has been unwound over several chapters, has permitted the volume to introduce and illustrate some of the nonobvious and more deeply running currents in module theory.

As an extension of the GRMPS idea, this chapter also treats MOMPS. Here the goal has been to exemplify the basic types of behavior which can and do occur in modules. First, a finitely generated module can be free on a basis; matrices result. This possibility has been carried through to the case of transfer function matrices. Second, a module can exhibit torsion, which is a phenomenon not observed in vector spaces. This type of situation has been pointed out by defining some of the algebraic elements involved in realization theory.

The chapter touches but lightly on vector spaces and algebraic realization theory. As explained in the Preface, this choice is a consequence of the fact that most readers are reasonably well acquainted with the theory of vector spaces and that text references are available for further reading on realization.

REFERENCES
AND FURTHER READING

1. Algebra

S. MacLane and G. Birkhoff, *Algebra*. Macmillan, New York, 1967.
The presentation of ideas in this general reference book is compatible with the approach we
have used here. A second edition appeared in 1979.

B. Hartley and T. O. Hawkes, *Rings, Modules, and Linear Algebra*. Chapman and Hall, London,
1974.
Readers wishing to add further background to the concepts of Chapter 7 should find this a
useful volume.

2. Set-Dynamical Systems

W. M. Wonham, Towards an Abstract Internal Model Principle, *IEEE Transactions on
Systems, Man, and Cybernetics*, **SMC-6**, 735–740 (1976).
This paper presents a more general version of the ideas occurring toward the end of Chapter 3.
The paper also offers special interpretations for the case of vector space structure, and thus
is useful followup reading for Chapter 7.

S. Eilenberg, *Automata, Languages, and Machines, Volume A*. Academic Press, New York,
1973.
The entire area of automata theory presents a possible interface to Chapter 2. This reference has
the added feature of presenting further reading on realization theory, as a sequel to
Chapter 7.

M. Gatto and G. Guardabassi, The Regulator Theory for Finite Automata, *Information and
Control*, **31**, 1–16 (1976).
This article takes a viewpoint related to that of Chapter 3.

M. K. Sain and J. J. Uhran, Jr., The Equivalence Concept in Criminal Justice Systems, *IEEE
Transactions on Systems, Man, and Cybernetics*, **SMC-5**, 176–188 (1975).
This paper was intended to illustrate the possibility of using algebraic ideas to shed light upon
social studies.

3. Group Morphic Systems

R. W. Brockett and A. S. Willsky, Finite Group Homomorphic Sequential Systems, *IEEE Transactions on Automatic Control*, **AC-17**, 483–490 (1972).
For the case of finite groups, this article explores topics related to, and in addition to, those of Chapter 4.

M. K. Sain and J. L. Massey, Invertibility of Linear Time-Invariant Dynamical Systems, *IEEE Transactions on Automatic Control*, **AC-14**, 141–149 (1969).
It is this paper which has been generalized in Chapter 5.

A. S. Willsky, Invertibility of Finite-Group Homomorphic Sequential Systems, *Information and Control*, **27**, 126–147 (1975).
This paper contains additional ideas related to Chapter 5, for the case of finite groups.

4. Interconnected Systems

M. K. Sain, The Quotient Signal Flowgraph for Large-Scale Systems, *IEEE Transactions on Circuits and Systems*, **CAS-25**, 781–788 (1978).
The material of Chapter 6 extends the basic notions of this paper.

M. K. Sain, E. W. Henry, and J. J. Uhran, Jr., An Algebraic Method for Simulating Legal Systems, *Simulation*, **21**, 150–158 (1973).
Historically, this was the study which eventually suggested the material of Chapter 6.

5. Modules and Realization Theory

R. E. Kalman, Algebraic Structure of Linear Dynamical Systems. I. The Module of $\sum$, *Proceedings National Academy of Sciences* (USA), **54**, 1503–1508 (1965).
This publication is often cited as the formal launching of algebraic system theory.

R. E. Kalman, P. L. Falb, and M. A. Arbib, *Topics in Mathematical System Theory*. McGraw-Hill, New York, 1969.
Chapter 10 contains discussion of algebraic realization theory.

M. L. J. Hautus and M. Heymann, Linear Feedback—An Algebraic Approach, *SIAM Journal on Control and Optimization*, **16**, 83–105 (1978).
The approach in this paper follows in much the same spirit as that of Section 7.9.

E. Sontag, Linear Systems over Commutative Rings: A Survey, *Ricerche di Automatica*, **7**, 1–34 (1976).
This work offers further reading along the lines of Section 7.9.

6. General

W. M. Wonham, *Linear Multivariable Control: A Geometric Approach* (Second Edition). Springer-Verlag, New York, 1979.
This is a very elaborate study of the application of algebraic concepts to solve complicated systems problems. It takes place in a vector space format and is must reading for persons wishing to see the extent to which algebra can aid in providing both insights and solutions to real tasks.

M. K. Sain, The Growing Algebraic Presence in Systems Engineering: An Introduction, *IEEE Proceedings*, **64**, 96–111 (1976).
This paper mentions some ideas on the solution of linear equations over rings, on the potential for using algebraic tensors in system theory, and on the relevance of exterior algebras to the abstract transfer function.

INDEX